85 Structure and Bonding

Springer-Verlag Berlin Heidelberg GmbH

Optical and Electronic Phenomena in Sol-Gel Glasses and Modern Application

Volume Editors: C. K. Jørgensen, R. Reisfeld

With contributions by
M. A. Aegerter, R. C. Mehrota, I. Oehme,
R. Reisfeld, S. Sakka, O. Wolfbeis, C. K. Jørgensen

With 73 Figures and 22 Tables

 Springer

In references Structure and Bonding is abbreviated
Struct. Bond. and is cited as a journal.

Springer WWW home page: http://www.springer.de

ISBN 978-3-662-14847-1   ISBN 978-3-540-49750-9 (eBook)
DOI 10.1007/978-3-540-49750-9

CIP-Data applied for

© Springer-Verlag Berlin Heidelberg 1996
Originally published by Springer-Verlag Berlin Heidelberg New York in 1996
Softcover reprint of the hardcover 1st edition 1996

Typesetting: Macmillan India Ltd., Bangalore-25, India
SPIN: 10509080   51/3020 - 5 4 3 2 1 0 Printed on acid-free paper

Volume Editors

Editorial Board

Preface

Transparent glasses can be prepared by a sol-gel process: Controlled hydrolysis and polycondensation of a variety of procursors at low temperature (e. g. 60 to 200 °C). The precursors are metal alcoxides $M(OR)_2$ ($M = Si$), Sn, Ti, Al, Mo, V, W, Ce...; OR being an alcoxy group OC_nH_{2n+1}.

The present volume is a sequel to volume 77 of Structure and Bonding published in 1992 (Chemistry, Spectroscopy and Applications of Sol-Gel Glasses).

In this new volume, S. Sakka discusses "Sol-Gel Coating Films for Optical and Electronic Applications" formed on surfaces of ceramics of conventional glasses.

O. S. Wolfbeis, R. Reisfeld, and I. Oehme write on "Sol-Gels and Chemical Sensors" of recent concern for short-distance or remote detection of haevy metals, gases, biological impurities and other reactive compounds ocurring as vapors, gases, or solids at low concentration, including droplets of fog or mist.

R. Reisfeld writes "New Materials for Non-Linear Optics", a review of efficient incorporation of nanoparticles of semiconductors, metals and organic colorants in transparent sol-gel glasses and their use for nonlinear optics.

M. Aegerter contributes "Sol-Gel Chromogenic Materials" which change from being colorless to being strongly colored by photochromic and/or electrochromic effects, or modify their color, as applied in the technology of "smart windows".

R. Reisfeld writes "Lasers in Sol-Gel Materials" and discusses high-yield, tunable laser emission from organic dye-stuffs (highly resistant to strong illumination of long duration) incorporated in sol-gel glasses either in bulk or wave guiding films.

C. K. Jørgensen writes "Luminescence of Cerium(III) Inter-Shell Transitions in Glasses" with high-energy particles (and gamma rays) providing high-yield scintillators. Comparison is made with other multi-dimensional Born-Oppenheimer potential surfaces.

Most of the chapters describe materials of great potential for technical application in the fields of optics, ionoelectronics, mechanics, the environment, and biology.

The editors would like to express their admiration of Dr. Helmut Dislich who started the modern technology of sol-gel glasses.

Renata Reisfeld
Christian Klixbüll Jørgensen

Table of Contents

Table of Contents of Volume 77
Chemistry, Spectroscopy and Applications of Sol-Gel Glasses

Sol–Gel Coating Films for Optical and Electronic Application

Sumio Sakka

Fukui University of Technology, Sakka Laboratory,
Kuzuha–Asahi 2-7-30, Hirkata, Osaka-Fu 573, Japan

Over the last few years, considerable progress has been made with sol–gel coating films for optical and electronic applications. Nonlinear optical films and ferroelectric films are among those that have been studied most extensively. Optical and electronic coating films based on inorganic-organic hybrid or composite films have attracted much attention. These and other recent developments are stressed in this article which deals with a wide variety of optical and electronic coating films.

Structure and Bonding, Vol. 85
© Springer-Verlag Berlin Heidelberg

1 Introduction

The sol–gel method for preparing solid materials is now regarded as a quite promising and important method along with solid-state reaction, melt-quenching and vapor-phase deposition methods.

In most sol–gel processes [1–3], we start from a solution containing source compounds for the target material, as shown in Fig. 1 [4]. The solution becomes a sol as a result of formation of fine colloidal particles or polymers and further reactions lead to gelation, i.e., wet gel formation. In the course of sol-to-gel conversion, which takes place at low temperatures, coating, fiber drawing and molding into bulky shapes can be achieved [5]. In many cases a target material is obtained by heating a shaped gel at higher temperatures.

Among these shapes, coating films are the most important products of the sol–gel method. Coating of glass, ceramic, metal and plastic substrates by the sol–gel method is very useful for modifying properties of substrates or providing substrates with new active properties which are needed for developing optical, electronic and chemical devices. The low temperature processing which can be

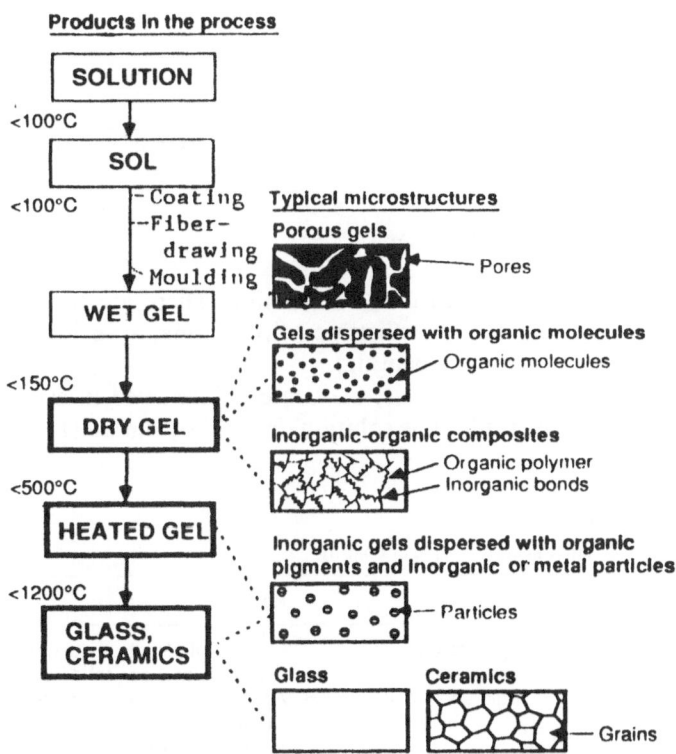

Fig. 1. Sol–gel process and microstructure of sol–gel products [4]

achieved by the sol–gel method is particularly important because it allows the application of the film to non-heat resistant substrates and the application of non-heat resistant films to substrate.

The sol–gel method of fabricating thin films offers potential advantages over traditional techniques as shown in Table 1. Considering that a substrate is used for coating, the low processing temperature is particularly important in the application to optical and electronic devices because the substrate and other active elements on the substrate are not necessarily highly heat resistant. Easy coating of large or small surfaces makes it possible to apply the sol–gel coating to wide display panels and windows as well as small parts of integrated circuits. Possibility of changing thickness may be advantageous for coating films for some optical and electronic devices. High optical quality films required for optical devices can also be provided. These advantages come from the fact that the sol–gel method does not need powder processing.

In this review article, a brief discussion is presented of the formation of sol–gel coating films in Sect. 2. The main parts of this article, Sects. 3–5, are devoted to optical and electronic coating films prepared by sol–gel methods. The present author published an article on Sol–Gel-Derived Coating Films and Application [6]. Compared with the previous article, the present article is characterized by a more detailed description of optical and electronic coating films and a description of new developments in these areas.

2 Formation of Coating Films by the Sol–Gel method

2.1 Coating Processes

Solutions or sols are used for coating. There are three kinds of methods for applying sols to the substrate: dip coating, spin coating and laminar flow coating (or meniscus coating).

In dip coating [7] the substrate is immersed in a dipping solution and is drawn up vertically. The solution dragged by the substrate is solidified into a gel

Table 1. Advantages of sol-gel coating

Term	Remarks
Low temperature processing	Coating of glasses, semiconductors, and integrated electronic and optical devices is possible
Easy coating of large and small surfaces	Application can be made to display and integrated circuits
Variable thickness	Very thin coating films and thick films are possible
High optical quality	Transparent films are possible

film. In spin coating [8] an amount of solution is dropped on the rotating substrate and the solution propagates outwards on the substrate, covering it. The sol film formed becomes solidified as gel film. The laminar flow method [9] was recently developed for sol–gel coating by Floch and Belleville [10]. In this method a substrate is coated in an upside-down position (the bottom surface is coated). The coating solution is pumped into a slot applicator tube, and flows out to the surface through the slot, forming a continuous liquid film on the outside. A substrate is placed in contact with the liquid film so that a narrow meniscus is created between it and the applicator tube. As the applicator is moved horizontally relative to the substrate, a liquid film is left on the substrate. In all these methods, a gel film is formed after evaporation of solvents. The resultant gel coating films are heat-treated when necessary to produce the required material.

Many of the precautions necessary for obtaining good coating films by dip coating apply to spin coating and laminar flow coating as well. Therefore discussion of the formation of coating films will concentrate on the dip coating method.

The typical sol–gel dip coating involves three steps – dipping, withdrawing and heating. A drawing apparatus is shown in Fig. 2. After dipping and withdrawal of the substrate, the film is heated to a certain high temperature of 200–500 °C for metal oxide films such as SiO_2, Al_2O_3 and $BaTiO_3$. This heating produces chemical bonds between the film and the substrate which are needed for the formation of coatings. It is assumed that chemical bonds –M–O–M′–, where M and M′ are metallic ions in the film and substrate, respectively, are formed on heating to a certain high temperature. For metal substrates, the bond is formed through a thin oxide layer. It has been shown that the formation of –M–O–M′– bonds at high temperatures may be easier when many –M′OH and –MOH groups are present at the contact surfaces of film and substrate [11]. Figure 3 shows the formation of siloxane bonds from two metal hydroxy groups.

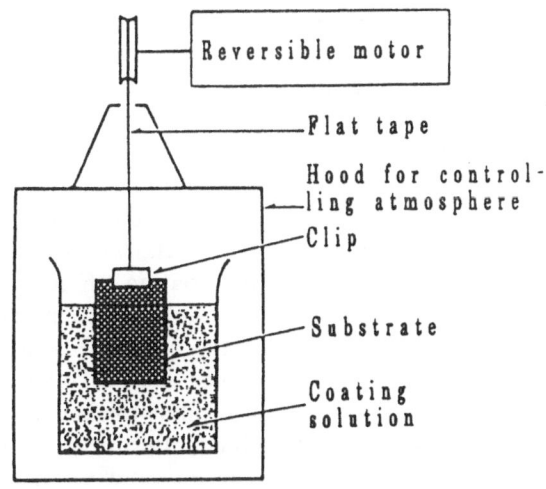

Fig. 2. Apparatus for dip coating

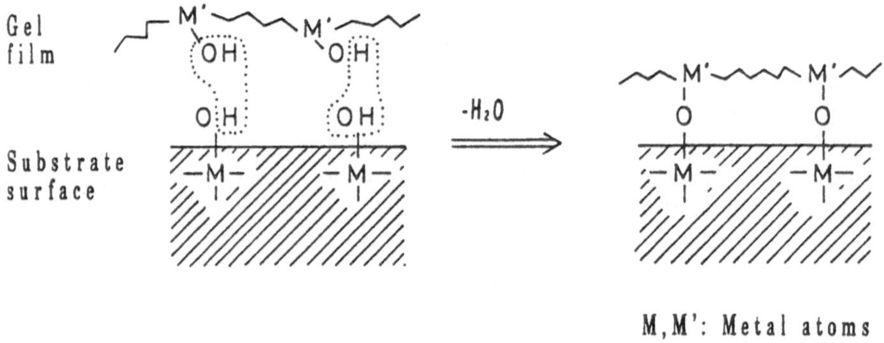

M,M': Metal atoms

Fig. 3. Formation of metalloxane bonding between the coating film and the substrate

2.2 Thickness of Coating Films

2.2.1 Critical Thickness

It was found in the formation of SiO_2 films on a glass substrate [11] that if one uses coating conditions (for instance, high viscosity of the coating solution) under which the film is very thick in a one-step coating procedure involving application of the gel film and heating to a certain high temperature, e.g. 500 °C, the film will completely peel off the substrate. When the film is thinner due to the use of lower viscosity solutions, the film stays on the substrate after the coating procedure, but cracks are formed when the film is thicker than a critical thickness. It was observed [11] that severe cracks are accompanied by peeling of local areas of the film and the film scatters light due to the occurrence of air spaces between the film and substrate.

With the same viscosity and accordingly, similar expected thickness, complete drying of the gel film at room temperature or lower temperatures than the temperature where the bonds between the gel film and substrate are formed is more likely to result in crack formation. This shows the importance of chemical bond formation accelerated by the presence of silanol groups or water. At the same time this indicates that the critical thickness is not unique but changes somewhat with processing conditions. For coating films which are expected to form some kind of chemical bond to the substrate, Brinker et al. [12] review the crack formation, the critical thickness and the method for avoiding cracks.

On heating the dip-coating film prepared from a $Si(OC_2H_5)_4$ $-H_2O-C_2H_5OH-HCl$ solution at high temperature, say 500 °C, it is assumed that the film adheres to the substrate and then the shrinkage due to the evaporation of the remaining solvent takes place essentially in a direction perpendicular to the film. In films thicker than the critical thickness, however, the coherent force within the film acting in the parallel direction may exceed the adhesion force, thereby causing cracks. The critical thickness experimentally found for Si(OR)$_4$ derived SiO$_2$ films and Ti(O–iCH$_3$)$_4$-derived TiO$_2$ films may be 0.2–0.5 µm [11].

The critical thickness is larger than 0.7 µm when silica films are prepared from a solution containing CH$_3$Si(OC$_2$H$_5$) [13]. This may be attributed to a lower stiffness of the film [12] or smaller shrinkage after the adhesion of the film to the substrate [13]. The critical thickness may be much larger (10 µm) for organic–inorganic composite films [14]. This is attributed to the small shrinkage and low stiffness of the film. Stress relaxation may be expected as a result of plastic deformation due to the presence of a considerable amount of organic polymer components.

The critical thickness is around 1.5 µm for brittle oxide coating films [15], when a starting solution containing a large amount of chelating molecules such as ethanolamine is used. In such a case, the decomposition and vaporization of organic materials occur, say, at 200–300 °C, when the film is very soft. The adhesion takes place at higher temperatures, say, at 500 °C, where essentially no shrinkage occurs.

It is to be noted that the thermal expansion difference between the film and substrate may be the cause of cracks in films thicker than 1 µm, for example.

2.2.2 Change of Thickness with Coating Condition

In dip coating the substrate is immersed in a dipping solution and is drawn upward. It is easily understood that the amount of the solution dragged by the substrate increases with increasing viscosity of the solution and increasing speed of drawing. Taking into account the effect of the surface tension, σ, Landau and Levich [16] derived a formula for the thickness, t, which was rewritten by Strawbridge and James [17] as

$$t = 0.944 \, (\eta v/\sigma)^{1/6} \, (\eta v/\rho g)^{1/2} \tag{1}$$

where ρ is the density of the solution and g is the acceleration of gravity. In the sol–gel coating, the evaporation of the solvent and development of condensation of the film-forming species take place immediately after drawing, which changes the viscosity of the solution. Basically, however, the thickness of the film is expressed by

$$t = (\eta v)^{2/3} \tag{2}$$

This relationship is valid [12, 17] when η and v are small as in most sol–gel dip coatings. When η and v are large, the thickness may be expressed by [17]

$$t \propto (\eta v)^{1/2}. \tag{3}$$

Generally,

$$t \propto \eta^m v^n. \tag{4}$$

Experimenal data relating to the dependence of t on η for SiO$_2$ films derived from Si(OC$_2$H$_5$)$_4$ solutions [11, 17] have shown that $m = 0.5$–0.6. As to the dependence of t on v, experimental observations [18–21] have shown that n changes in the range 1/2–2/3. According to Guglielmi and Zenezini [22], different investigators report different n values, although n is close to 1/2 in most cases.

In order to increase the thickness without hurting the film quality, one can repeat the coating procedure. An example for $BaTiO_3$ films is shown in Fig. 4 [19]. Similar results have been obtained for TiO_2–GeO_2 films [23]. It is assumed that for every subsequent coating procedure one uses a new substrate consisting of a combination of the original substrate and previously applied films. A possible stress induced by the thermal expansion difference between the film and substrate may be negligibly small in the present thickness range.

2.3 Microstructure of Coating Films

As seen from Fig. 1, it is possible to produce many kinds of micro structures by sol–gel methods. The microstructures which the coating film can have are

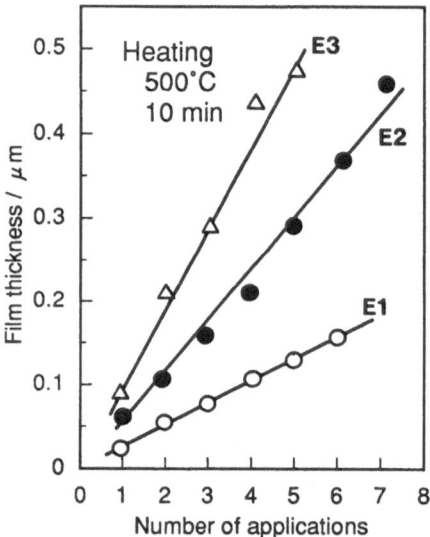

Fig. 4. Film thickness vs number of applications for $BaTiO_3$ films by the sol–gel dip-coating technique [19]. Solutions E1, E2, and E3 contain 1.72, 4.28 and 6.34 wt% $BaTiO_3$, respectively

Table 2. Possible microstructures for sol-gel coating films

Material	Gels with specific microstructure
Gels	Inorganic porous gel
	Gel with dispersed organic molecules
	Inorganic–organic hybrid
Heated gels	Amorphous oxide
	Metal (or semiconductor) colloid–dispersed oxide
Glasses and ceramics	Dense glass
	Ceramics with randomly dispersed crystals
	Preferentially crystal–oriented ceramics

summarized in Table 2. Most of the microstructures listed in the table can be shared by bulk body and fibers. The only exception is preferential crystal-orientation. It is expected that the preferential crystal orientation will produce high performance coating films.

3 Coating Films with Optical Functions

3.1 Films as Planar Waveguides

Slab or channel planar waveguides are important components of optical integrated circuits and devices for transmitting light. An example is taken from the work of LaSerra et al. [24]. Figure 5 shows the $PbTi_4O_9$ planar waveguide on a glass substrate prepared by sol–gel coating from solutions containing lead acetate $Pb(C_2H_3O_2)_2$, titanium iso-propoxide $Ti(OC_3H_7)_4$, methoxymethanol $CH_3OCH_2CH_2OH$ and water. The refractive index of the $PbTi_4O_9$ film is 2.15, which is much higher than those of the glass substrate (1.517) and air. This makes it possible for the light to propagate through the film by total reflection at both surfaces. This figure also shows that, with a rutile prism coupler, a He–Ne laser is coupled to a thin planar film 0.82 μm thick on the glass substrate.

According to considerations based on papers by Uhlmann et al. [25, 26], sol–gel synthesis is a very useful method for preparing planar waveguide due to the ability to form films which satisfy the requirements. Besides the low temperature nature which is beneficial to most of the coating films, the following characteristics can be cited for the planar waveguides.

(1) Low optical loss – sol–gel techniques give coating films of high optical quality, that is, films of high optical quality which are transparent and homogeneous, leading to low optical loss. In crystalline films, a small crystal size suppresses the loss due to scattering of light at grain boundaries. Amorphous films which are characterized by the absence of grain boundaries can be prepared.

(2) Varying refractive index – in the sol–gel method, the composition, and thereby the refractive index of films can be continuously varied over a wide range.

(3) Patterning – the sol–gel process inherently includes the change of stiffness or viscosity during the conversion of soft gel to hard gel. Patterning is possible for soft gels by stamping and for harder gels by heating.

(4) Thickness of the film – very thin films are usually obtained for inorganic oxides. The thickness can be increased by adding organic groups, which are not so easily decomposed, or by employing inorganic-organic hybrid materials.

Some examples of sol–gel prepared optical planar waveguides produced so far will be given below.

SiO_2–TiO_2 coatings have been studied widely, because the refractive index of the film can be varied over a wide range by changing the TiO_2 content, and the film has good chemical, thermal, and mechanical stability and low optical losses of about 1 dB/cm [26]. It has been shown that the optical loss increases with the amount of residual carbon and roughness of the substrate surface [27].

Densification by laser heating and embossing by stamping can be used for preparing channel waveguides in gels. Araujo et al. [28] showed that optical waveguides 50 μm wide can be produced by CO_2 laser densification of porous gel-silica matrices with pore radii in the region of 30 Å. The CO_2 laser is suitable for generating heat without affecting the deeper parts of the silica matrices in porous silica and the substrate in the case of deposited films. Also the processing time is very short. Laser processing has been applied to patterning of SiO_2–TiO_2 films [29, 30].

Mechanical stamping has also been applied to SiO_2–TiO_2 films [31]. Laminar-shape embossed waveguide gratings with a 0.52 μm width and 100–200 nm peak-to-trough depth have been prepared.

Recent trends in the development of sol–gel processed optical waveguides have involved the use of inorganic-organic nanocomposite films [26]. This is because, with nanocomposite materials, thicker films than inorganic films like TiO_2–SiO_2 or TiO_2–SiO_2 films can be easily prepared without occurrence of cracks in the film. Schmidt et al. [32] showed with slab waveguides obtained by copolymerization of methacrylate-substituted silanes with methacrylate acid and Zr components that slab waveguide films up to 20 μm thick with low losses of 0.1 dB/cm can be obtained.

With these inorganic-organic composites, planar waveguides can be formed by laser patterning; photoinitiators are incorporated in order to control the polymerisation reaction [33].

3.2 Films with Optical Absorption and Coloring

Coating films which have optical absorption bands in the UV-visible-IR range are very useful for providing substrate glasses with desirable color, UV shielding ability or optical filter property.

Sheet glasses are colored with thin coating films of transition metal oxides [34, 35] or thin coating films consisting of gel network-forming oxides and coloring materials such as transition metal oxides [36–41] and organic pigments [42].

FeO [34, 35], NiO and Cr_2O_3 [34] coating films about 0.1 μm thick show pretty intense color. Yamamoto et al. [36] and Duran et al. [37] made SiO_2–R_mO_n (R = Cr, Mn, Fe, Co, Ni, and Cu) coating films from tetra-ethoxysilane-metal nitrate-water-acid solutions. Most metal nitrates are soluble in water. In this method very high concentrations up to 30–50 mol% of metal oxides could be dissolved in the starting solution and coating films. Some of the transition metal elements, for example Co and Ni, give very thin films, about

0.3 μm thick, with an unexpectedly strong color [36]. The optical absorption is much higher than that calculated from the extinction coefficient of the coloring element, assuming that the element is dissolved in gel or oxide matrix as ion. It is then suggested that the strong color is caused by colloidal particles of the oxide of the coloring element.

Transparent, yellow-colored coating film of the TiO_2–CeO_2 system [38–40] can be obtained by sol–gel coating using titanium isopropoxide and cerium chloride as starting materials. This film has a bright yellow color and is also highly UV-absorbing [40], and so is applied to the shielding of UV light. Blue-colored GeO_2–V_2O_5 coating films have been made from germanium and vanadium alkoxides [41]. The coating films have been heated at 500 °C in reducing atmosphere to reduce the V^{5+} ions to V^{4+} ions. The optical absorption of the film measured as a function of temperature shows a hysteresis in the temperature range of 40–80 °C. This might be related to semiconductor-metal transition of crystalline VO_2, although no such change in electrical conductivity Vis seen. TiO_2–SiO_2 coating films are used for UV shielding [42], since they have a strong optical absorption in the UV region of the spectrum.

Organic pigments can be used for coating films which show desirable selective absorption. Organic dye molecules are very sensitive to heat, and so more heat-resistant pigments which are aggregations of dye molecules are used [43]. Sol–gel prepared ZrO_2–SiO_2 coating films which contain organic pigment particles have been applied to the glass face plate of commercial color television sets [44]. The organic pigments selectively absorb the light, enhancing the color contrast of the television. Melpolder [45] made composite color filter materials consisting of organic dyes, polymers and inorganic skeletons.

The color caused by interference of light is favored by some people, because it is soft-tuned like that of rainbow. This interference color is achieved with sol–gel coating films consisting of high refractive index oxide. The change of optical absorption spectrum [46] of the TiO_2 film as a function of the film thickness indicates that the number of maxima and minima of light intensity in the visible range increases with increasing thickness of the film, as the theory of interference shows.

3.3 Reflecting Coating Films

The most important application of the reflecting coating film is the preparation of sunlight shielding window glasses which serve in cutting the thermal radiation of sunlight coming into a room of a building, lowering the power required for cooling the room during summer time. For this purpose, coating films have to be applied to a large area of glass. Sol–gel dip-coating is quite suitable for such coatings. Thus In_2O_3–SnO_2 [47] and VO_2–SiO_2 [48] films are made for this purpose by sol–gel coating. In_2O_3–SnO_2 films reflect thermal radiation in the wavelength region of 2–12 μm and absorb the high temperature thermal radiation in the region of 0.3–2 μm.

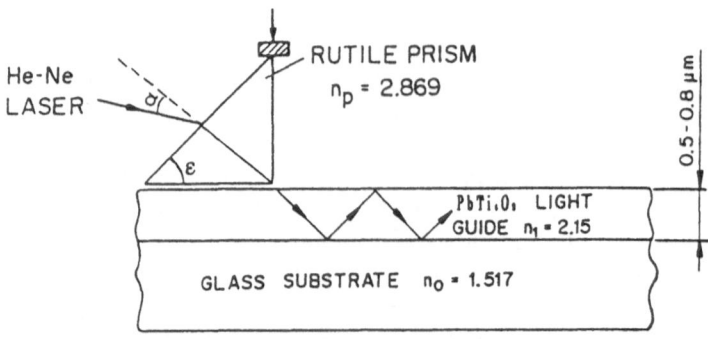

Fig. 5. $PbTi_4O_9$ planar waveguide prepared by sol–gel coating. He–Ne laser light beam is coupled to the waveguide by prism coupler method. After LaSerra et al. [49]

LaSerra et al. [49] have proposed $PbO–TiO_2$ and $Bi_2O_3–TiO_2$ coating films for this purpose. Coating films of these systems have practically no optical absorption in the visible region. Therefore, control of $n \times t$ values, that is, the refractive index times the film thickness will provide almost flat transmission and reflection characteristics. Accordingly, the films are suitable for solar-energy-reflecting windows. The cutting of solar energy by reflection is more advantageous than by the sun-shielding glass window which absorbs sunlight. The calculation of $n \times t$ value for reflection of the light is made by taking the interference of the lights reflected from both surfaces of the film.

$PbTiO_3$ ceramics can be prepared from a few different combinations of the starting compounds. Gurkovich and Blum [50] used the reaction of lead acetate $Pb(CH_3COO)_2$ dissolved in methoxyethanol $CH_3OCH_2CH_2OH$ with titanium isopropoxide $Ti(OC_3H_7)_4$ in order to make a $PbTiO_3$ precursor. LaSerra et al. [49] used $Ti(iso\text{-}OC_3H_7)_2$ $(acac)_2$ prepared by modifying $Ti(iso\text{-}OC_3H_7)_4$ by acetylacetone Hacac and lead acetate $Pb(CH_3COO)_2$. For compositions of the $Bi_2O_3–TiO_2$ system, a bismuth sol prepared from $Bi(NO_3)_3 \cdot 5H_2O$ in acetic acid and lead acetate were mixed.

Another interesting application of reflecting films is found in the reflecting coating film on the windshield which is part of a head up display (HUD) apparatus [51]. The HUD system makes it possible for a driver of an automobile to see the speed and other driving information on the running car in his field of view without looking down to the instrument panel. $TiO_2–SiO_2$ films, about 2000 Å thick, were applied to the wind shield by a sol–gel coating method using titanium and silicon alkoxides. The coating film of about 10 cm × 10 cm in size selectively reflects the light signal to show driving information. The signal is generated from a high intensity fluorescence tube. As the HUD for the automobile, the film has to satisfy the following conditions [52]:

(1) transmittance of visible light should be higher than 70%;

(2) the film shows the selective reflectance peaking at 530 nm with more than 25% reflectivity for the better contrast;

(3) the film has to have the chemical and mechanical resistance equal to or higher than the substrate windshield glass.

These requirements are satisfied by the TiO_2–SiO_2 film. The design of optical characteristics can be made based on the change of refractive index with TiO_2 content and the change of thickness. Figure 6 shows the reflectance spectrum of a TiO_2–SiO_2 coating film which is practically applied as HUD for some Japanese automobiles [52].

3.4 Antireflecting Coating Films

Antireflecting films applied to glasses prevent loss of light due to reflection at the surfaces of the glass, providing clear vision of pictures through a glass sheet or window. They also reduce stray light in optical systems. So far, multilayer films have been prepared by vapor-phase deposition techniques, such as sputtering, vacuum deposition and vapor phase chemical deposition. The sol–gel coating technique enables easy application of single layer and multilayer antireflecting films on large surfaces.

Two types of single-layer antireflecting coating films with gradient index have been designed for the purpose of protecting optical systems from damage by laser light and increasing the threshold of a glass laser element in high intensity optical lasers for nuclear fusion. Figure 7 shows a model of non-reflection glass [53]. The gradient change from the surface of the glass toward the surface of the film results in prevention of reflection.

Mukherjee and Lowdermilk [54] developed a technique in which a Na_2O–B_2O_3–SiO_2 film is applied to glass by sol–gel processing. The borosilicate glass film is phase-separated by heating and then etching with hydrofluoric

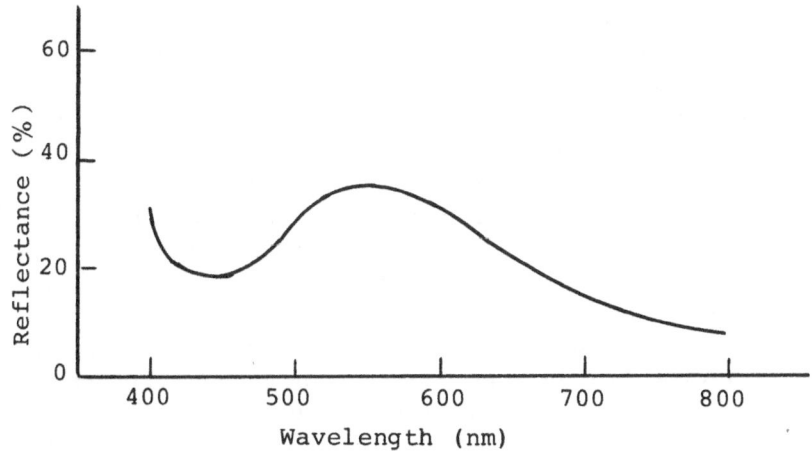

Fig. 6. Reflectance spectrum of a TiO_2-SiO_2 coating film on automobile windshield used as combiner for head up display (HUD) apparatus [52]

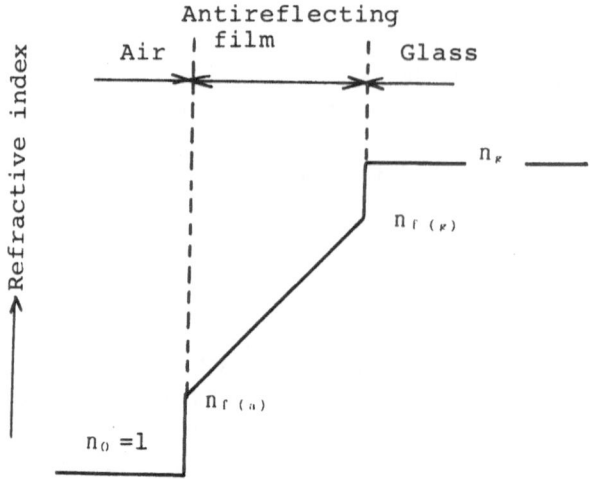

Fig. 7. Model of non-reflection glass with antireflecting monolayer film of gradient refractive index. After Asahara and Izumitani [53]. $n_g > n_{f(g)} > n_{f(a)} > n_0$. n_g: refractive index of glass. $n_{f(g)}$ and $n_{f(n)}$: refractive index of the film at the film-glass and film-air interfaces, respectively. $n_0 = 1$: refractive index of air

acid (HF) or ammonium fluoride (NH$_4$F). This etching dissolves the B$_2$O$_3$–Na$_2$O phase and SiO$_2$ skeletons to a lesser extent, producing silica film of high porosity. The porous silica film thus prepared has a refractive index gradient such that the refractive index becomes small towards the free surface of the film. This antireflecting film raises the damage threshold of optical laser glass lenses for the laser nuclear fusion systems from 4–5 to 20 J cm^{-2} for multilayer antireflecting coating, increasing the efficiency of the laser.

The hydrofluoric acid etching of silica film prepared on the glass from a tetraethoxysilane solution makes a film having a similar refractive index gradient [55] to that mentioned above. The etching solution enters the pores in the film from the free surface, starting to etch the silica skeleton near the entrances of pores and making the porosity in the position near the free surface larger than that for the deeper positions. This results in the antireflecting film refractive index becoming lower towards the free surface. This coating provides antireflectivity over the entire spectral range with better than 99% transmission, as shown in Fig. 8, and increases the laser damage resistance to 9 J cm^{-2}.

Floch and Belleville [10, 56] prepared scratch-resistant single-layer anti-reflecting coating films on both plastic and inorganic substrates from a composite material made from silica as the discontinuous phase and polytetrafluoroethylene-derived organic polymer as the continuous phase. The substrate was coated with a colloidal silica suspension containing TEOS-derived 20 nm particles and binders, and heated to 120 °C. The porous silica film was then impregnated by the fluorocarbon polymer. Figure 9 shows the transmission curves of fused silica substrate coated with the scratch resistant composite film

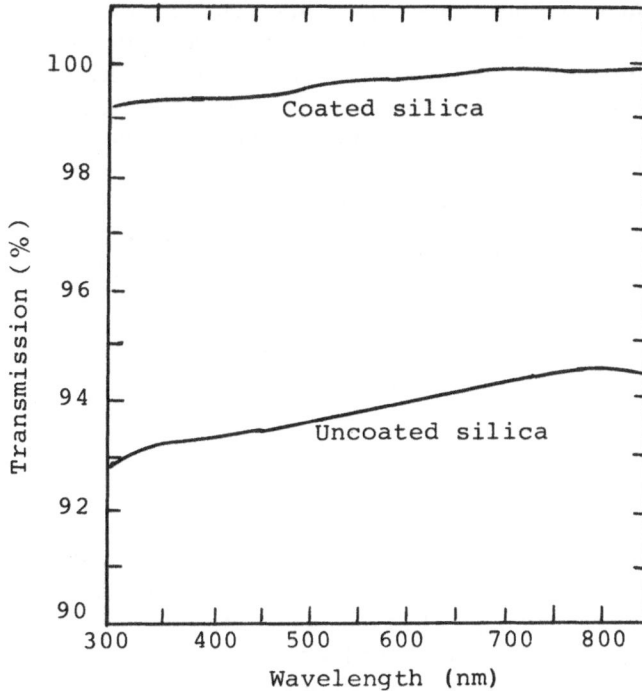

Fig. 8. Typical spectral transmission curves for a fused silica substrate before and after application of the anti-reflecting coating. After Yoldas and Partlow [55].

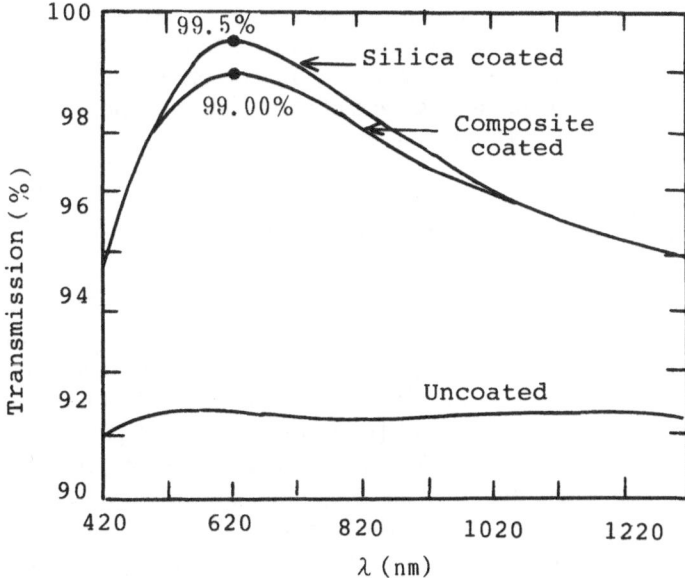

Fig. 9. Transmission curves of fused silica substrate coated with the scratch resistant composite film and the sol–gel silica film. After Floch and Belleville [56]

and the silica (sol-siloxane) film. The slightly higher transmission for the silica coating can be explained by the difference of the refractive index of the film: 1.26 for the porous silica film and 1.30 for the composite film. The laser damage threshold of the fused silica substrate is improved by the antireflecting coatings.

Besides single-layer antireflecting films, multiple-layer anti-reflecting films are produced by the sol–gel method. Multiple-layer anti-reflecting films consisting of triple interference layers $(TiO_2–SiO_2)$–TiO_2–SiO_2 are applied to plate glass [57]. This film makes it possible to see the picture covered by the glass sheet very clearly. Also, show-cases constructed from antireflecting glasses will give a clear sight of goods inside the case.

3.5 Fluorescent and Laser Films

3.5.1 Nd-Doped Silica

There have been many attempts to introduce fluorescent and lasing ion and molecules into coating films by the sol–gel method. These coating films are important as active elements in microoptic circuits, display panels and other light utilizing panels. Preliminary attempts to prepare silica coating films containing neodimium have been made [58, 59] in order to obtain heat-resistant lasers. The problem encountered in SiO_2 glass doped with Nd atoms is the clustering of Nd atoms in SiO_2 [60], which lowers the fluorescence lifetime. In order to avoid this problem, SiO_2 glass was codoped with Nd_2O_3 and Al_2O_3 [60] and was prepared by the sol–gel method [61, 62]. The clustering of Nd atoms is still seen, however, in these cases. Almeida et al. [59] prepared SiO_2 and $90SiO_2–10TiO_2$ films doped with Nd in a Nd/Si atomic ratio up to 15% by the sol–gel method using $NdCl_2$ as precursor. The resultant films, subjected to various heat treatments, were of good quality, although whether the problem of clustering was eliminated or not is not clear.

3.5.2 Organic Molecules in Gels and Composites

Organic molecules with fluorescent properties and dye laser organic molecules have been incorporated into coating films of silica gels, alumina gels and inorganic–organic composites. It is expected that the incorporation of active organic molecules into inorganic and composite matrices may increase their photostability. Avnir et al. [63, 64] and Tani et al. [65] started the incorporation of organic molecules, such as rhodamine B and 6G, coumarin 1 and 4, acridine, crystal violet and oxazine 170, into sol–gel silica gels, including thin films. Photo-decomposition of rhodamine 6G is slower in SiO_2 gels than in water [63]. Oxazine 170, a well-known laser dye in the region 600–700 nm, was incorporated into silica gel coating films [66] and it was found that laser action in the sol–gel glass is possible in the range 640–730 nm for a moderate pumping power. Kurokawa's group prepared amorphous Al_2O_3 films doped with optically active organic materials [67–69] by the sol–gel method. Laser action was observed upon excitation by N_2 laser of 10^{-3} mol l^{-1} rhodamine 6G molecules in Al_2O_3 gel films of 0.1 mm thickness.

3.5.3 Solar Collector

The film flat plate LSC (luminescent solar concentrator) [70] has a coating film containing fluorescent dye molecules, as shown in Fig. 10. A LSC absorbs solar radiation in the flat coating film containing one or more fluorescent species. The fluorescent emission is trapped in the coating film by total reflection and concentrated at the edge of the collector which constitute photovoltaic cells.

A LSC has advantages [70] over a conventional solar concentrator in that both direct and diffuse lights are collected, tracking the sun is not necessary and the luminescent species can be chosen to allow matching of the concentrated light to the maximum sensitivity of the photovoltaic cell. The advantage of the configuration of the doped thin film with the plate over the luminescent plates in which the dye is incorporated in the entire bulk is that the luminescent light emitted from the thin film is trapped in the plate with reduced self-absorption and scattering.

Reisfeld [71] showed, on the basis of the experimental data of absorption and emission of rhodamine 6G in sol–gel derived gel, that the quantum efficiency of rhodamine 6G is 0.75 and its molar extinction coefficient is 82 000. The calculation based on the overlap of the absorption of rhodamine 6G with the solar spectrum using a 50 μm thick film deposited on a plate having refractive index of 1.5 and area of 1 m^2 gives 1.5% as the optical efficiency. A larger efficiency can be obtained by using the photostable dye and combination of dyes which increases the overlap of absorption with the solar spectrum.

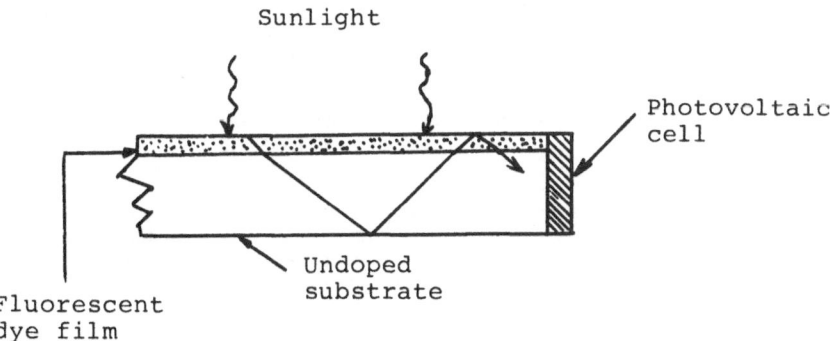

Fig. 10. Diagram of a flat plate luminescent solar concentrator with a sol–gel thin film. After Reisfeld [70]

3.6 Nonlinear Optical Films

Nonlinear optical materials have been attracting much attention due to their important applications to optoelectronics and photonics. For example, they are used for ultra-high speed optical switching, optical memory and frequency-conversion devices based on the optical Kerr effect, optical bistability and harmonic generation [72–74].

The films are especially useful for nonlinear optics. It is known that the nonlinear effect is greatly enhanced as the intensity of light becomes large. With thin films, the incident light can be concentrated in the thin film when the light is passed through the film in the direction of the film plane. For uses requiring a wide area, the incident light can enter the film in the direction perpendicular to the film plane. It is noted that the sol–gel method facilitates the formation of thin films of various types.

Table 3 shows the type of nonlinear optical coating films prepared by the sol–gel method. Since many studies have been made, the examples shown in Table 3 are only a fraction of the total.

3.6.1 Second-Order Nonlinear Optical Films

When ferroelectric oxide or efficient second-order nonlinear oxide crystals are highly oriented in the film, the film in noncentrosymmetric and is important for its second-harmonic generation (SHG) and electrooptic properties. Among such films, there are $LiNbO_3$ single crystal films grown on $LiNbO_3$ [73] and $LiTaO_3$ [74] single crystal substrate, $PbTiO_3$ films on fused silica [75] and $SrTiO_3$

Table 3. Types of materials for nonlinear optical coating films prepared by the sol-gel method

Nonlinearity	Materials	Examples
Second-order optical nonlinearity	Single crystalline oxide and oriented crystalline oxide films	$LiNbO_3$ $PbTiO_3$ $KTiOPO_4$ β–BaB_2O_4
	Films with organic molecules by poling	Silica doped with N, N–diethylamino–(β)–nitrostyrene
Third-order optical nonlinearity	Randomly crystalline oxide film Semiconductor particles–doped oxide film	α–Fe_2O_3 SiO_2 doped with CdS colloid
	Metal particles–doped oxide film Inorganic–organic nano–composites containing fine particles of metal or semi-conductor	SiO_2 doped with Au colloid Composites doped with Au colloid
	Organic molecules–doped gel film	SiO_2 and Al_2O_3 doped film

single crystal [76] substrates and $KTiOPO_4$ films [77, 78]. These single crystals and crystal-oriented coating films are obtained by selecting the precursor molecules and composition of the starting solution, controlling the temperature and time of heating and using the appropriate substrate. The formation of β-BaB_2O_4 film, which has a very high SHG coefficient, is being studied [79, 80].

Films of inorganic oxide-organic oxide composite materials containing a large concentration of second-order chromophore molecules show second-order nonlinear optical properties when the chromophores are aligned by poling. Poling is achieved by corona discharge or application of d.c. voltage. Zieba et al. [81] prepared the composite films consisting of polyvinylpyrrolidone (PVP) and silica composite doped with a second-order chromophore N,N-diethylamino-(β)-nitorostyrene (DEANST). The resultant film containing 35% SiO_2, 18% DEANST and 47% of PVP showed that $\chi^{(2)} = 1.4 \times 10^{-7}$ e.s.u. Kim et al. [82] used silylated second-order chromophore to increase the stability of orientation formed by poling.

3.6.2 Third-Order Nonlinear Optical Films

From their microstructures, sol–gel coating films showing high third-order optical nonlinearities are classified into randomly crystalline oxides, semiconductor particle-dispersed oxides, metal particle dispersed oxides, inorganic-organic composites containing semiconductor metal particles and oxides or composites containing organic molecules. It is noted that the mechanism of nonlinearity is different from one material group to another. Since there is a chapter devoted to the sol–gel nonlinear optical materials in this monograph, only a limited number of examples of third-order nonlinear optical coating films are cited here.

α-Fe_2O_3 coating films [83] were prepared from $Fe(NO)_3 \cdot 9H_2O$–$CH_3OCH_2CH_2OH$(2-methoxyethanol)-$CH_3COCH_2COCH_3$ (acetylacetone)-H_2O solution on SiO_2 substrate. TiO_2 thin films [84] were prepared from $Ti(OC_3H_7^i)_4$-i-C_3H_7OH–H_2O–$HN(CH_2CH_2OH)_2$ solution (for rutile film) and $Ti(OC_3H_7^i)_4$-i-C_3H_7OH–H_2O–HNO_3 solution (for anatase film). Figure 11 shows the plot of nonlinear optical susceptibilities $\chi^{(3)}$ of the resultant films vs refractive index. It is seen that α-Fe_2O_3 film shows the highest $\chi^{(3)}$ value (5.8×10^{-11} e.s.u.) among metal oxides. The rutile film has a higher $\chi^{(3)}$ than the anatase film. This difference may be attributed to the higher refractive index (larger polarizability) for the former.

SiO_2 glasses containing fine particles of semiconductors such as CdS [85], Cd(S, Se) [86] and CuCl [87] were prepared by the sol–gel method. The semiconductor compounds were synthesized from the compounds containing pertinent components in the state of sol or gel. Some of the glasses were obtained in bulk form, but it is easy to apply the results to the preparation of coating films. Alumina coating films containing CdS particles were also obtained [88].

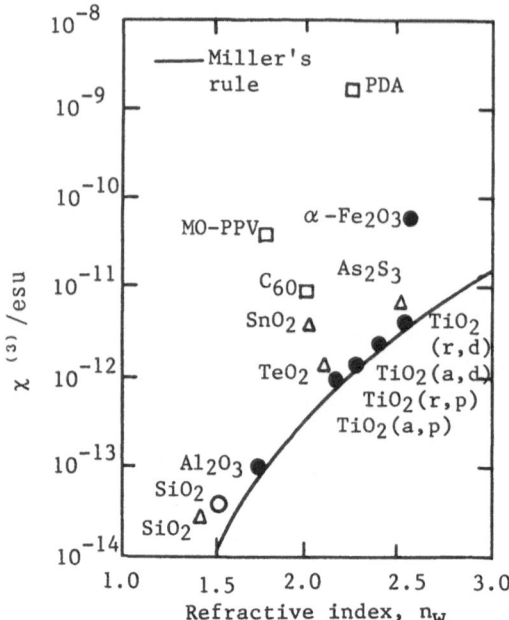

Fig. 11. Relationship between $\chi^{(3)}$ and refractive index, n_w, at 1900 nm for an α-Fe_2O_3 thin film and other nonlinear optical materials [84]. *Circles, triangles* and *squares* denote inorganic crystal, inorganic amorphous material, and organic material, respectively. *Closed circles* are the data by the present authors. The letters r, a, d, and p in parenthesis denote rutile, anatase, dense, and porous, respectively. PDA: poly-diacetylene, MO-PPV: poly(2,5-dimethoxy *p*-phenylene vinylene)

Yeatman et al. [89] discussed pore size control for the silica coating films containing CdS particles.

Paying attention to high third-order nonlinear susceptibilities $\chi^{(3)}$ of glasses embedded with gold particles [90], Matsuoka et al. [91] and Kozuka and Sakka [92] have made SiO_2 glasses with dispersion of about 1 vol.% of Au small crystals by the sol–gel method using $Si(OC_2H_5)_4$–C_2H_5OH–$HAuCl_4 \cdot 4H_2O$ solutions. The average size of Au particles could be changed from 40 to 200 A by controlling the HCl content of the starting solution, ripening period of gel, and heating temperature and time. Innocenzi et al. [13] found that the partial replacement of $Si(OC_2H_5)_4$ by $CH_3Si(OC_2H_5)_3$ in the starting solution produced smaller and more spherical gold particles with a more uniform size distribution. As shown in Fig. 12, the change of matrix from SiO_2 to TiO_2 ceramic with higher dielectric constant shifts the peak due to the surface plasmon enhanced absorption band to longer wavelengths. The Au/SiO_2 and Au/TiO_2 films have absorption peaks at 538 and 644 nm, respectively [93].

Similarly, it was confirmed that Pt and Pd particles, which in conventional silicate glasses show optical absorption in the ultra-violet region, can exhibit absorption in the visible region when embedded in TiO_2 crystalline film [93, 94].

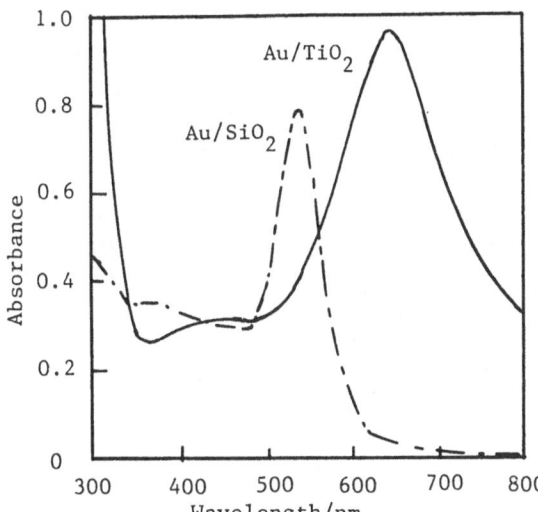

Fig. 12. Optical absorption spectra of the Au/SiO$_2$ and Au/TiO$_2$ composite films containing round-shaped Au particles of about 20 nm in size [93]

Reliable experimental data of $\chi^{(3)}$ on the above metal particle-dispersed coating films have not been obtained.

Reisfeld and Minti [95] showed that CdS nanoparticles embedded in inorganic-organic composite films show similar strong third harmonic generation (THG) signals like silica films with dispersed CdS nano-particles. The signals were observed at 0.355 µm when the film were irradiated at 1.06 µm.

There are many organic molecules with large $\chi^{(3)}$ values [96]. In order to increase optical property stability, organic molecules are incorporated into silica gel or inorganic-organic matrix [74]. Nakamura et al. [97] incorporated an organic dye 4'-dimethylamino-N-methyl-4-stilbazolium iodide [DMSI] into SiO$_2$ gel by the sol–gel method and measured $\chi^{(3)}$ of the resultant dye-doped gels. The material showed much higher $\chi^{(3)}$ values than the pure SiO$_2$ matrix. These materials can be easily formed into coating films.

3.7 Photochromic Films

Photochromic materials become colored upon exposure to UV or short wavelength visible light and the transparent or non-colored state comes back when the colored material is kept in the dark or is irradiated with longer wavelength visible light. Photochromic eye glasses based on silver halide particles are the most important application so far. In the area of information processing, photochromic materials may be used for memory, recording and switching. Generally, only the surface layer of the material is colored by UV light irradiation, because the UV light is strongly absorbed by the surface layer. This suggests that thin or thick coating films are very effective for photochromism.

There are two kinds of active photochromic species: one is silver halide particles embedded in glass and the other is spiropyranes.

Mennig et al. [98] prepared AgCl photochromic glass coating films of about 1.5 μm in thickness on microscopic slide glasses by sol–gel method. The gel film of the $Na_2O–B_2O_3–Al_2O_3–SiO_2$ system was formed on the glass substrate by dip-coating using the sol prepared from $Si(OC_2H_5)_4–CH_3Si(OC_2H_5)_3–Al(O-isoC_3H_7)_3–B(OCH_3)_3–NaOCH_3$ solution. AgCl colloids of about 40 nm in diameter were formed in the film by infiltration of Ag^+ ions, heating for formation of Ag colloids and exposing to HCl gas for the chlorination of Ag colloids. Darkening and fading of the photochromic coating films could be repeated in numerous cycles without decay.

It is noted that AgCl-doped silica glasses of 1–3 mm thickness prepared by the sol–gel method show darkening on exposure to UV light [99].

It was known that spiropyranes show different photochromic behavior in a silica gel and an inorganic-organic composite gel [100]. In silica gel, normal photochromic behavior, i.e., the change from colorless to colored state upon UV exposure, was observed, while in silica-polydimethylsiloxane composite gel, reverse photochromism, i.e., the change from colored to colorless state occurred. This indicates [100] that the property of the cage in which a spiropyrane molecule is trapped affects the state and behavior of the spiropyrane dye.

Levy et al. [101] prepared coating films aluminosilicate and ORMOCER (inorganic-organic) containing photochromic dyes, 1,3-dihydro-1,3,3-tri-methylspiro-[2H-indole-2,3′-[3H]-naphth-[2,1-b] [1,4]-oxazine) (SO) and 1′, 8a′-di-hydro-2′,3′-dimethoxycarbonyl-spiro [fluorene-9,1′-indolzine] (DHJ), comparing their photochromic behaviors. It was found that the ORMOCER coating film retains a reasonably high photochromic activity (coloring on UV exposure and fading in the dark), while aluminosilicate film does not show the photochromic behavior. It was also found that the photostability of photo-chromic dyes is considerably improved when they are incorporated in OR-MOCER coating films.

It is noted, however, that photochromic coating films containing organic dyes have to be further improved in stability in order for them to be employed as photochromic materials.

3.8 Electrochromic Films

In electrochromic devices, the electrochromic film is colored when an electric field is applied and the color fades on reversing the field. Therefore, electroch-romic devices can be applied to optical displays and windows controlling the light intensity in buildings and cars.

Figure 13 shows the cross-section of the all solid-state transmissive elec-trochromic cell prepared by Baudry et al. [102]. It is seen that this electroch-romic cell consists of multiple layers sandwiched between two glass sheets and has the configuration:

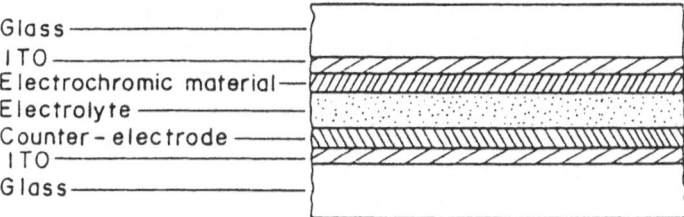

Fig. 13. Cross-section of a transmissive electrochromic plate. After Baudry et al. [102]

glass/transparent electron conductor (ITO)/
anode (electrochromic material)/ionic conductor (electrolyte)/
counterelectrode (ion storage layer)/
transparent electron conductor (ITO)/glass.

In a conventional electrochromic cell, sulfuric acid solution or lithium perchlorate solution is used as electrolyte which serves as ion storage layer as well. This indicates that an all solid-state electrochromic plate has been realized by replacing the liquid electrolyte by solid electrolyte and ion storage films. It is also seen that the sol–gel coating is very important for this replacement. Furthermore, it is easily understood that the application of the sol–gel coating to other layers that have been prepared by the vapor phase deposition method so far may be possible. Baudry et al. [102] studied the sol–gel preparation of TiO_2–CeO_2-films for transparent counter-electrode. It was found that this material shows a reversible lithium insertion.

The most important constituent of the electrochromic cell is the electrochromic film. Amorphous WO_3 film is an important candidate as an electrochromic film. In an electrochromic cell, both sides of a WO_3 film contacted with an electrolyte have electrodes. The electric field applied between the two plane electrodes causes the following coloring reaction accompanied by incorporation of electrons and ions into the film:

$$WO_3 + xe^- + xM^+ = M_xWO_3 \tag{5}$$

where M^+ represents H^+ or Li^+ and M_xWO_3 is a colored tungsten bronze. The color is blue for H_xWO_3. The use of H^+ ions gives a fast-responding element and the use of Li^+ gives a chemically inactive, and stable element. Livage [103] prepared WO_3 films by the sol–gel method using inorganic tungstates, showing that the resultant films color and fade quickly in an electrochromic cell using H_2SO_4 as electrolyte.

Unuma et al. [104] prepared colorless and transparent amorphous WO_3 thin films by the sol–gel method using tungsten hexaethoxide as starting substance. Oxygen plasma treatment decomposed and eliminated organic matter without change of amorphous nature of the film. The system consisting of an amorphous WO_3 film deposited by dip-coating on a Nesa film (electroconducting transparent SnO_2-Sb_2O_3 film), an electrolyte of propylene carbonate containing lithium perchlorate and a Pt counter electrode showed electrochromic properties i.e., blue coloring and bleaching.

In_2O_3-SnO_2 films (ITO) deposited on glass are used as electroconducting transparent films for applying an electric field. These films might be prepared by the sol–gel method. At present, however, the electric conductivity and stability of the sol–gel prepared ITO films are somewhat inferior to vapor-deposited ITO films, as described in Sect. 4.2. Therefore, the latter films are being used at present.

Macedo and Aegerter [105] prepared two kinds of all solid-state electrochromic cells. One of them has the configuration:

glass/ITO/WO_3/TiO_2/TiO_2-CeO_2/ITO/glass.

The three internal layers, i.e. the electrochromic layer of WO_3, the ion-conducting layer of TiO_2 and the counter electrode of TiO_2-CeO_2 are prepared by the sol–gel method. The counter electrode works as the ion storage layer for H^+ ions. The three intermediate layers are mixed electron-ion conductors. When the electric field is applied, the ions stored in the ion storage layer diffuse into the electrochromic layer, and as a result the optical transmittance of the electrochromic layer changes. The lifetime of the cell is still not enough and more studies have to be made for practical applications.

Judeinstein et al. [106] made all layers by the sol–gel method for the electrochromic cell of the configuration:

Conducting electrode (ITO) /WO_3 gel/electrolyte gel/
conducting electrode (ITO).

The electrolyte gel is a mixed organic-inorganic gel. It is said that the response time of this cell is 40 s under an applied voltage of ± 2.5 V and the cell can be cycled more than 4×10^5 times without failure.

Sol–gel synthesis for electrochromic oxide films other than WO_3 has been attempted for V_2O_5 [107], TiO_2 [107] and V_2O_5-TiO_2 [108].

3.9 Photovoltaic and Photocatalytic Films

Utilization of solar energy is necessary for meeting future energy demands. Conversion of solar energy into electrical energy based on photovoltaic effects and chemical energy based on photo electrochemical decomposition of water into hydrogen and oxygen are two important methods for utilizing solar energy. Sol–gel coating films used for these purposes will be discussed in this section.

3.9.1 Photovoltaic Coating Films

Grätzel and his collaborator [109, 110] prepared a new type of photovoltaic device of which overall light to electric energy conversion efficiency is 10%

under simulated solar radiation. This efficiency is comparable to that of the photovoltaic device using amorphous silicon semiconductors.

Figure 14 shows the schematic view of the structure of the photovoltaic cell module constructed by Grätzel [109, 110]. The cell consists of two glass plates which are coated with a thin, transparent electrically conducting tin oxide layer, a nanocrystalline titanium dioxide film deposited on one plate by the sol–gel method and a sensitizer dye layer of *cis*-dithiocyanato bis(2,27-bipyridyl-4, 4-dicarboxylate) ruthenium (II) on the titanium film. The titanium dioxide film acts as a light trap. The sensitizer absorbs visible light, injecting an electron in the excited state to the conduction band of the TiO_2 film [11]. The electrons injected in the conduction band travel across the membrane, reaching the external current circuit where the work is done. The electrons then return to the cell through a counter electrode. The redox electrolyte containing a redox couple of iodine and iodide placed between the counter electrode and the TiO_2 membrane allows for charge transport between the two electrodes. The electrons reduce iodine to iodide ions which diffuse from the counter electrode to the TiO_2 film, where they regenerate the sensitizer by electron transfer to the sensitizer cations, while simultaneously the iodide is oxidized back to iodine. This redox cycle works in converting light into electrical current.

In the TiO_2 film, the fine colloidal titanium dioxide particles of 10–20 nm diameter touch each other as a result of sintering at 500 °C. This film gives a very large surface area, attaining a high efficiency of conversion of light to electricity, reaching 10% in this device.

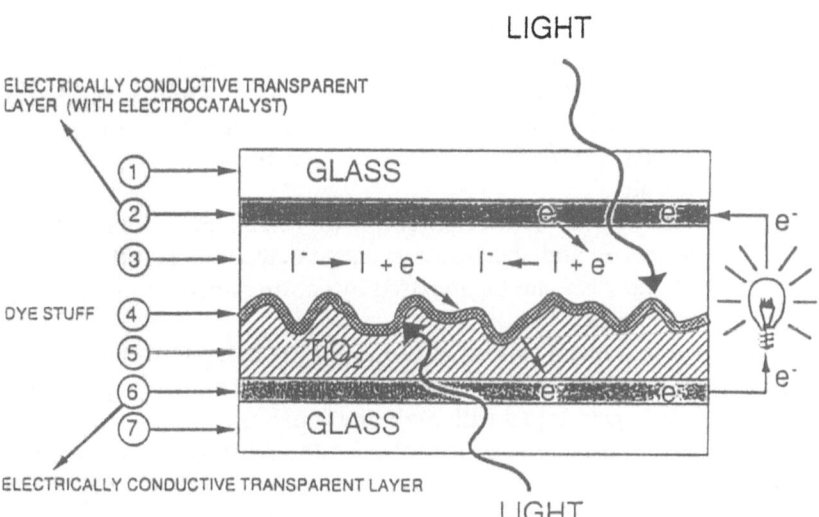

Fig. 14. Construction of a cell module with transparent glass electrode. *1.* and *7*: glass sheets, *2* and *6*: transparent conductive layer of fluorine doped tin oxide, *3*: electrolyte, *4*: dye layer, *5*: colloidal TiO membrane (about 10 μm thick). After Grätzel [110]

3.9.2 Coating Films and Photocatalytic Effect

Another method of utilizing solar energy is to convert light energy directly into chemical energy using a semiconductor electrode [111]. Yoko et al. [112, 113] applied the sol–gel derived TiO_2 films to a photoanode to decompose water into oxygen and hydrogen. Figure 15 shows the photocurrent-bias potential curves obtained with a TiO_2 film electrode of various thicknesses. The TiO_2 film is n-type semiconductor. It is found that the TiO_2 film electrodes heated at 500 °C for 20 min show the saturated photocurrent of 14 mAcm^{-2} which is comparable to or even better than that of single crystal TiO_2. Since the sol–gel derived oxide films are porous in nature and characterized by a large surface area at which the electrode reaction takes place under illumination [113], they are very suitable as photoelectrodes. The effects of the solvent of the starting solution for preparation of TiO_2 has been studied [114].

The same workers also reported the preparation and photoelectrochemical properties of the sol–gel derived $NiFe_2O_4$ [115] film electrode, which may become a p-type semiconductor, forming a photocathode. The future objectives in this field are to develop photoelectrochemically stable n- and p-type semiconductor film electrode materials with an optical band-gap at around 1.6 eV and high photoelectrochemical stability. Also, development of stable sensitizers is expected.

3.10 Optical Chemical Sensors

High resolution and sensitivity are the most important aspects of the sensors. Sol–gel coating films containing sensing materials satisfy these requirements. The group of Avnir [116, 117] proposed that the sol–gel method would be quite suitable to produce chemical sensor elements for the following reasons:

  (1) sol–gel prepared gels can be doped with non-leachable organic molecules;
  (2) the analytes can diffuse into the pore network of the gel, reacting with reagent molecules;
  (3) the gel matrix can be transparent down to 250 nm and makes quantitative spectrophotometric and spectrofluorimetric detections possible;
  (4) very small probes can be prepared by coating fibers or planar substrates or making small pieces.

It was shown that metal cations, protons (for pH indication), anions and organic molecules in gases and liquids can be detected by color tests [116]. It was also shown that SiO_2 gels doped with pyranine, a pH sensitive fluorophore, placed in a micropipette can be used as a probe for the fluorescent pH meter [117].

Kubecková et al. [118] prepared SiO_2 (0.26 μm thick) and TiO_2 (0.09 μm thick) coating films doped with phenolphthalein on flat and fiber silica substrates. These films were sensitive to pH changes when the optical attenuation was measured at 570 nm for SiO_2 and at 530 nm for TiO_2. It was found that TiO_2 films are better in their stability and reversability of the color change and that

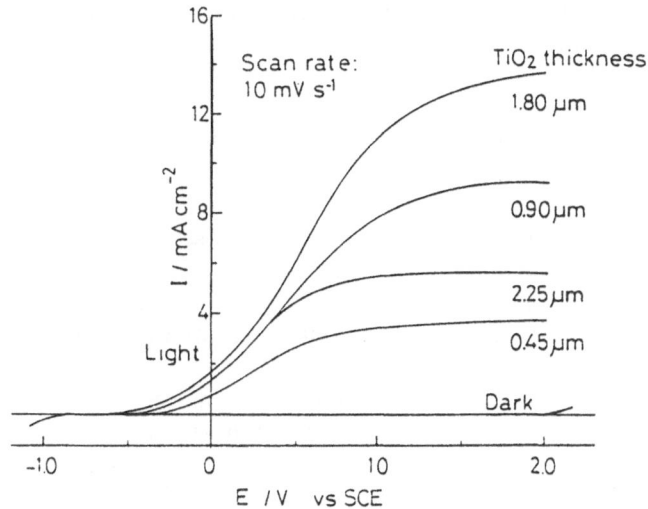

Fig. 15. Photocurrent-bias potential curves for TiO$_2$ film electrodes with different film thicknesses. After Yoko et al. [112]

the thicker films are desirable for decreasing the effect of the refractive index of solution on the propagation of evanescent waves.

MacCraith et al. [119] fabricated fiber-optic chemical sensors based on side-coating an unclad portion of an optical fiber with a sol–gel prepared porous gel film containing analyte-sensitive dyes. This technique employs evanescent wave interactions, i.e., evanescent wave absorption and evanescent wave excitation of fluorescence and gives rise to control of sensor performance.

The evanescent wave is a fast decaying wave, which penetrates a short distance (less than the wavelength of light) into a low index medium when total internal reflection occurs. The evanescent wave of guided light interacts with the analyte-sensitive dye in the coating film prepared on the unclad part of the fiber. In the example of a pH sensor using evanescent wave excitation [119] of fluorescence from fluorescein, the entrapped fluorescein is excited by the evanescent wave of argon-ion laser radiation, and a fraction of the fluorescence is captured by the fiber and detected at $\lambda = 530$ nm. The plot of fluorescence intensity vs pH was found to be fully reversible and the response time is less than 5 s.

As an example of the sensor employing evanescent wave absorption, an ammonia sensor has been constructed [119].

Kiernan et al. [120] have proposed that analysis of optical decay time is more useful than fluorescence intensity, in order to monitor oxygen by ruthenium complexes, because decay times are not affected by bleaching of the chromophore. For this purpose, sol–gel silica films of about 300 nm thickness were produced on glass slides by dip-coating and impregnated with Ru

(Ph₂phen)$_3^{+}$. The optical decay time was measured with the films pumped at 532 nm.

3.11 Magnetic and Magneto-Optic Films

Magnetic and magneto-optic thin films are important for magnetic recording devices and microwave devices. So far several attempts have been made to form thin films of magnetic and related oxides Fe_2O_{3-x} [121], α-Fe_2O_3 [83], $NiFe_2O_4$ [122, 123], and Fe_3O_4 [124, 125] by the sol–gel method. Some of them were prepared for use as nonlinear optical material or electrodes for photo-catalysis.

The Fe_3O_4 film prepared from iron acetylacetonate on silica glass substrate showed a magnetization curve with hysteresis loop which is characteristic of ferromagnetism [124], as shown in Fig. 16. The coercive force is estimated to be 230 Oe which is larger than that of bulk Fe_3O_4. This is ascribed to the smaller Fe_3O_4 crystallite size of about 50 nm in the sol–gel derived film. The Fe_3O_4 film was also made from iron nitrate solution. The above result indicates that the sol–gel coating films are promising materials for magnetic films.

The sol–gel derived oxide films are usually transparent due to small crystal-lite size, and so are promising materials for magneto-optical devices. It is known that iron garnet type compounds may be suitable as magneto-optical recording materials for their excellent magnetic properties. The large Faraday rota-tion serves in making sensors out of these oxides. The thin coating films

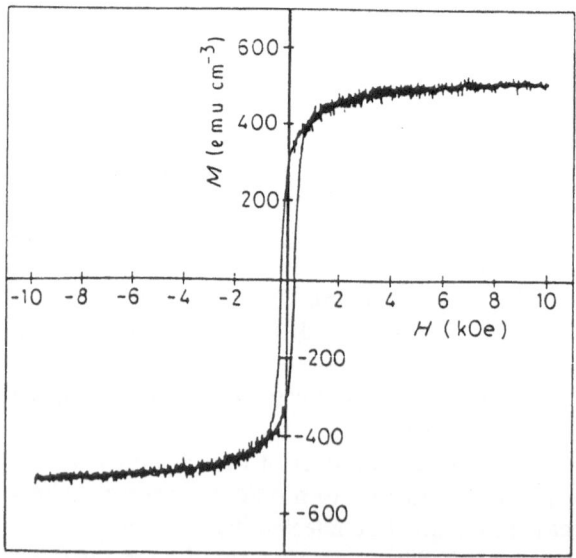

Fig. 16. Magnetization curve of sol–gel-derived Fe_2O_3 film. Tanaka et al. [124]

of $Y_{3-x}Bi_xFe_5O_{12}$ (bismuth-substituted yttrium iron garnet) [III], $Y_{3-x}Bi_xFe_{5-x}Al_yO_{12}$ (bismuth and aluminium-substituted yttrium iron garnet) [128] and Cu doped Bi, Al: DyIG (bismuth and aluminium-substituted dysprosium iron garnet doped with copper) and Cu-doped Bi, Ga: DyIG (bismuth and gallium-substituted dysprosium iron garnet doped with copper) were prepared by the sol–gel method. Preparation and properties of those coating films are shown in Table 4.

$Y_{3-x}Bi_xFe_5O_{12}$ and $Y_{3-x}Bi_xFe_{5-x}Al_yO_{12}$ coating films [126, 127] were prepared from starting solutions containing corresponding metal nitrates as starting materials and ethylene glycol or ethyl acetyl acetate and water as solvents. Silica glass plate and gadolinium gallium garnet single crystals were used as substrate. After dip-coating, the films were heated to various temperatures between 500 and 800 °C for crystal formation, and magnetic properties were measured. These films showed very large Faraday rotation; the maximum values were $50 \sim 57°$ μm^{-1} at the wavelength of 520–550 nm for the composition $x = 1.0$ in $Y_2BiFe-5O_{12}$ and $Y_2BiFe_{4.6}Al_{0.4}O_{12}$.

Cu-doped Bi, As: DyIG and Cu-doped Bi, Ga: DyIG thin films were prepared on glass substrate from solutions consisting of metal nitrates and alkoxides as starting materials and ethanol and water as solvents [128]. The films prepared by spin coating were heated to 300–400 °C (amorphous film formation) and to 600–700 °C (crystalline film formation). The films showed Faraday rotation of $10–12°$ μm^{-1} at maximum. The coercive force H_c increased with Cu content from 1 kOe to 7 kOe, respectively, for Al- and Ba-containing systems.

4 Coating Films with Electronic Functions

Dielectric and conducting coating films are described in this section.

4.1 Ferroelectric and Related Dielectric Coating Films

Ferroelectric and related dielectric coating films play an important role in electronics and optoelectronics. Besides the application as electro-optic and nonlinear optical film in photonics [129], ferro-electric coating films are applied as capacitor, non-volatile IC memories in microelectronics, piezoelectric transducers, pyroelectric temperature sensors, surface acoustic wave (SAW) elements and so on [130, 131] in electronics. It will be easily understood that the sol–gel method gives thin coating films much more easily than conventional sintering of powder compacts. Accordingly, the sol–gel method has been applied to produce coating films of many kinds of existing ferroelectric materials. Among these are $BaTiO_3$ [132–135], $PbTiO_3$ [130, 136, 137], $PbZr_{1-x}Ti_xO_3$(PZT) [137–143],

Table 4. Preparation and properties of magnetic coating films of iron garnet type compounds [127, 128]

No. Composition	Starting material	Film thick. (μm)	Magneti- zation	H_c (kOe)	Faraday rot. (deg/μm)	Curie point ($^\circ$C)
1. $Y_{3-x}Bi_xFe_5O_{12}$	Nitrates	0.3	75 emu/g (sat. mag.)		50 (520 $^\circ$C, $x = 1$)	293 (YIG)
2. $Y_{3-x}Bi_xFe_{5-y}Al_yO_{12}$	Nitrates	0.44	100–130 emu/cc (mag.)		57 ($x = 1$. $y = 0.4$)	
3. Bi, Al : DyIG (Cu-doped)	$Al(NO_3)_3$ $Dy(NO_3)_3$ $Fe(OC_3H_7)_3$ $Bi(OC_2H_5)_3$			1–7 (kOe)	5–12	
4. Bi, Ga : DyIG (Cu-doped)	$Ga(NO_3)_3$ $Dy(NO_3)_3$ $Fe(OC_3H_7)_3$ $Bi(OC_2H_5)_3$			2–13	8–10	

Pb–La–Zr–Ti–O (PLZT) [130], $Pb(Mg_{1/3} \cdot Nb_{2/3})O_3$ (MN) [144, 145], $Pb(Fe_{0.5}Nb_{0.5})O_3$ (PFN) [146, 147], $LiNbO_3$ [148], $KNbO_3$ [149] and $Sr_{1-x}Ba_xNb_2O_6$ (SBV) [150].

The following four items are important for promoting the performance of ferroelectric coating films;

(1) precipitation of perovskite single phase;
(2) preferred orientation of crystals in the film;
(3) deposition of multilayers including ferroelectric and insulating films;
(4) amorphous ferroelectric films.

4.1.1 Perovskite and Pyrochlore Phase

In the preparation of ferroelectric materials, which consist of complex oxides, we often encounter the formation of pyrochlore phase. The pyrochlore phase interferes with the ferroelectric performance of perovskite phase. This is also the case for sol–gel coating films. Precipitation of single phase perovskite compounds can be made by forming precursor compounds in the solution corresponding to the final oxides [151], controlling the composition of the starting solution (contents of water and catalyst) and adjusting conditions of heat treatment [146].

4.1.2 Preferred Crystal Orientation

It is known that the coating films in which ferroelectric crystals are preferentially oriented may show better ferroelectric properties (for example, pyroelectricity) or novel properties (for example, second-order optical nonlinearity). Hirano and Kato [151] prepared homogeneous and stoichiometric $LiNbO_3$ films with preferred orientation by crystallizing the sol–gel coating film at 400 °C on α-Al_2O_3 single crystal (sapphire) substrate by using double alkoxide solutions as coating solution. Recently, many authors [152–155] reported the formation of lead zirconate titanate (PZT) coating films with preferentially oriented crystals. Also, $Pb(Mg_{1/3}Nb_{2/3})O_3$ (PMN) [144] and $Pb(Fe_{1/2}Nb_{1/2})O_3$ (PFN) [144] crystalline films with preferred crystal orientation have been prepared on platinum and single crystalline MgO substrates by the sol–gel method.

Preferred orientation has been achieved in the sol–gel coating films of $Li_2B_4O_7$ crystalline phase applied on Si single crystal substrates [131]. Figure 17 shows the models and X-ray diffraction patterns of three kinds of $Li_2B_4O_7$ coating films of different microstructures. It should be noted that the preferred orientation can be produced by controlling the concentrations of water and acid catalyst and selecting the crystallographic plane of the single crystal substrate.

4.1.3 Multilayer Coating

Multilayer films involving ferroelectric films show improved dielectric properties which may lead to new applications. Sol-gel preparation of coating films is very useful for preparing a multilayer film consisting of nanometer thick films. Wu et al. [156] constructed an apparatus for preparing multilayered thin films. With this apparatus they made a multilayered film consisting of either a combination of many different ferroelectric oxides or a repetition of two different oxide layers.

Ohya et al. [157] showed that the multilayered film consisting of $PbTiO_3$ and $PbZrO_3$ layers exhibits a very clear interface between them, while the interface between $PbTiO_3$ and $SrTiO_3$ layers is not clear due to the diffusion of lead component from the former layer into the latter, leading to the formation of a solid solution. It was shown that the temperature dependence of the dielectric constant is flat in multilayered films, but the dielectric constant of the multilayered film could not be explained by the series connection of the component films. Multilayered films consisting of $Pb(Zr, Ti)O_3$ films and very thin $PbTiO_3$ or $(Ba_{0.5}Sr_{0.5})TiO_3$ buffer layers have been prepared by the sol–gel method [158]. It was found that the insertion of 2–4 nm thick buffer layers yields good ferroelectric properties, such as increased remanent polarization, small coercive force and good leakage current characteristics.

4.1.4 Amorphous Ferroelectrics

Using the sol–gel method, Xu et al. [159] prepared amorphous thin coating films of compositions $LiNbO_3$ and $Pb(ZrTi)O_3$, which show ferroelectric-like behavior, i.e., hysteresis in polarization-electric field measurements, pyroelec-

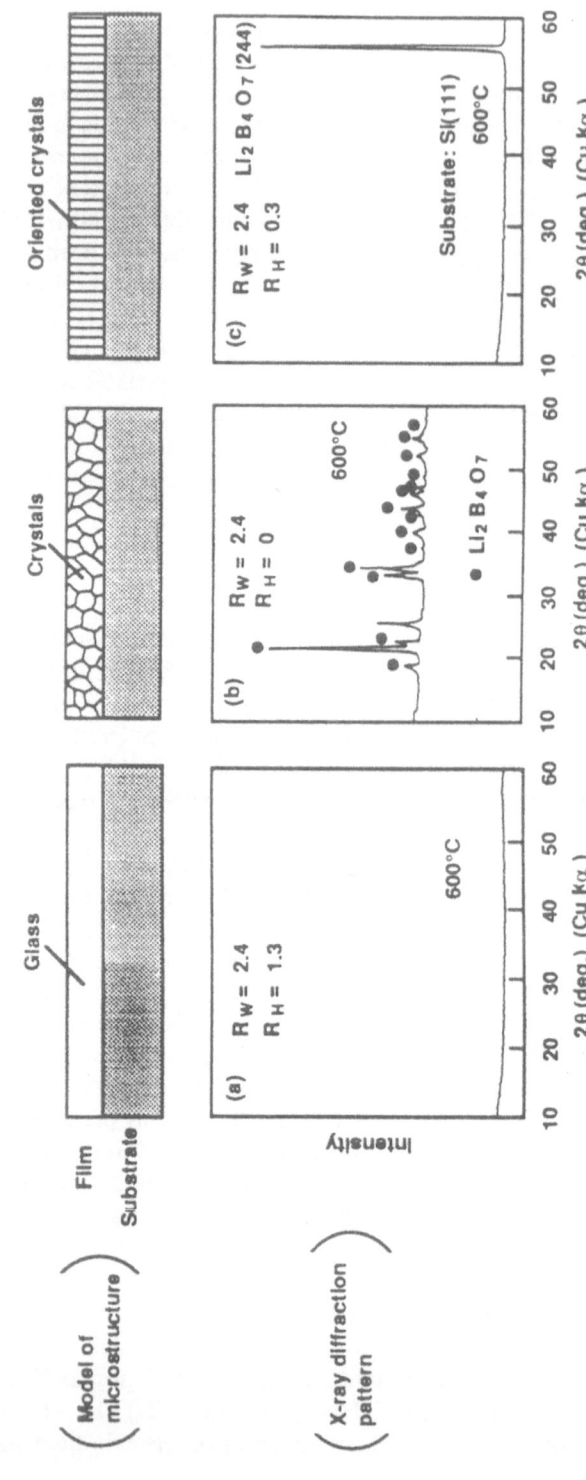

Fig. 17. Microstructure models and X-ray diffraction patterns of coating films of the composition $Li_2B_4O_7$ prepared from $LiOCH_3$–$B(OC_4H_9)_3$–H_2O–HCl solution. $R_W = ([H_2O]/[LiOCH_3] + [B(OC_4H_9)_3])$; $R_H = [HCl]/([LiOCH_3] + [B(OC_4H_9)_3])$

tricity after being poled and piezoelectric resonance. It is interesting to note that these amorphous thin films are obtained by heating at much lower temperatures than those for crystallization. Amorphous $LiNbO_3$ ferroelectric film is obtained by heating at temperatures lower than 350 °C, while its crystallization counterpart is obtained by heating at temperatures higher than 450 °C. The heating temperature for amorphous $Pb(Zr_{0.52}Ti_{0.48})O_3$ film is lower than 350 °C, which is much lower than 600 °C required for obtaining the crystalline counterpart.

The marked difference between amorphous and crystalline ferro-electric films is caused by the absence of grain boundaries in the former films. For example, the amorphous film lacks the dielectric loss at low frequencies which is caused by grain boundaries in crystalline ferroelectric materials. Xu et al. [159] suggest that the absence of grain boundaries indicates that optical loss due to the light scattering at grain boundaries is absent and this makes the amorphous ferroelectric films useful as low-loss optical waveguides.

4.2 Electronic Conductor Films

The most important electron-conducting coating films are those of $SnO_2–Sb_2O_3$ (SnO_2 : Sb, antimony-doped tin oxide) and $In_2O_3–SnO_2$ (In_2O_3 : Sn, tin-doped indium oxide). In this section the main emphasis is laid on these two oxides prepared by the sol–gel method. SnO_2 : Sb films, also called Nesa films and In_2O_3 : Sn films, also called ITO films are transparent and highly electron-conductive, being employed as transparent electrodes for liquid crystal displays, electrochromic devices, solar cells, etc. and as heating elements for windows of cars and aircraft.

Commercially, SnO : Sb films are prepared by chemical vapor deposition, while ITO films are prepared by vacuum deposition or sputtering. ITO films are employed for liquid crystal display and electrochromic devices because of lower resistivity compared with SnO_2 : Sb films. SnO_2 : Sb films are characterized by less expensive raw materials, and higher heat-resistance and are used for devices which need high temperature treatment in manufacturing [160].

4.2.1 SnO_2 : Sb Films

Preparation and electrical properties of SnO_2 : Sb coating films prepared by the sol–gel method are summarized in Table 5 [161–164]. In the column of electrical resistance, lower resistivity values are cited from a series of values obtained in a particular paper because lower resistivity is desirable.

In the pioneering work of Prof. Matsushita's group (No. 1 in Table 5), tin caprylate $Sn(C_7H_{15}COO)_2$ and di-iso-amyloxyethoxyantimony $C_2H_5O \cdot (C_5H_{11}O)_2$ Sb were the source for tin and antimony, respectively. The butanol solutions of these compounds were dripped onto glass substrates, dried and heated at 200–600 °C for 20 min. The resultant films showed the metallic

Table 5. SnO$_2$-based electronic conductor coating films prepared by the sol–gel method

No.	Composition of film	Starting solution	Coating	
			Method	Substrate
1.	SnO$_2$:Sb	Butanol solution of (C$_7$H$_{15}$COO)$_2$ Sn and C$_2$H$_5$O·(C$_5$H$_{11}$O)$_2$Sb	Dripping	Slide glass
2.	SnO$_2$:Sb	Butanol solution of Sn(OC$_4$H$_9$)$_4$ and Sb(OC$_4$H$_9$)$_3$	Dip-coat	Soda-lime glass, borosilicate glass, silica glass
3.	SnO$_2$:Sb	NH$_4$OH-added solution of SnCl$_4$ ·nH$_2$O, SbCl$_2$ and water	Dip-coat	Silica glass
4.	SnO$_2$:Ti	Alcoholic solution of SnCl$_z$ and Ti (OC$_4$H$_9$)$_4$	Dip-coat	Soda-lime glass

No.	Thickness of film	Electrical properties		Remark	Reference (year)
		Resistance of film	Resistivity		
1.	$\sim 1\ \mu m$	$5 \times 10^3 \sim$ $5 \times 10^4\ \Omega/cm^2$	–	Minimum resistivity at 6–12% Sb$_2$O$_3$	[161] (1972)
2.	247 nm	300 Ω/cm^2	$2.5 \times 10^{-3}\Omega$ cm	Possible orientation of crystals	[162] (1986)
3.	140 nm		$3 \times 10^{-3}\ \Omega$ cm	Minimum resistivity	[163] (1990)
4.			$2 \times 10^{-2}\ \Omega$ cm		[164] (1990)

conduction and a resistance of $5 \times 10^3 \sim 5 \times 10^4\ \Omega\ cm^{-2}$. Later, this research group [165] modified its method by employing Sn(C$_7$H$_{14}$COO)$_2$ and Sb(OC$_4$H$_9$)$_3$ as sources for Sn and Sb, respectively. Heat treatment of the resultant gel films produced coating films of low specific resistivity at $2.1 \times 10^{-2}\ \Omega$ cm. This resistivity is still one or two orders of magnitude higher than that ($10^{-4} \sim 10^{-3}\ \Omega$ cm) of SnO$_2$:Sb films obtained by spraying or CVD methods. This study indicates, however, that the sol–gel coating is promising for preparing electronic conductor films.

Employing Sn(OC$_4$H$_9$)$_3$ and Sb(OC$_4$H$_9$)$_3$, Gonzalez-Oliver and Kato [162] obtained a low specific resistivity of $2.5 \times 10^{-3}\ \Omega$ cm (No. 2, Table 5). It is interesting to see that the coating film obtained by spraying the same starting solution on the high temperature substrate shows a specific resistivity of $7.5 \times 10^{-3}\ \Omega$ cm.

Chlorides of tin and antimony were employed in the method No. 3 in Table 5 [163]. In this method, a transparent water solution containing SnCl·nH$_2$O ($n = 4$–5) and SbCl$_5$ is first obtained. Adjusting pH of the solution to 7 by addition of NH$_4$OH hydrolyzes the metal chlorides, precipitating metal hydrates, which are rinsed with water to remove chloride ions. The coating solution consists of water, 5% metal hydrates and 0.3 wt% polyvinyl alcohol (bonding agent). Coating gel films on quartz glass obtained by dip coating were heated to 550–800 °C, resulting in transparent crackless SnO_2:Sb films of 90 –900 nm thickness. Only a crystalline phase consisting of crystallites of 30 nm diameter is found. The specific resistivity of the film as a function of Sb content showed minimum at the Sb content of 3–7 wt % and the value was $3 \times 10^{-3} \Omega$ cm for the coating film heated at 800 °C. It can be said that this specific resistivity is as low as that of SnO_2:Sb films prepared by spraying or CVD.

Maddalena et al. [164] examined the effect of addition of TiO_2 and ZrO_2 on the conductivity of SnO_2 coating films on soda-lime glass sheet prepared from alcoholic solutions of $SnCl_2 \cdot 2H_2O$. For Ti and Zr doping, $Ti(OC_4H_9)_4$ and $Zr(OC_3H_7)_4$ were added to the starting solution. Relatively low specific resistivity of $2 \times 10^{-2} \Omega$ cm was obtained with Ti doping. This indicates that Ti doping is not so effective as Sb doping in lowering the specific resistivity of SnO_2 film.

Recent studies [166, 167] are directed toward the mechanism of formation of SnO_2:Sb compound in the sol–gel coating process.

4.2.2 In_2O_3–Sn Films (ITO Films)

Preparation and properties of ITO films are summarized in Table 6. In the coating film of No. 1 in Table 6, indium octanoate $(C_7H_{17}COO)_3$ In and tin octanoate $(C_7H_{17}COO)$ Sn are employed as sources of In and Sn, respectively. The coating solution consisting of 10% metal octanoate mixture, 10% linoleic acid and 80% benzene was dripped on soda-lime glass substrate to form a film. After drying and heating to 400–600 °C, the film 200–500 nm thick showed the resistance of 350–850 Ω/cm. The resistivity was lowest at $1.7 \times 10^{-2} \Omega$ cm when 10% SnO_2 was added.

For No. 2 in Table 6, $In(NO_3)_3 \cdot 3H_2O$ and metallic tin were dissolved in acetylacetone and a mixture of HNO_3 and acetylacetone, respectively [169]. The mixture of these solutions was employed to coat soda-lime glass substrate by dip coating. Heating of the film to 550 °C produced 100–200 nm thick ITO films of low specific resistivity at $1.3 \times 10^{-3} \Omega$ cm. Ogiwara and Kinugawa [169] coated the substrate with 100–200 nm thick SiO_2 film before applying ITO film in order to avoid deterioration of the ITO film due to diffusion of Na^+ ions from soda-lime glass substrate into the ITO film.

Furusaki et al. [170] prepared coating film No. 3 in Table 6 by modifying No. 1. The coating solution was prepared by adding benzene to a mixture of indium 2-ethylhexanoate and tin 2-ethylhexanoate and by dissolving the resultant mixture in a mixed solvent of butanol and butylacetate. After heating to 550 °C, the In_2O_3:Sn coating film showed a low resistivity of $6 - 8 \times 10^{-3} \Omega$ cm. The subsequent annealing in vacuum of 10^{-3} Torr at 550 °C lowered the resistivity to $3 - 5 \times 10^{-4} \Omega$ cm.

Table 6. In_2O_3–based electronic conductor coating films prepared by the sol–gel method

No. Composition Starting solution of film		Coating	
		Method	Substrate
1. In_2O_3:Sn	Linoleic acid-benzene solution of $(C_7H_{15}COO)_3In$ and $(C_7H_{15}COO)_2Sn$	Dripping	Soda-lime glass
2. In_2O_3:Sn	Indium-nitrate–acetylacetonato[a] and tin solution[b]	Dip-coat	Soda-lime-glass (Precoated with 100–200 nm SiO_2 film)
3. In_2O_3:Sn	Benezene solution of indium–2-ethyl-hexanoate and tin–2-ethylhexanoate	Dip–coat	Soda-lime glass, Silica glass
4. In_2O_3:Sn	Water solution of indium sulfate and tin sulfate	Dip–coat	Silica glass

[a] Prepared by dissolving $In(NO_3)_3 \cdot 3H_2O$ in acetylacetone
[b] Metallic tin is dissolved in a mixture of acetylacetone and nitric acid

No.	Thickness of film	Electrical property		Remark	Reference (year)
		Resistance of film	Resistivity		
1.	200–500 nm	350–850 Ω/cm^2	$1.7 \times 10^{-2} \, \Omega\,cm$	Minimum specific resistivity at10% Sn.	[168] (1978)
2.	100–200 nm		$5.3 \times 10^{-3} \, \Omega\,cm$	Heated at 550 °C	[169] (1982)
3.			$6–8 \times 10^{-3} \, \Omega\,cm$	Resistivity decreases on heating in vacuo	[170] (1986)
4.	120–140 nm		$2–4 \times 10^{-3} \, \Omega\,cm$	Resistivity decreases on heating in vacuo	[171] (1991)
			(After vacuum heating) $6–8 \times 10^{-4}$		

Coating film No. 4 in Table 6 is prepared from inorganic compounds $In_2(SO_4)_3 \cdot nH_2O$ and $SnSO_4$. Gel coating films were converted to transparent ITO films of 120–140 nm thickness. The films showed a resistivity of $2 - 4 \times 10^{-3} \, \Omega\,cm$ for the film containing 14% Sn. Annealing in vacuum at 300 °C further lowered the resistivity to $6 - 8 \times 10^{-4} \, \Omega\,cm$.

4.2.3 Other Kinds of Electronic Conductor Coatings

Cd_2SnO_4 shows electronic conduction like SnO_2- and In_2O_3-based oxides. Cd_2SnO_4 coating films prepared by the sol–gel method using solutions containing $Cd(OAc)_2$ and $Sn(n-OC_4H_9)_4$ showed the film resistance of $100–1000 \, \Omega\,cm^{-2}$ [174].

It is known that transition metal oxides such as TiO_2, V_2O_5 and WO_3 are electronic conductors. They also show ionic conduction when protons or alkali metal ions are incorporated in their structure. It is easily seen that their conductivities vary greatly with preparation conditions. This is especially the case when they are prepared by the sol–gel method at low temperatures [175–179]. Due to the dual conduction mechanism (electronic and ionic), they are useful as coating films for electrochromic devices.

V_2O_5 coating films prepared by Livage [175, 178] were amorphous and highly conductive (the conductivity value is similar to that of V_2O_5 crystal), and proved to be suitable as antistatic films for plastics and photographic films.

Recently, molecular design of organic-inorganic materials with electronic conduction has been studied for polydimethylsilane units linked with vanadium oxo-species [179], although the conductivity is pretty low. This type of hybrid material may easily form coating films, and further studies are expected.

4.3 Ionic Conductor Films

Solids with high ionic conductivities (called fast ion conductors or superionic conductors) can be applied as solid electrolytes for solid state batteries and electrochromic display devices and as materials for gas sensors and capacitors. Among solid materials with room temperature conductivity close to $10^{-2} \Omega^{-1} cm^{-1}$, there are α-AgI and β-alumina ($NaO \cdot 11Al_2O_3$) crystals, NASICON (crystalline solids of compositions $Na_{1+x}Zr_2Si_xP_{3-x}O_{12}$ or Na_{1+x} $Zr_{2-x/3}Si_xP_{3-x}O_{12-2/3}$ ($1.5 < x < 2.2$)), Li_3N and $AgI-Ag_2O-P_2O_5$ glasses.

Some of the highly ion-conducting coating films prepared by sol–gel methods are introduced in this section. Electronic and ionic mixed conductors such as $V_2O_5-H_2O$ gels are not discussed here.

β-alumina coating films were prepared from alkoxides of sodium and aluminium by dip coating [180]. Fine grained β-alumina crystals were precipitated at 1300 °C in the film of about 1 μm thickness. It has been said that the low heating temperature is useful in avoiding vaporization of sodium and abnormal grain growth.

Li^+ ion is one of the ion species which easily diffuse in solids. Tatsumisago and Minami [181] prepared Li^+-conducting Li_2O-SiO_2 amorphous films by the sol–gel method using alcoholic solutions of Li and $Si(OC_2H_5)$ as coating solution. The gel film showed relatively high conductivity of $10^{-4} - 10^{-3} \Omega^{-1} cm^{-1}$ at 200 °C after being heated at 300–500 °C. The conductivity is caused by Li^+ ions. The conductivity is suppressed by the formation of Li_2CO_3, however, when too high Li_2O content is aimed at.

Li^+ ion conducting lithium aluminosilicate films were prepared in the lithium aluminosilicate system. Thick films of eucryptite composition (LiAl-SiO_4) were made by screen printing using fine powder of $LiAlSiO_4$ composition prepared from $Al(OC_4H_9)_3$, $Si(OC_2H_5)_4$ and $LiNO_3$ [182]. The ionic conductivity was $\sim 10^{-8} \Omega^{-1} cm^{-1}$ at room temperature and $\sim 4 \times 10^{-5} \Omega^{-1} cm^{-1}$ at 300 °C. Ogasawara and Klein [183] prepared $Li_2O-Al_2O_3-SiO_2$ gel films from solutions consisting of $Si(OC_2H_5)_4$, methanol, water, aluminum nitrate, HCl and $LiNO_3$ to obtain electrolytes in lithium batteries. The good gels showed the ionic conductivity of $10^{-7} \Omega^{-1} cm^{-1}$ at room temperature.

$(LiCl)_2$–Al_2O_3–SiO_2 thin films prepared by the sol–gel method with spin coating show the ionic conductivity of $2.5 \times 10^{-4} \, \Omega^{-1} \, cm^{-1}$ at 300 °C [184] after heat treated at 300 °C for 30 min. The conductivity data of the heat-treated thin film was represented by an Arrhenius plot with the activation energy of 0.75 eV.

Besides those mentioned above, sol–gel preparation was studied for NASI-CON-like materials [185, 186], LISICON-like materials [185], new high ionic conductor materials of the Na_2O–Re_2O_3–P_2O_5–SiO_2 system, where Re is a rare earth element [187], and Li^+ ion conducting ORMOSILS [188]. The results of those studies may be easily applied to the preparation of coating films.

4.4 High T_c Superconducting Films

Immediately after the discovery of high temperature superconducting oxides in 1986 by Bednorz and Müller [189], application of the sol–gel method to preparation of superconducting films and fibers was started. Both these forms aim at constructing wires or electric current pathways. Considering that high temperature superconductors are oxides and thus brittle in nature, it is easily seen that the coatings on mechanically strong substrates of fiber or ribbon shape may work well as wires.

Great efforts have been made in preparing yttrium cuprate and bismuth cuprate type superconducting coating films by the sol–gel method. The former type of superconductors are also called "Y–system", "Y–Ba–Cu–O system", "$YBa_2Cu_3O_{7-x}$", etc. The latter are called "Bi-system", "Bi–Sr–Ca–Cu–O", "Bi–(Pb)–Sr–Ca–Cu–O", etc. Tables 7 and 8 show some of the examples of such superconducting coating films prepared by the sol–gel method, together with their superconducting properties.

Principally, the sol–gel method has the advantage over conventional solid state sintering of powder compacts in producing materials of high homogeneity. Low processing temperature is another significant advantage. Both these advantages have to be realized more fully and strictly in superconducting oxides. Coatings are accompanied by other problems due to the presence of the substrate. Including the above aspects and demands unique to superconductors, the following problems emerge in preparing oxide superconducting coating films by the sol–gel method.

(1) The film must be sintered to maximum density without being heated to high temperatures so that the aimed compounds may not be decomposed.

(2) The film must be free from undesired impurities. Especially, carbon particles which originate in organic residues from starting compounds have to be minimized.

(3) The film must consist of the single superconducting oxide phase. Small amounts of non-superconducting phase will result in the deterioration of superconducting properties. However, incorporation of minor amounts of pinning species may be desirable.

Table 7. Sol–gel coating films of high temperature superconductors of Y–Ba–Cu–O system

No.	Starting materials	Solvent	Substrate
[$YBa_2Cu_3O_{7-x}$ superconductor]			
1.	Stearate (Y), naphthenate (Ba, Cu)	n-Butanol	YSZ
2.	2–Ethylhexanoate (Y, Ba, Cu)	Toluene	YSZ
3.	2–Ethylhexanoate (Y, Ba, Cu)	Chloroform	MgO
4.	Neodecanoate, octanoate	Xylene	Si
5.	Acetate (Y, Ba, Cu)	Methoxyethanol, triethanolamine	YSZ
6.	n–Butoxide (Y), methoxide (Ba, Cu)	Xylene, triethanolamine	YSZ
7.	Methoxyethoxide (Y, Ba), ethoxide (Cu)	Methoxyethanol, toluene	Si single (100) plane
8.	n–Butoxide (Y, Ba, Cu)	n–Butanol	YSZ
9.	Ethoxide (Y, Ba, Cu)	Organic solvent	YSZ
10.	Acetylacetonate (Y),	Ethoxyethanol	
11.	Propoxide (Y), ethoxide (Ba, Cu)	Diaminoalcohol	YSZ
12.	Nitrate (Y, Ba, Cu)	Water, ethanol	YSZ
13.	Nitrate (Y, Ba, Cu)	Toluene	YSZ
[$YBa_2Cu_4O_8$ superconductor]			
14.	iso–Propoxide (Y), Methoxyethoxide (Ba), Methoxide (Cu)	2–Methoxyethanol, ethylacetoacetate	MgO Ag

No.	Film thick. (μm)	Heating temp. (°C)	T_c (K)		Orientation	Reference Year
			Onset	End (zero)		
[$YBa_2Cu_3O_{7-x}$ superconductor]						
1.	1.3	800	90	60		[190] 1987
2.	10	900	100	82		[191] 1987
3.	1.4	990	89	77	(001)	[192] 1988
4.	0.1–1	800	98	80	(110)	[193] 1988
5.		800–850		84	C axis-oriented	[194] 1992
6.	1.5–7	800	98	56		[195] 1988
7.	0.15–1	700	(40)	–		[196] 1988
8.	5–10	920		80	(001)	[197] 1988
9.	5–10	950	96	90		[198] 1988
10.				75		[199] 1990
11.	10	850–950		54		[200] 1992
12.		930–950	95	89		[201] 1988
13.		900	93	63		[202] 1992
[$YBa_2Cu_4O_8$ superconductor]						
14.		800 (on MgO)	75	45		[203] 1992
		800 (on Ag)	80	37	High oriented	

(4) The grain size should be relatively large. Too fine grain size will produce many weak bonds, which deteriorate the superconducting characteristics.

(5) The chemical reaction of the film with the substrate must be suppressed. The reaction products are usually non-superconducting oxides, which makes the part of the film contacting the substrate non-superconducting.

Table 8. Sol–gel coating films of high temperature superconductors of Bi–(Pb)–Sr–Ca–Cu–O system

No.	Starting materials	Solvent	Subtrate
1.	2–Ethylhexanate (Bi), cyclohexane butyl (Sr), 2–ethyl hexanoate (Ca, Cu)	Organic solvents	MgO (single crystal
2.	2–Ethylhexanoate (Bi, Sr, Ca),	Toluene	YSZ
3.	iso–Propoxide (Bi), n–butoxide (Sr, Ca), ethoxyethoxide (Cu)	ethoxyethanol, toluene, 2-dimethyl-amino-ethanol	
4.	Ethoxide (Bi, Sr, Ca, Cu)	Ethanolamine	MgO (single crystal)
5.	Ethoxide (SR, Ca), iso–propoxide (Bi)	Ethanol, 2-dimethyl-amino-ethanol	
6.	Acetate (Bi, Sr, Ca, Cu)	Acetic acid, water	YSZ

No.	Film thick. (µm)	Heating temp. (°C)	T_c (K)		Orientation	Reference Year
			Onset	End (zero)		
1.	2	800–900	100	84	c-axis	[206] 1988
2.		850–930	97	79		[207] 1988
3.				65		[208] 1990
4.		850–870		80	c-axis	[209] 1990
5.			115			[210] 1991
6.		825	115	79		[211] 1990

(6) Preferred orientation of the precipitated superconducting crystals is desirable. Superconducting oxides are highly anisotropic and the current flows only in the CuO_2 planes and does not flow in the direction of the c-axis. Therefore, preferred orientation of the crystals for c-axis perpendicular to the coating plane permits the highest current density in the direction along the substrate planes.

The efforts to overcome some of these problems or achieve the desired properties will be discussed in the following comments on the examples shown in Tables 7 and 8.

4.4.1 Yttrium System Superconducting Films

As seen from Table 7, organic acids (Nos. 1–5), metal alkoxides and acetylacetonato (Nos. 6–11) and inorganic compounds (metal nitrates) (Nos. 12 and 13) are employed as source compounds for metal oxides. Various kinds of solvent are used for the starting solutions. It is noted that chelating agents, such as methoxyethanol and ethanolamine are employed. It is expected that the chelating solvents may stabilize the starting solution, suppressing the precipitation of metal compounds.

In order to precipitate the pertinent superconductor phase, the gel films have to be heated to 800–950 °C for the yttrium system superconductors. In the heat treatment the film may react with the substrate, precipitating non-superconductor phase. This is known to deteriorate markedly the superconducting characteristics of the film. In order to avoid this reaction, less-reactive yttria-stabilized zirconia (YSZ), MgO single crystals and $SrTiO_3$ single crystals are usually employed as substrates. Thicker films, 1–10 μm thick, may decrease the deteriorating effect of the reaction, because in thicker films more of the film may remain unreacted. It is possible to lower the rate of reaction between the film and substrate by lowering the heat treatment temperatures. However this would lead to insufficient formation of superconductor phase. For No. 4 in Table 7, $BaZrO_3$ was formed at the interface when the film was heated at 900 °C. Lowering the heating temperature to 850 °C, however, led to the insufficient sintering of crystals in the film and, accordingly, low T_c of the film.

For No. 5 in Table 7, the heating temperature was lowered below 850 °C by using Ar atmosphere instead of oxygen atmosphere. This reduced the reaction of the film with the substrate. The c-axis orientation of $YBa_2Cu_3O_{7-x}$ crystals was also observed. It is assumed that these improvements gave rise to the high $T_{c\,(end)}$ of 84 K.

In Table 7, Nos. 3 [192], 4 [193], 5 [194] and 8 [197] show the orientation of $YBa_2Cu_3O_{7-x}$ crystals and quite high $T_{c\,(end)}$. It is seen, however, that some other coating films which have no crystal orientation show higher $T_{c\,(end)}$. This suggests that the reaction taking place in sol–gel coating films during gelation and heating processes are complicated and so many factors affect the superconducting properties of the resulting films. Further studies are needed on this point.

Recently, it was shown [204] that the sol–gel method can produce $YBa_2Cu_4O_8$ (also called 124 compound) superconductor, which is more chemically stable than $YBa_2Cu_3O_{7-x}$ (also called 123 compound), at one atmospheric oxygen pressure. This is very significant for the application of high temperature superconductors, and a brief explanation will be given. It was known that $YBa_2Cu_3O_{7-x}$ superconductor is chemically weak [205], while $YBa_2Cu_4O_8$ superconductor is chemically stable. This makes the latter more useful for practical application, although T_c of the latter is several degrees lower than that of the former. There was a difficulty in synthesis of $YBa_2Cu_4O_8$; 400–500 atmospheric oxygen pressure was required in solid state reaction or sintering in order to avoid decomposition of this composition. Later it was found that, in sol–gel preparation, the synthesis of this compound needs only one atmospheric oxygen pressure. Katayama et al. [203] prepared the coating film of $YBa_2Cu_4O_8$ by the sol–gel method, although the superconducting characteristics of the product (No. 14 in Table 7) has to be improved.

4.4.2 Bismuth System Superconducting Films

Preparation and superconducting properties of sol–gel coating films of the Bi–(Pb)–Cu–Sr–Cu–O system are shown in Table 8. The superconductors of this system are characterized by T_c higher than 100 K. Pb is added in order to promote the precipitation of high T_c phase in this system. It is seen that coating films with crystal orientation (Nos. 1, 4 and 6 in Table 8) have higher T_c, although the value is not so high as expected for this superconductor. Figure 18 shows the resistance-temperature relationship for Bi–(Pb)–Ca–Sr–Cu–O coating films [211]. It is to be noted that Sei et al. [212] indicate that Pb and Bi remain only at the surface of the coating film. It is assumed that a similar phenomenon may have occurred more or less in coating films prepared by other groups.

5 Coating Films for Fabricating and Protecting Optical and Electronic Devices

In this section, sol–gel coating films employed for creating fine patterns on optical and electronic memory disks will be described. Also coating films applied for protecting active elements of optical and electronic devices will be introduced.

5.1 Films for Patterning

Optical and electronic disks have submicron scale fine patterns, i.e., pregrooves for tracking and random access. Fine patterning for preparing complex optical

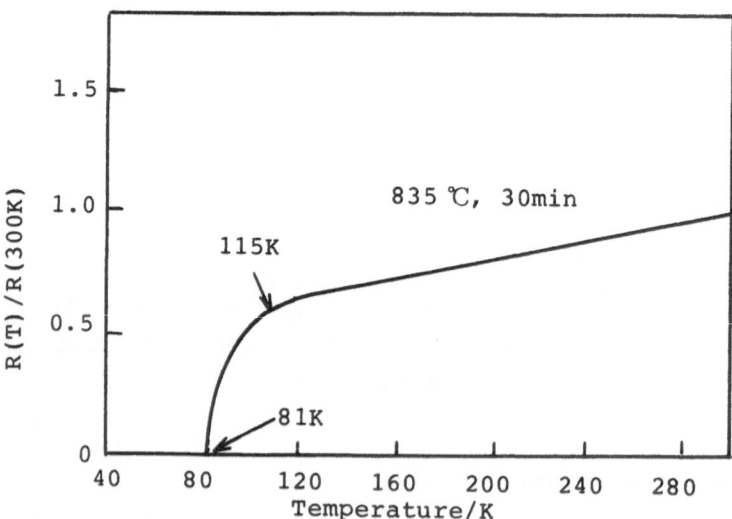

Fig. 18. D.C. electrical resistance of Bi–Ca–Sr–Cu–O coating film as a function of temperature [194]

waveguides have been described in Sect. 3.1. It was shown that fine patterns are created in the sol–gel coating films on glass and metal substrate by the three methods – stamping, laser densification and UV-assisted chemical etching.

Here, only the films for patterning on glass disks are described. It should be noted that glass is one of the most promising substrates for optical memory disks, because it is transparent and mechanically stable and a microscopically smooth surface is easily produced.

Based on sol–gel processing, Tohge et al. [213] and Matsuda et al. [214] developed the stamping method to form submicron fine patterns on soda-lime-silica glass substrate. It was shown that this method is applicable to fabrication of optical ROM disks and diffraction gratings with high reliability. As shown in Fig. 19, the whole process consists of (1) coating of a glass substrate with a gel film, (2) patterning of the gel film by pressing a stamp and (3) drying and heating. It was found that this method makes it possible to prepare uniform formation of pregrooves on the whole surface of 130 mm diameter glass disks.

The most important point in the process is to control the hardness of gel film just prior to stamping. This is achieved by adding a specific amount of organic substances such as polyethylene glycol (PEG) to a sol. Addition of PEG makes a soft gel which is easy to mold. Embossing by stamping of PEG-doped SiO_2–TiO_2 film results in replication of the 1.6 μm pitch of the master.

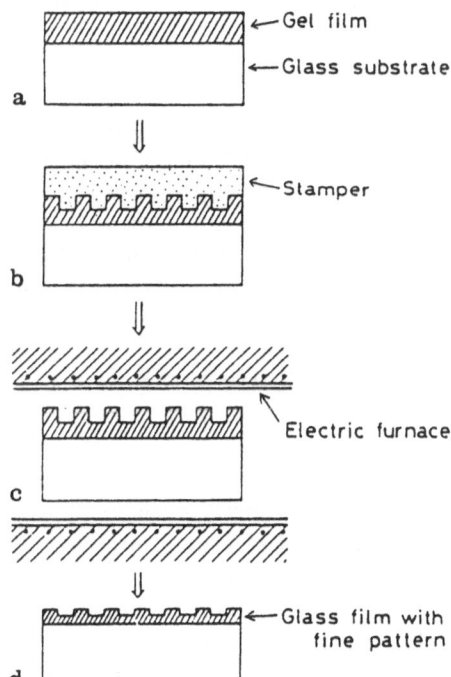

Fig. 19a–d. Fine patterning process by the sol–gel method: a formation of gel film on a substrate; b patterning on the gel film; c heat treatment; d glass film with fine patterning. After Tohge et al. [213]

The patterned films are exposed to various vapors and solvents for further processing and the chemical durability of the gel has to be high. The chemical durability to moisture is considerably improved by choosing the glass composition from TiO_2–SiO_2 [215] and ZrO_2–SiO_2 systems.

5.2 Films for Protecting Optical and Electronic Elements of Devices

Sol–gel coatings are suitable for protecting active elements of optical and electronic devices, because coating films can be applied at low temperatures [216].

Passivation of silicon semiconductors can be achieved by SiO_2 and Al_2O_3 coatings [217]. It is known that introduction of nitrogen into SiO_2 film increases the protection ability of SiO_2 films [218, 219]. In this case SiO_2 coating films are heated in NH_3 atmosphere to form a nitrogen-rich layer at the outer surface of the coating film. Sigorov et al. [216] prepared SiO_2 isolating and passivating film of 0.4 µm thickness on GaAs by the sol–gel method in the process of fabricating GaAs devices. For high-voltage devices, thicker films doped with phosphorus were developed.

There are many examples in which sol–gel coating films are used as barrier layers. Sol–gel $ZrO_2(Y_2O_3)$ films on silicon and sapphire were used as barriers for protection of high temperature superconductor films [216]. Direct contact of a superconductor film with silicon and sapphire substrate markedly deteriorates its superconducting properties due to formation of a non-superconducting phase as a result of reactions. Ogiwara and Kinugawa [169] coated a soda-lime-silica glass substrate with SiO_2 film before applying ITO coating film. The SiO_2 film protects the ITO film by suppressing the diffusion of Na^+ ions from the substrate into the ITO film.

Inorganic-organic composite thick films can be used for passivation of thin film capacitors. Popali et al. [220] developed inorganic-organic composite coatings with barrier properties against water vapor, excellent electrical properties (low dielectric constant, high resistivity and high dielectric strength) and good adhesion to various substrates. It was shown that these films of less than 10 µm thickness can replace several mm thick, expensive encapsulations.

6 Concluding Remarks

As shown in the preceding section, considerable developments have been seen in preparation, property measurements and applications of sol–gel coating films for optical and electronic use. Some are already commercially produced and many are close to practical application. This is partially attributed to the low processing temperature nature of the sol–gel method. Easy formation of the film may be another reason. It is expected that sol–gel coating films have a wide variety of applications and will continue to grow.

7 References

1. Sakka S (1982) In: Tomozawa M, Doremus RH (eds) Treatise on materials science and technology, Vol 22, Glass III. Academic Press, New York, p 129
2. Sakka S (1988) Science of sol-gel method. Agne-Shofu-Sha, Tokyo, p 221 (in Japanese)
3. Brinker CJ, Scherer GW (1990) Sol–gel science, Academic Press, San Diego, p 908
4. Sakka S (1994) J Sol–Gel Sci Tech 3: 69
5. Klein LC (ed) (1988) Sol–gel technology for thin films, fibers, preforms, electronics, and specialty shapes. Noyes Publications, Parkridge, p 407
6. Sakka S, Yoko T (1992) In: Reisfeld R, Jorgensen CK (eds) Structure and bonding 77, Springer-Verlag, p 89
7. Brinker CJ, Scherer GW (1990) Sol–gel science, Academic Press, San Diego, p 787
8. Scrivan LE (1988) In: Brinker CJ, Clark DE, Ulrich DR (eds) Better ceramics through chemistry III, Mater Res Soc Symp. Proc. vol 121, Materials Research Society, Pittsburgh, p 717
9. Thomas IM (1994) In: Klein LC (ed), Sol–gel optics, Kluwer Academic Publishers, Boston: p. 141
10. Floch HG, Belleville PF (1994) J Sol–Gel Sci Tech 2: 695
11. Yamamoto Y, Kamiya K, Sakka S (1982) J Ceram Soc Japan 91: 328
12. Brinker CJ, Hurd AJ, Shunk PR, Frye GC, Ashley CS (1992) J Non-Cryst Solids 147 & 148: 424
13. Innocenzi P, Kozuka H, Sakka S (1994) J Sol–Gel Sci Tech 1: 305
14. King H, Merl N, Schmidt H (1992) J Non-Cryst Solids 147 & 148: 44
15. Monde T, Kozuka H, Sakka S (1988) Chem Lett 1988: 287
16. Landau LD, Levich BG (1942) Acta Physicochim USSR 17: 42; Yang CC, Josefowicz JY, Alexandru L (1980) Thin Solid Films 74: 117
17. Strawbridge I, James PF (1986) J Non-Cryst Solids 86: 381
18. Schroeder H (1969) Phys Thin Films 5: 87
19. Sakka S, Kokubo T (1983) Jap J Appl Phys 22, Supplement 22–2: 3
20. Strawbridge I, James PF (1986) J Non-Cryst Solids 63: 394
21. Yoldas BE (1980) Appl Opt 19: 1425
22. Guglielmi M, Zenezini S (1990) J Non-Cryst Solids 121: 303
23. Makishima A, Asami M, Wada K (1990) J Non-Cryst Solids 121: 310
24. LaSerra ER, Charbouillot Y, Baudry P, Aegerter MA (1990) J Non-Cryst Solids 119: 21
25. Uhlmann DR, Zelinski BJJ, Teowee G, Boulton JM, Koussa A (1991) J Non-Cryst Solids 129: 76
26. Uhlmann DR, Motakef S, Suratwala T, Wade R, Teowee G, Boulton JM (1994) J Sol Gel Sci Tech 2: 335
27. Bahat M, Mugnier J, Lou L, Serughetti J (1992) SPIE 1758 Proc Sol–Gel Optics III: 173
28. Aráujo FG, Chis T, Hench LL (1994) J Sol–Gel Sci Tech 2: 729
29. Zaugg TC, Fabes BD, Weisenbach L, Zelinski (1991) SPIE Proc 1590: 26
30. Guglielmi M, Colombo P, Mancinelli deglu Esposio L, Righini GC, Pelli S, Rigato V (1992) J Non-Cryst Sol 147–148: 641
31. Rancone RL, Weller-Brophy LA, Weisenbach L, Zelinski BJJ (1991) J Non-Cryst Solids 128: 111
32. Schmidt H, Krug H, Kasemann R, Tiefensee F (1991) SPIWE Proc 1590: 36
33. Krug H, Tiefensee F, Oliveira PW, Schmidt H (1992) SPIE Proc 1758: 448
34. Geotti-Bianchini F, Guglielmi M, Polato P, Soraru GD (1984) J Non-Cryst Solids 63: 251
35. Atta AK, Biswas PK, Ganguli D (1990) J Non-Cryst Solids 121: 315
36. Yamamoto Y, Makita K, Kamiya K, Sakka S (1983) J Cersm Soc Jpn 91: 222
37. Duran A, Fernandez Navarro JM, Mazen P, Joglar A (1988) J Non-Cryst Solids 100: 494
38. Makishima M, Kubo H, Wasa K, Kitami Y, Shimohira T (1986) J Am Ceram Soc 69: C127
39. Mohallen NDS, Aegerter MA (1988) J Non-Cryst Solids 100: 526
40. Sains MA, Duran A, Fernandez Navarro JM (1990) J Non-Cryst Solids 121: 315
41. Hou LS, Sakka S (1988) J Non-Cryst Solids 100: 526
42. Dislich H, Hussmann E (1981) Thin Solid Films 77: 129
43. Nakazumi H (1991) Preprints Meet Kinki Chem Assoc, Tokyo: 1–11
44. Ito T (1990) Toshiba Review 45: 831

45. Melpolder S (1994) Paper A24, Inter Symp on Sol–Gel Sci Tech in Pacific Coast Region Meet, Oct 20–22, Los Angeles
46. Sakka S, Kamiya K, Yoko T (1988) In: Zeldin M, Wynne KJ, Allcock HR (eds) ACS Symposium Series 360, American Chemical Society, Washington DC: p 345
47. Arfsten NJ, Kaufmann R, Dislich H, In: Hench LL, Ulrich DR (eds) Ultrastructure Processing Of Ceramics, Glasses and Composites, Wiley, New York, p 189
48. Hutter F, Schmidt H, Scholze H (1986) J Non-Cryst Solids 82: 373
49. LaSerra ER, Charbouillot Y, Baudry P, Aegerter MA (1990) J Non-Cryst Solids 121: 323
50. Gurkovich SR, Blum JB (1984) In: Hench LL, Ulrich DR (eds) Ultra-structure Processing of Ceramics, Glasses and Composites, Wiley, New York, p 152
51. Hattori A, Makita K, Okabayashi S (1989) Abstract of Meeting of the Society of Photo-Optical Instrumentation Engineers, USA, Aug, 1989
52. Makita K, (1990) New Glasses 5: 186
53. Asahara Y, Izumitani T (1982) J Non-Cryst Solids 42: 269
54. Mukherjee SP, Lowdermilk WH (1982) J Non-Cryst Solids 48: 177
55. Yoldas BE, Partlow DP (1985) Thin Solid Films 129: 1
56. Floch HG, Belleville PF (1994) J Sol–Gel Sci Tech 1: 293
57. Hinz P, Dislich H (1986) J Non-Cryst Solids 82: 411
58. Hong G, Sakka S (1991) In: J Rare Earth, Special Issue, Proc of 2nd Inter Conf on Rare Earth Development and Application: 681
59. Almeida RM, Orignac X, Barbier (1994) J Sol–Gel Sci Tech 2: 465
60. Arai K, Namikawa H, Kumata K, Honda T, Ishii Y, Handa T (1986) J Appl Phys 59: 3430
61. Pope EJA, Mackenzie JD (1988) J Non-Cryst Solids 106: 236
62. Thomas IM, Payne SA, Wilke GD (1992) J Non-Cryst Solids 151: 183
63. Avnir D, Levy D, Reisfild R (1984) J Phys Chem 88: 5956
64. Levy D, Avnir D (1991) J Photchem. Photobiol A, Chem 57: 41
65. Tani T, Namikawa H, Arai K, Makishima A (1985) J Appl Phys 58: 3559
66. Gvishi R, Reisfeld R (1991) J Non-Cryst Solids 128: 69
67. Kobayashi Y, Imai Y, Kurokawa Y (1988) J Mater Sci Lett 7: 1148
68. Tanaka H, Takahashi J, Tsuchiya J, Kobayashi Y, Kurokawa Y (1989) J Non-Cryst Solids 109: 169
69. Kobayashi Y, Kurokawa Y, Imai Y, Muto S (1988) J Non-Cryst Solids 105: 198
70. Reisfeld R (1990) J Non-Cryst Solids 121: 254
71. Reisfeld R (1989) In: Aegerter MA, Jafelicci M, Souza ED, Zanotto ED (eds) Sol–gel science and technology, World Scientific, Singapore, p 323
72. Mackenzie JD (1993) J Sol–Gel Sci Tech 1: 7
73. Partlow DP, Greggi J (1987) J Mater Res 2: 595
74. Mackenzie JD, Xu R, Xu Y (1992) In: Hench LL, West JK (eds) Chemical processing of advanced materials, John Wiley & Sons, New York p 365
75. Boulton JM, Teowee G, Bommersbach WM, Uhlmann DR (1992) SPIE Vol 1758 Sol–Gel Optics II: 292
76. Gan F, Xian X (1992) SPIE Vol 1758 Sol–Gel Optics II: 310
77. Livage J, Schmutz C, Griesmar, Barboux P, Sanchez C (1992) SPIE Vol 1758 Sol–Gel Optics II; 274
78. Barbé CJ, Harman MA, Scherer GW (1994) J Sol–Gel Sci Tech 2: 507
79. Nie W, Lurin C, Paz-Pujalt GR (1992) SPIE Vol 1758 Sol–Gel Optics II; 284
80. Hirano S: Private communication
81. Zieba J, Zhang Y, Prasad PN (1992) SPIE Vol 1758 Sol–Gel Optics II: 403
82. Kim J, Plawski JL, Lapuerta R, Korenowski GM (1992) Chem Mater 4: 249
83. Hashimoto T, Yoko T, Sakka S (1993) J Ceram Soc Jpn 101: 64
84. Hashimoto T, Yoko T, Sakka S (1993) Bull Chem Soc Jpn 67: 653
85. Nogami M, Yamada K, Watanabe M, Nagasaka K (1981) 99: 625
86. Bagnall CM, Zarsycki J (1991) J Non-Cryst Solids 135: 182
87. Nogami M, Zhu YQ, Tohyama Y, Nagasaka K, Tokizaki T, Nakamura (1991) J Am Ceram Soc 74: 238
88. Kawaguchi H, Miyakawa T, Tanno N, Kobayashi Y, Kurokawa (1991) Jpn J Appl Phys 30: L280
89. Yeatman EM, Green M, Dawnay EJC, Fardad MA, Horowitz F (1994) J Sol–Gel Sci Tech 2: 711
90. Hache F, Ricard D, Flythanis C, Kreibig K (1988) Appl Phys A 47: 347

91. Matsuoka J, Mizutani R, Nasu H, Kamiya K (1992) J Ceram Soc Japan 100: 599
92. Kozuka H, Sakka S (1993) Chem Mater 5: 222
93. Kozuka H, Zhao G, Sakka S (1994) J Sol–Gel Sci Tech 2: 741
94. Zhao G, Kozuka H, Sakka S (1995) J Sol–Gel Sci Tech 4: 37
95. Reisfeld R, Minti H (1994) J Sol–Gel Sci Tech 2: 641
96. Reisfeld R (1990) Proc SPIE 1328: 29
97. Nakamura M, Nasu H, Kamiya K (1991) J Non-Cryst Solids 135: 1
98. Mennig M, Krug H, Fink-Straube C, Oliveira PW, Schmidt H (1992) SPIE Vol 1758 Sol–Gel Optics II : 387
99. Oohira T, Yokota Y, Iseki Y (1991) In: Sakka S, Soga N (eds) Proceedings of the international conference on science and technology of new glasses: 364
100. Hou L, Hoffmann B, Mennig M, Schmidt H (1994) J Sol–Gel Sci Tech 2: 635
101. Levy D, Einhjorn S, Avnir D (1989) J Non-Cryst Solids 113: 137
102. Baudry P, Rodrigues ACM, Aegerter MA (1990) J Non-Cryst Solids 121: 319
103. Livage J (1983) In: Solid state chemistry, Elsevier, Amsterdam: p 17
104. Unuma H, Tonooka K, Suzuki Y, Furusaki T, Kodaira K, Matsushita T (1986) J Mater Sci Lett 5: 1248
105. Macedo MA, Aegerter MA (1994) J Sol–Gel Sci Tech 2: 667
106. Judeinstein P, Livage J, Zarudiansky A, Rose R (1988) Solid State Ionics 28: 1722
107. Nabavi M, Doeuff S, Sanches C, Livage J (1989) Mater Sci Eng B3: 203
108. Nagase K, Shimizu Y, Miura N, Yamazoe N (1983) J Ceram Soc Japan 101: 1032
109. O'Regan B, Grätzel M (1991) Nature 335: 737
110. Grätzel (1994) J Sol–Gel Sci Tech 2: 673
111. Fujishima A, Honda K (1972) Nature 238; 37
112. Yoko T, Kamiya K, Sakka S (1986) Dinki Kagaku 54: 284
113. Yoko T, Yuasa A, Kamiya K, Sakka S (1988) J Electrochem Soc 138: 2279
114. Hu LL, Yoko T, Kozuka H, Sakka S (1992) Thin Solid Films 219: 18
115. Yoko T, Inagaki Y, Sakka S (1992) Res Rep Asahi Glass Foundation 56: 13 [in Japanese]
116. Zusman R, Rottman C, Ottolenghi M, Avnir D (1990) J Non-Cryst Solids 122: 107
117. Samuel J, Strinkovski M, Avnir D, Lewis A (1994) Mater Lett 21: 431
118. Kubecková M, Pospisilová M, Matejec V (1994) J Sol–Gel Sci Tech 2: 591
119. MacCraith BD, McDonagh C, O'Keeffe G, Butler T, O'Kelly, McGilp JF (1994) J Sol–Gel Sci Tech 2: 661
120. Kiernan P, McDonagh C, MacCraith BD and Mongey K (1994) J Sol–Gel Sci Tech 2: 513
121. Kordas G (1986) Mat Res Soc Symp Proc 73: 685
122. Chen KC, Janah A, Mackenzie JD (1986) Mat Res So Symp Proc 73: 731
123. Yoko T, Nakanishi K, Sakka S (1992) Res Rep Asahi Glass Foundation 61: 17 [in Japanese]
124. Tanaka K, Yoko T, Atarashi M, Kamiya K (1989) J Mater Sci Lett 8: 83
125. Sedlar M, Matejec V (1994) J Sol–Gel Sci Tech 2: 587
126. Tsuchiya T, Sei T, Kanda H (1991) In: Sakka S, Soga N (eds) Proc Intern Conf Sci Tech New Glasses, p 406
127. Tsuchiya T, Sei T, Kanda H (1992) SPIE Vol 1758 Sol–Gel Optics II 2: 304
128. Gan F, Zhou Y (1994) J Sol–Gel Sci Tech 2: 461
129. Uhlmann DR, Motakef S, Surutwala T, Wade R, Teowee G, Boulton JM (1994) J Sol–Gel Sci Tech 2: 335
130. Payne DA (1994) J Sol–Gel Sci Tech 2: 311
131. Yamashita H, Yoko H, Sakka S (1989) J Ceram Soc Japan 98: 913
132. Sakka S, Kokubo T (1983) Jpn J Appl Phys Vol 22 Supplement 22–2: 3
133. Yanovskaya MI, Turova NY, Turevskaya EP, Novoselova AV, Venetseva YN, Sagitov SI (1981) Inorg Mater 17: 221
134. Campion JF, Payne DA, Chae HK, Xu Z (1991) Ceramic Trans 22: 477
135. Hayashi T, Oji N, Maiwa H (1994) Jpn J Appl Phys 33: 5277
136. Ogawa T, Senda A, Kasami T (1991) Jpn J Appl Phys 30: 2147
137. Budd KD, Dey SK, Payne DA (1985) Brit Ceram Proc 36: 107
138. Budd KD, Dey SK, Payne DA (1986) Mat Res Soc Symp Proc 73: 711
139. Yi G, Wu Z, Sayer M (1988) J Appl Phys 64: 2717
140. Coffman PR, Dey SK (1994) J Sol–Gel Sci Tech 1: 251
141. Lakeman DE, Payne DA (1992) 75: 3091
142. Swlvaraj U, Brooks K, Paradarao AV, Komarneni S, Roy R, Cross LE (1993) J Am Ceram Soc 76: 1441

143. Livage C, Safari A, Klein LC (1994) J Sol–Gel Sci Tech 2: 605
144. Okuwada K, Imai M, Kakuno K (1989) Jpn J Appl Phys 28: L271
145. Hirano S, Yogo T, Kikuta K, Sakamoto W (1994) J Sol–Gel Sci Tech 2: 329
146. Kang J, Yoko T, Sakka S (1992) Proc of SPIE Vol 1758 Sol–Gel Optics II: 249
147. Okuwada K, Nakamura S, Imai M, Kakuno K (1990) Jpn J Appl Phys 29: 1153
148. Hirano S, Kato K (1987) Adv Ceram Mater 2: 142
149. Xu Y, Chen CJ, Xu R, Mackenzie JD (1990) J Appl Phys March 15
150. Xu R, Xu Y, Chen CJ, Mackenzie JD (1990) J Mater Res 5: 916
151. Hirano S, Kato K (1988) Adv Ceram 3: 503
152. Barlingay CK, Dey SK (1990) Appl Phys Lett 61: 1278
153. Tuttle BA, Voigt JA, Goodnow DC, Lamppa DL, Headley TJ, Eatough MO, Zender G, Nasby RD, Headley TJ, Rodgers SM (1993) J Am Ceram Soc 76: 1537
154. Nashimoto K, Nakamura S (1994) Jpn J Appl Phys 33: 5147
155. Aoki K, Fukuda Y, Numata K, Nishimura A (1994) Jpn J Appl Phys 33: 5155
156. Wu Q, Xu Y, Mackenzie JD (1994) SPIE Sol–Gel Optics III, July, 1994, Los Angeles
157. Ohya Y, Ito T, Takahashi Y (1994) Jpn J Appl Phys 33: 5272
158. Doi H, Arsuki T, Soyama N, Sakaki G, Yonezawa T, Ogi K (1994) Jpn J Appl Phys 33: 5159
159. Xu R, Xu Y, Mackenzie JD (1992) SPIE Vol 1758 Sol–Gel Optics II: 261
160. Mizuhashi M, (1992) New Ceramics 5 [12] : 25 [in Japanese]
161. Matsushita T, Sekiya T, Yamai I (1972) J Japan Chem Soc 1972: 880
162. Gonzalez-Oliver CJR, Kato I (1986) J Non-Cryst Solids 82: 400
163. Kodaira K, Sohma M, Furusaki T (1990) Ceram Tranactions 11, Thin and thin films, The American Ceramic Society p 301
164. Maddalena A, DalMaschio R, Dire S, Raccanelli A (1990) J Non-Cryst Solids 121: 365
165. Tsunashima A, Yoshimizu H, Kodaira K, Shimada S, Matsushita T (1986) J Mat Sci 21: 2731
166. Furusaki T, Takahashi J, Takaha H, Kodaira K (1993) J Ceram Soc Japan 101: 451
167. Lada W, Deptula A, Olczak T, Torbicz W, Pljanowska D, DiBartolomes A (1994) J Sol–Gel Sci Tech 2: 551
168. Tsunashima A, Asai T, Kodaira K, Matsushita T (1978) Chem Letters 1978: 835
169. Ogiwara S, Kinugawa K (1982) Bull Ceram Soc Japan 90: 157
170. Furusaki T, Kodaira K, Yamamoto M, Shimada S, Matsushita (1986) Mat Res Bull 21: 803
171. Furusaki T, Kodaira K (1991) In: Vincenzini (ed) High performance ceramic films and coatings, Elsevier Science Publisher, p. 241
172. Furusaki T, Takahashi J, Kodaira K (1994) J Ceram Soc Japan 102: 200
173. Harrison PG, McGieron JK, Harrison CC (1994) J Sol-Gel Sci Tech 2: 295
174. Dislich H, Hinz P (1982) J Non-Cryst Solids 48: 11
175. Livage J (1984) Mat Res Soc Symp Proc Vol 32: 125
176. Nabavi M, Dourff S, Sanches C, Livage J (1989) Materials Science and Engineering B3: 203
177. Bullot J, Cordier P, Gallais O, Gauthier M, Livage J (1984) J Non-Cryst Solids 68: 123
178. Livage J (1983) In: Proceedings of the Second European Conference, Veldhoven, The Netherlands, June 7–9, 1982, Elsevier, Amsterdam: 17
179. Sanchez C, Alonso B, Chapusot F, Ribot F, Audebert P (1992) J Sol–Gel Sci Tech 2: 161
180. Yoldas BF, Partlow DP (1980) Am Cer Soc Bull 59: 640
181. Tatsumisago M, Minami T (1987) J Chem Soc Japan 87: 1958
182. Perthuis H, Colomban Ph (1985) J Mat Sci Lett 4: 344
183. Ogasawara T, Klein LC (1994) J Sol–Gel Sci Tech 2: 611
184. Wang B, Greenblatt M, Yan J, Wu Y (1994) J Sol–Gel Sci Tech 2: 323
185. Boilot JP, Colomban P (1985) J Mat Sci Lett 4: 22
186. Garrido FMS, Aleves OL (1994) J Sol–Gel Sci Tech 2: 421
187. Yamashita K, Tanaka M, Najiri T, Umegaki T (1991) In: Sakka S, Soga N (eds) Science and Technology of New Glasses, Tokyo, p 269
188. Ravaine D, Seminel A, Charnbouillot Y, Vinces M (1986) J Non-Cryst Solids 82: 210
189. Bednorz JG, Müller KA (1986) Z Phys B–64: 189
190. Kumagai T, Yokota H, Kawaguchi K, Kondo W, Mizuta S (1987) Chem Lett 1987: 1645
191. Nasu H, Makida S, Kato T, Ihara Y, Imura T, Osaka Y (1987) Chem Lett 1987: 2403
192. Cross ME, Hong M, Liou SH, Gellagher PK (1988) Mat Res Soc Symp Proc 99: 731
193. Davison WW, Shyu SG, Buchanan RC (1988) Mat Res Soc Symp Proc 99: 289
194. Yang YX, Howson MA, Milne SJ (1992) J Non-Cryst Solids 147 & 148: 715
195. Monde T, Kozuka H, Sakka S (1988) Chem Lett 1988: 287
196. Kramer S, Wu K, Kordas G (1988) Mat Res Soc Symp Proc 99: 323

197. Shibata S, Kitagawa T, Okazaki H, Kimura T, Murakami (1988) Jpn J Appl Phys 27: L53
198. Tatsumisago M, Saato H, Minami T (1988) Chemistry Express 3: 311
199. Hirao S, Hayashi T, Miura M (1990) J Ceram Soc Jpn 73: 885
200. Masuda Y, Matsubara K, Tateishi T, Sakka S, Kawate Y (1992) J Mater Res 7: 819
201. Cooper EL, Frisch MA, Giess EA, Gupta A, Hussey BW, O'Sullivan, Raider SI, Scilla GJ (1988) Mat Res Soc Symp Proc 99: 165
202. Nasu H, Makida S, Imura T, Osaka Y (1988) In: Capone DW, Butler WH, Batlogg B, Chu CW (eds) High Temperature Superconductors II, Materials Research Society, Pittsburgh, p 101
203. Katayama S, Sekine M, Fudouzi H, Kuwabara M (1992) J Appl Phys 71: 2795
204. Fujihara S, Zhuang H, Yoko T, Kozuka H, Sakka S (1992) J Mater Res 7: 2355
205. Komori K, Kozuka H, Sakka S (1989) J Mater Sci 24: 1889
206. Agostinelli JA, Paz-Pujalt GR, Mehrotra AK (1988) Physica C156: 208
207. Nasu H, Makida S, Ibara Y, Kato T, Imura T, Osaka Y (1988) Jpn J Appl Phys 27: L536
208. Hirano S, Hayashi T, Tomonaga H (1990) Jpn J Appl Phys 29: L40
209. Tohge N, Tatsumisago M, Minami T (1990) J Non-Cryst Solids 121: 443
210. Katayama S, Sekine M (1991) J Ceram Soc Jpn 99: 345
211. Zhuang H, Kozuka H, Yoko T, Sakka S (1990) J Appl Phys 29: L1107
212. Sei T, Okano T, Tsuchiya T (1992) J Non-Cryst Solids 147 % 148: 711
213. Tohge N, Matsuda A, Minami T, Matsuno Y, Katayama S, Ikeda Y (1988) J Non-Cryst Solids 100: 501
214. Matsuda A, Matsuno Y, Kataoka S, Katayama S, Tsuno T, Tohge N, Minami T (1990) SPIE Vol 1328: 62
215. Matsuda A, Kogure T, Matsuno Y, Katayama S, Tsuno T, Tohge N, Minami T (1993) J Am Ceram Soc 76: 2889
216. Sigorov AS, Vorontilov KA, Valeev AS, Yanovskaya MI (1994) J Sol–Gel Sci Tech 2: 563
217. Schlichtung J, Neumann S (1982) J Non-Cryst Solids 48: 185
218. Martinsen J, Figat Ra, Shafer MW (1984) Mat Res Soc Sympo Proc 32: 145
219. Brow RK, Pantano CG (1984) Mat Res Soc Symp Proc 32: 361
220. Popali M, Kappel J, Pilz M, Schulz J, Feyder G (1994) J Sol–Gel Sci Tech 2: 157

Sol-Gels and Chemical Sensors

Otto S. Wolfbeis[1,2], Renata Reisfeld[3], and Ines Oehme[1]

[1] Karl-Franzens Universität Graz, Institute for Organic Chemistry, Heinrich St. 28, A-8010 Graz, Austria
[2] University of Regensburg, Institute for Analytical Chemistry, Chemo- and Biosensors D-93040 Regensburg, Germany
[3] The Hebrew University of Jerusalem, Department of Inorganic and Analytical Chemistry, Jerusalem 91904, Israel

We review the principles of optical and electrochemical sensors based on the use of the sol-gel technique, in particular their fabrication, working principles, and various configurations. We also report on potential applications, e.g. to environmental and clinical analysis, to gas sensing, and to bioprocess monitoring. Methods are critically reviewed for making such sensors, how to encapsulate organic, inorganic and biological matter, and how to control the properties of the resulting materials. Specifically, sensors for gases, ions, and organic as well as bioorganic molecules are discussed in some detail.

1 Introduction

1.1 Sol-Gels

The sol-gel process allows monoliths and glass films to be prepared into which specific reagents can be incorporated [1–3]. One particular example of a sol-gel process involves the system tetraethylorthosilicate of chemical structure $Si(OC_2H_5)_4$ (TEOS), ethanol, and water. This is a one-phase solution that undergoes sol-gel transition to a rigid two-phase system of solid silica (SiO_2) and solvent-filled pores. Silica gels may be prepared from the sol-gel polymerization of silicon alkoxides (e.g., TEOS) or other alkoxides. Hydrolysis occurs when TEOS and water are mixed in a mutual solvent, generally ethanol.

The intermediates that exist as a result of partial hydrolysis possess free silanol (–Si–OH) groups. Complete hydrolysis of TEOS to form $Si(OH)_4$ would give silicic acid, but this does not normally occur. Instead, either two silanols or one silanol and an ethoxy group condense to form a bridging oxygen in a siloxane group (Si–O–Si). Representative examples of a hydrolytic reaction and the subsequent condensations between either a silanol and an alkoxysilane (with the elimination of ethanol) or between two silanols (with the elimination of water) are given below:

$$Si(OEt)_4 + H_2O \rightarrow Si(OEt)_3OH + EtOH \qquad \text{(hydrolysis)}$$

$$Si(OEt)_3OH + Si(OEt)_4 \rightarrow Si(OEt)_3-O-Si(OEt)_3 + EtOH \qquad \text{(condensation)}$$

$$R-Si-OH + HO-Si-R \rightarrow R-Si-O-Si-R \qquad \text{(condensation)}$$

These reactions proceed as long as ethoxy groups (which can be hydrolyzed) or hydroxy groups (which may condense) are available. As has been discussed in great detail in reviews on the fabrication and properties of sol-gels [1–3], the hydrolysis and polycondensation reactions initiate at numerous sites within the TEOS/water solution.

Both hydrolysis and condensation may occur by acid- or base-catalyzed bimolecular nucleophilic substitution reactions. Mineral acids and ammonia are the most generally used catalysts in sol-gel processing [4]. At low pH levels – where hydrolysis is slow – the silica tends to form linear molecules that are occasionally cross-linked. These molecular chains entangle and form additional branches resulting in gelation. Under basic conditions, where hydrolysis is faster, highly branched clusters are formed that are not interpenetrable before drying and thus behave as discrete species. Gelation occurs by linking of the clusters.

Once a sufficient quantity of interconnected Si–O–Si bonds has been formed in a region, they respond cooperatively to form colloidal particles (a "sol"). With time the colloidal particles and condensed silica species link to form a 3-dimensional network. At the subsequent onset of gelation, the viscosity increases sharply, and the result is a solid object in the shape of the mold. After the sol-gel transition, the solvent phase is removed from the interconnected pore network. If removed by conventional drying such as evaporation, so-called xerogels are obtained, if removed via supercritical evacuation, the product is an aerogel.

During the first steps of the reaction, a high surface area microporous gel-glass is formed by a complex sequence of polymerization, sol formation, gelation and gel drying. Gels tend to crack during drying due to the stress produced by capillary forces associated with the gas–liquid interfaces. This can be avoided to a large extent by adding drying control chemicals that modify the surface tension of the interstitial liquid and the pore size, resulting in a crack-free dry gel (xerogel) after heat treatment.

The second step, which has been studied in great detail [3] consists in annealing the porous gel-glass at elevated temperature, which results in a dense non-porous glass. When the intention is to form mixed-oxide matrices using alkoxides of different metals (e.g. Al or Ti) with different hydrolysis rates, prehydrolysis of the alkoxysilane is preferred. A review [5] discusses the chemical aspects of the molecular precursors, the aggregation phenomena involved in the sol-gel to material transformation, and the physical properties and applications of transition metal oxide gels.

Materials other than alkoxysilanes that have been used to fabricate sol-gels include ethoxyaluminates, ethoxytitanates [6], alkoxy-zirconates and -tantalates, and the like [7]. While chemically related to the alkoxysilanes, the respective aluminates, titanates and zirconates are different in many respects including reaction rates of hydrolysis and condensation, and the refractive index and pore size of the resulting glasses.

1.2 Chemical Sensors

Chemical sensors and biosensors are small devices capable of continuously and reversibly recording the concentration of a (bio)chemical species. It is quite common, though, to refer to irreversible detection devices as "sensors" as well, provided they are small and give a fast response. In fact, the majority of sol-gel-based chemical sensors act irreversibly. Such sensors are the chemical equivalents to physical sensors which, for instance, measure light intensity, temperature, pressure, or acceleration. Typical chemical species, for which an extraordinary interest exists in terms of sensing, include pH, oxygen, pollutants; and typical species for which biosensors have been developed include glucose, cholesterol, phenytoin, immunoglobulins, and pollutants such as atrazine.

Given the unique advantages of sensors over conventional assays, numerous sensing schemes have been developed in the past 20 years, most of them based on optical or electrochemical methods which form the majority of sensors, although techniques such as calorimetric sensors, quartz-microbalance sensors, and a number of less common sensing schemes have found impressive special applications. Notwithstanding the variety of sensing schemes, what they all have in common is that, in order to achieve a certain selectivity, a specific interaction takes place between a receptor – in a wide sense – and the target analyte. This usually occurs at a membrane or coating or a thin film located at the sensor – sample interface. Typical receptors include organic molecules such as immobilized chelating agents, but also more complex species such as enzymes, antibodies, DNA, as well as natural or synthetic receptors, valinomycin and crown ethers being typical examples.

In all kinds of sensors, the receptor is present in immobilized form. Aside from covalent immobilization which is the preferred but tedious method, embedding the receptor into an inert matrix is another frequently applied method. In order to be useful as a matrix material, the polymer must be permeable to the target analyte. Numerous polymers have been used for incorporating or chemically immobilizing indicators, chelators, biomolecules, and the like. The choice of polymer strongly depends on the kind of analyte and has a decisive effect on the performance of a sensor. In fact, the proper choice of the immobilization matrix is a most crucial step in designing a sensor and can have a significant effect in terms of whether a sensor works or not.

Chemically sensitive coatings may be deposited on electrodes, optical fibres, or other sensing elements by numerous methods including dip- and spin-coating, casting, photo-polymerization, and covalent immobilization. Spreading, printing or ink-jetting the chemically sensitive layers on a planar support is the method of choice in manufacturing disposable (single-shot) sensor tests and test strips.

The two most important kinds of electrochemical sensors are based on amperometry and potentiometry, respectively. In the former, the quantity of current flowing through a sensing electrode is measured, whilst in potentiometry it is the potential formed at the sample/electrode interface. However, up to now sol-gels have rarely been used in electrochemical sensors (see below). There is a number of examples, though, where sol-gels have been used to coat conductivity-based sensors, e.g. the well-known Tagushi ("Figaro") sensors [8]. It should be kept in mind that sol-gels by themselves are non-conductive, but may become conductive using appropriate additives.

Optical sensors are based on measurement of light [9]. Hence, the support must be optically transparent in the wavelength range of interest which can be from the ultraviolet to the infrared. In the most common arrangement, a sensor membrane – or a test strip – is placed in front of an opto-electronic system comprising a light source such as an LED which shines light onto the sensing membrane in contact with the sample. The sensing membrane changes its color in accordance with the concentration of the analyte to be measured. Reflected, diffracted, or otherwise modulated light then hits a photodetector and after amplification is processed and displayed in the desired form. Fig. 1 shows a schematic of a module-type of arrangement with a sensing layer over which a sample is passed.

In other cases, optical fibers are employed in order to guide light from the source to the sensor head (which may be placed in a hardly accessible site such as an artery), and to guide it back to an opto-electronic detection system.

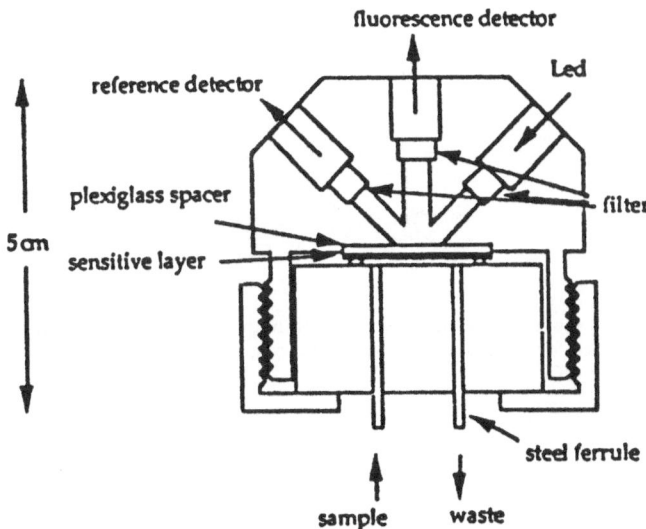

Fig. 1. Optical module including a flow-through cell and a chemically-sensitive layer for continuous monitoring of chemical species such as pH, oxygen, or pollutants (via fluorescence detection)

Depending on the wavelength used, fibers transmissable to the UV, VIS, NIR, or IR have to be used, the last-mentioned still being rather expensive. Fig. 2 shows the tip of an optical-fibre chemical sensor with a threefold indicator placed at the distal end; it is used for in-vivo monitoring of blood gases and blood pH. Apart from coating the tip of an optical fibre, the coating may be deposited on the fibre core as well.

A third group of optical sensors comprises integrated optical waveguide structures, usually embedded in glass (Fig. 3). This is another field of application where sol-gels hold great promise since applications of planar integrated optical waveguide (IOW) technology to problems in surface spectroscopy and optical chemical sensing have been partly limited by the difficulty of producing high-quality glass IOWs.

A final group of optical sensors comprises the capillary-type sensors, i.e. small capillaries coated with a sensitive chemistry in the interior or on the inner wall. For example, glass tubes have been used [11] which were filled with porous sol-gel powder doped with a complexing agent which gave a color with the analyte. The length of the stained section of the tube was found to be a measure for the concentration of the analyte. Schemes of tube detectors using alternative sampling modes can be seen in Fig. 4, namely (a) capillary force; (b) hydrostatic pressure; and (c) external pumping.

Alternatively, the inner surface of a capillary may be covered with a plastic sensing chemistry whose color changes in accord with the concentration of the analyte, which in this case was carbon dioxide [12] or oxygen [13] (Fig. 5). Others have used two glass plates to form a capillary gap, and have deposited

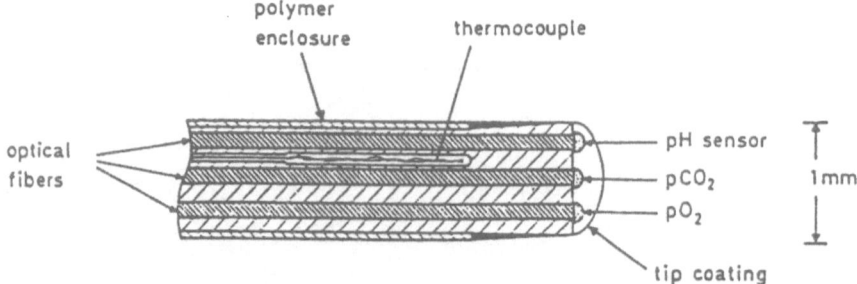

Fig. 2. Optical fibre triple sensor with a chemically sensitive coating at the tip of each fibre and designed for in-vivo monitoring of oxygen, carbon dioxide and pH [9]

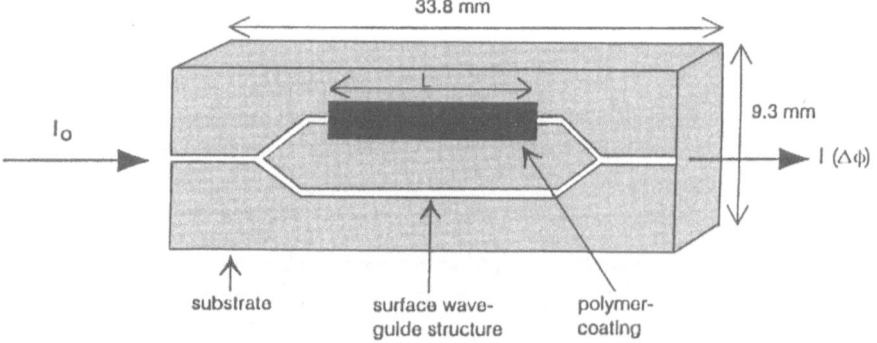

Fig. 3. Optical sensing using a bifurcated integrated optical waveguide with a chemically sensitive layer deposited on one arm [10]

a pH-sensitive coating on the inner wall of one plate to obtain a small pH-sensing element [14]. In both cases, evanescent wave spectroscopy is the preferred method of optical interrogation.

Although sol-gels were first used mainly in sensors other than optical, it is obvious now that they have their greatest potential in optical sensors. They may be used in both reagentless sensors and – in particular – in sensors based on the use of immobilized indicator dyes, reagents, and biomolecules. Consequently, the number of sol-gel based optodes (optical chemical sensors; from the Greek οπτοδε, "the optical way") [15] greatly increased in the past 10 years. In this chapter, we intend to give an account on the use of sol-gels in chemical sensors and biosensors. Rather than by method, the article is structured according to the kind of analyte detected with a sol-gel-based sensor chemistry.

A rather wide patent [16] covers the preparation of sol-gel glasses doped with organic or inorganic compounds, any materials of biological origin including enzymes, and the application of the resulting materials to analytical tests, as a chromatographic medium, in sensors, as a catalyst or biocatalyst, in electrodes and enzyme electrodes, and in any other detection device.

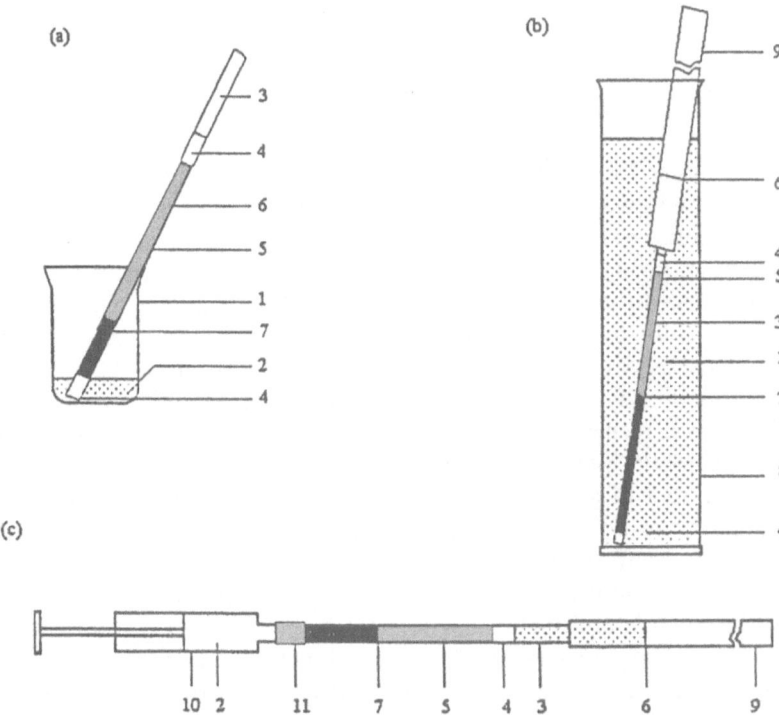

Fig. 4a–c. Schemes of tube detectors using alternative sampling modes: **a** by capillary force; **b** by hydrostatic pressure; **c** via external pumping. In all figures, (*1*) is a beaker, (*2*) the sample solution, (*3*) the capillary tube, (*4*) a paper plug, (*5*) the doped sol-gel powder, (*6*) the front of the solution, (*7*) the color front, (*8*) a cylinder, (*9*) an auxiliary (plastic) tube, (*10*) the syringe, and (*11*) the rapid connector [11]

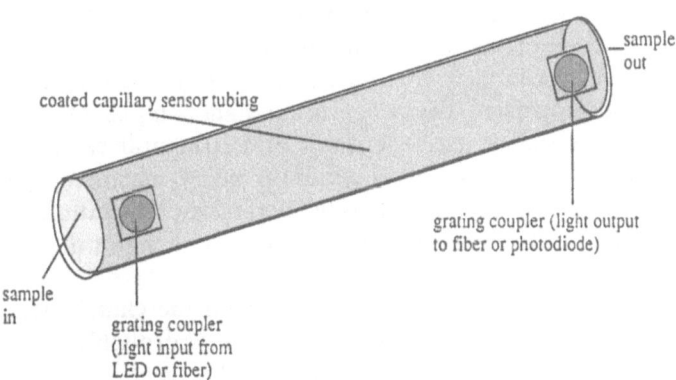

Fig. 5. Capillary sensor with an inner coating that responds to a chemical parameter by a change in its optical properties and which can be measured via the evanescent wave of the light beam coupled in and out at the two grating couplers

2 Techniques

2.1 Preparation of Sol-Gels for Sensors

Techniques for the preparation of sol-gels have been described in numerous books and reviews [1, 2, 17, 18]. The sol-gel process provides a great variety of possibilities to *modify* and to *taylor* mechanical, physical and optical properties of the resulting material. This includes pore size, refractive index, permeation and permeation selectivity, adhesion to other materials, chemical modification of the bulk glass (by co-condensing organic components or other tetraalkoxymetals), and chemical modification of the surface.

Organically modified silica (*ormosils*) and organically modified ceramics (*ormocers*), respectively, were obtained by use of organofunctional alkoxysilanes as one of the precursors. Organic modification results in reduction of cross-linking and of surface silanol groups. Furthermore the film adhesion to the support can be improved, polarity and hydrophilicity can be fine-tuned and partition coefficients can be altered. Reactive functional groups, such as amino- or vinyl groups may be introduced and offer the possibility of anchoring organic or bioorganic molecules.

The sol-gel process, with its associated mild conditions, offers a new approach to the synthesis of composite materials with domain sizes approaching the molecular level. Transparent organic-inorganic composites can be prepared by dissolving performed polymers into sol-gel precursor solutions, and then allowing the tetraalkyl orthosilicates to hydrolyze and condense to form glassy phases of different morphological structures. Alternatively, both the organic and inorganic phases can be simultaneously formed through the synchronous polymerization of the organic and inorganic components, the phase morphology, the degree of interpenetration, and the presence of covalent bonds between phases, the properties of these composites can vary greatly and range from elastomeric rubbers to high-modulus materials [19].

The use of nanocomposites in materials processing can lead to monophasic or multiphasic ceramics, glasses or porous materials, with tailored and improved properties. The properties of nanocomposites such as sol-gel, intercalation, entrapment, electroceramic and structural ceramic types were reviewed [20] and shown to be superior in many instances when compared to the monophasic or microcomposite alternatives. The utilization of nanocomposites in materials processing is expected to have a major impact on catalytic, sensor, and optical materials.

The effects of various parameters on the characteristics of sol-gel-derived dip-coated thin silica films have been thoroughly investigated [21, 22]. In essence, film thickness is a function of dip speed, aging time of the sol, and the water precursor ratio R. In the case of acidic catalysis (pH 1), increasing R leads to increasing film thickness. High R values (> 3) produce more stable films. While for these films the thickness was stabilized after 20 days, the thickness of films fabricated at R = 2 decreased gradually over a period of 60 days. Silica and silica/titania liquicoat films (commercial sol-gel solutions) also require ~20 days to stabilize, regardless of titania content [22]. Numerous films with refractive indices ranging from ~ 1.42 for a porous silica film to ~ 2.0 for a porous, undensified titania film were fabricated.

2.2 Entrapment of Foreign Molecules

All sol-gels used for sensing purposes contain various entrapped foreign materials such as ions, indicators and dyes, proteins, carriers, and the like. With respect to sensor performance, the method of entrapping foreign species is a most critical step because the sol-gel microstructure strongly affects (a) the diffusion of entrapped materials, (b) diffusion of the analyte penetrating the sol-gel or interacting with its surface, (c) the interaction between entrapped species and diffusing analyte; (d) the activity and thermodynamic properties of entrapped materials, and (e) aggregation.

The three most common methods for immobilization of foreign molecules are doping, impregnation, and covalent bonding. With respect to sensing applications, the resulting materials are expected to be resistant to leaching, especially for applications which require continuous monitoring. In many cases it is impossible to predict whether a dopant trapped by standardly used techniques will leach out or not, and at what rate. By trial-and-error fine-tuning of the preparation procedure, it has been possible to overcome this difficulty. However, a standardization of such procedures is badly needed for the efficient and rational design of sensors although a generally valid protocol is unlikely to exist in view of the variety of dopants with their differences in size, shape, and charge. Hence, encapsulation in a tightly-closed cage always will be a trade-off between leaching out and retainment of full activity.

2.2.1 Entrapment of Organic and Inorganic Dopants

Inorganic ions and organic molecules or ions having characteristics absorption and/or luminescence spectra can be incorporated into glasses formed by the sol-gel process from appropriate precursors. The typical spectra of the dopants change during the sol-gel transformation and may serve as optical probes for the process. Reisfeld & Jörgensen [23] have reviewed the variety of new materials that can be designed from the doped glass bulks or films. Some examples include nonlinear materials, tunable lasers, luminescent solar concentrators, and optical sensors. Typical glasses are doped by Co(II), Eu(III), Tb(III), Ru(bipyridyl), CdS, fluorescein and its derivatives, oxazine, 2,2'-bipyridyl-3,3'-diol, methyl orange, methyl red, acridine orange, and acridine yellow. Composite materials (either formed from porous inorganic glasses impregnated with a dissolved monomer, subsequently polymerized in the pores; or ormosils formed by simultaneous treatment of alkylsilicon alkoxides, or a precursor containing reactive glycidyloxopropyl substituents) can dissolve (otherwise almost insoluble) perylimide dyes, which are exceedingly photostable, and used in tunable lasers in the visible.

Sol-gel matrices doped with organic or bioorganic molecules are used as either *optically* active materials or as *chemically* active materials. The first group includes filters and lightguides, luminescent materials and laser components, and a variety of materials for information processing and recording purposes. The latter covers sensors and bioactive glasses, photoactive materials, and catalysts including enzymes [17, 24]. Finally, the advantages and the current limitations of the sol-gel immobilization procedure were also demonstrated for organic photometric reagents such as 1-nitroso-2-naphthol (for detection of cobalt ions) and o-phenanthroline (for determination of divalent iron) [25]. Similarly, various porphins [26, 27] and porphyrins [28] were immobilized in sol-gel matrices.

Organic reagents may be physically adsorbed, chemisorbed, or caged in porous supports [17]. Typical impregnation procedures are carried out by exposing the porous support to a concentrated solution of the reagent in organic solvent and discarding the solvent after equilibration, thus, after drying, obtaining a reagent-coated support. The technology is versatile in that the same solvents, manufacturing apparatus and impregnation protocol may be used to encapsulate different reagents in a variety of matrices. However, the adhesion of reagents to the support is rather weak, thus excluding in-vivo applications and limiting the practical operation and shelf-life of the sensors. Only when the leaching driving force is low (e.g. in case of water exclusion or gas analysis) can such a device maintain long operation life. Otherwise, impregnation methods are restricted to disposable or renewable devices.

Traditionally, imobilization through molding or chemical doping has been used to entrap organic compounds during the molding or polymerization of *organic* polymers. The advent of low temperature sol-gel synthesis of metal oxides paved the road to the development of similar inorganic/organic combinations. Generally, doped glasses exhibit properties which are intermediate between physical impregnation and chemical bonding. Doping technology is similar to impregnation techniques in its general application: the same procedure can be applied with little modification to encapsulate a plethora of organic reagents. As in impregnation technology, no derivatization of the chemical reagent is required prior to the encapsulation. Thus, reactivity and specificity of the dopant are generally maintained, except for effects of chemisorption and intraparticle microenvironment.

Since the cavities produced by the chemical doping are not hermetically sealed, at least a part of the doped reagent is free to move between them. This limited mobility is of value when the formation of a chromophoric group requires the participation of several ligands to form multiple ligand chelates. For example, iron(II) forms a three-ligands orange-red chelate with o-phenanthroline (ferroine), but chemical bonding of o-phenanthroline to silica glass was reported to produce only a two ligand chelate [29]. Observations of the relative mobility of dopants are supported by ESR (electron spin resonance) studies [30], demonstrating that polyamine copper(II) chelates entrapped in wet sol-gel alumina exhibit almost free tumbling motion. The mobility of reagents in the silica support also brings about better performance of the doped reagent and thus contributes to the increase in the practical capacity and detection range of silica sensors.

The penalty for these advantages is paid by increased leaching of organic reagents from doped glasses. Free motion is not restricted to the ability to form multiple ligands but also implies an ability to reach the solid–liquid interface and leak out. Fig. 6 demonstrates the leaching problem in the case of silica glass doped with 1-nitroso-2-naphthol [25]. This is a considerable drawback since even a small leak excludes in vivo medical applications. The long-term stability of doped silica sensors is expected to be intermediate between impregnated and chemically immobilized sensors. However, it would be premature at this early stage to draw conclusions regarding long-term stability and shelf life, although it is likely to be high.

Nevertheless, there is increasing effort to overcome the leaching problem [31]. Tetramethoxysilane polymerization at high acidity and low water content was found to result in non-leachable yet reactive matrices. A second possibility is the covalent bonding to pre-*prepared matrices*. Direct chemical bonding of organic reagents to silica supports gained much interest in parallel with the proliferation of chromatographic applications. For example, a siloxane bond (–Si–O–Si–C–) is formed by a condensation reaction of a chlorosilane reagent with surface silanol groups. This has been applied to modify the surface of sol-gels and has resulted in the covalent immobilization of pH probes and other indicators. This type of encapsulation is highly specific so that reaction conditions and precursors have to be tailored for each case. This limits the application of chemical derivatization to expensive devices or to general purpose tools amenable to mass production.

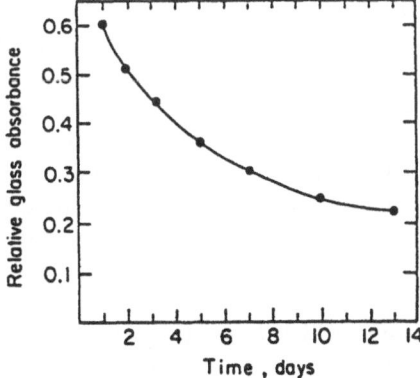

Fig. 6. Leaching of an indicator dye as evidenced by the decrease in the relative absorbance at 386 nm of a sol-gel glass plate doped with 1-nitroso-2-naphthol and immersed into water solution for 14 days [25]

Surface modification is another option. For example, the surface of a pre-prepared sol-gel was modified [32] with 3-aminopropyltriethoxysilane, the amino group of which formed a covalent bond with fluorescein isothiocyanate. Chemical bonding of the reagent can also be performed by a covalent link of the respective reagent to a sol-gel precursor such as an alkoxysilane monomer. Methylred was bound to trimethoxysilylpropylamine via dicyclohexylcar-bodiimide coupling [31], while aminofluorescein was covalently linked to 3-thiocyanatopropyl-triethoxysilane [33], and the resulting derivatives were used in a conventional sol-gel protocol with tetramethoxysilane (TMOS).

A promising method of immobilization by molding makes use of mixed organic/inorganic sol-gel precursors [34]. Rhodamine B, for example, was immobilized in a 50/50% mixture of TMOS and phenyl-trimethoxysilane under base catalysis to result in a non-brittle but highly permeable material that was used for sensing ammonia dissolved in water. The best results were obtained when the counterion of the rhodamine cation was an organic anion. The resulting sensor films, while having a response time in the order of several minutes, firmly retained the dyes, resulting in a constant fluorescence intensity even after prolonged exposure to water-sample solutions. The limits of detection were 10 ppb ammonia in water. In an alternative approach, the sol-gel layers were covered with an 8-μm film of either teflon or polydimethylsiloxane, both of which are gas-permeable but ion-impermeable. Dye leaching was completely suppressed, while the response to ammonia in aqueous solution was uncompromised, except for much longer response times which are in the order of 5 min for the forward and 10–15 min for the reverse response.

2.2.2 Entrapment of Biomaterials

Introductions to sol-gel encapsulation techniques for biosensors have been given in competent reviews [18, 35]. It is mainly the mild conditions under which sol-gels may be formed that have awakened interest in their use as immobilization matrices for biomolecules. It has been demonstrated [36, 37] that biomolecules (in this case enzymes) can be entrapped in sol-gel while retaining their activity. It turned out that enzymes were even stabilized by the sol-gel against thermal inactivation, and did not leach out [38, 39]. Entrapped biomolecules are retained in an aqueous microenvironment inside the pores of the silica glass which support retention of reactivity. However, conventional sol-gel protocols involve conditions, such as extremes of pH and high concentrations of alcohols, that promote aggregation and denaturation of proteins. In order to avoid denaturation and aggregation of the proteins due to low pH, the conventional procedures were refined [40] by first making the sol, then (i.e., after the first reaction has occurred after acid-catalyzed hydrolysis of TMOS) adding buffer which raises the pH to biological values, and finally adding the enzyme which then becomes entrapped.

The methoxy precursor tetramethyl orthosilicate (TMOS) is often used to encapsulate biomolecules because its methanol byproduct has a polarity closer to water than the less polar higher homologs. TMOS generally causes less denaturation. However, the large quantities of alcohol, (which is frequently used as a solvent) invariably prove harmful to protein stability. Transparent xerogels containing various enzymes were obtained by mixing a solution of an enzyme with TMOS at room temperature followed by gelation and drying [39]. Effective immobilization was usually obtained at initial pH values > 7, where there is a change in the gelation mechanism from predominant hydrolysis/condensation to predominant direct polymerization of silicate precursors.

Glasses with alkaline phosphatase showed 30% retention of enzymatic activity and improved stability to thermal deactivation compared to solutions [36]. During a storage period of two months at room temperature, no loss in activity was observed and a non-Michaelis-Menten-kinetic was found. Chitinase, aspartase and β-glucosidase have been entrapped as well.

Trypsin- and acid phosphatase-containing silica-gel glasses were obtained by mixing a solution of enzyme with polyethylene glycol 6000 and TMOS [38]. The activity toward small substrates equaled that of the enzyme in solution. Polylysine ($M_r < 13,000$) and aprotinine (M_r 6,500) inhibited, while larger polylysines as well as soybean trypsin inhibitor (M_r 20,100) were ineffective. The firm interaction between the protein molecules and the silica matrix stabilized the enzyme. Thus, the half-life of sol-gel-entrapped acid phophatase at 70 °C (pH 8.0) was two orders of magnitude larger than that of the enzyme in solution.

The immobilization of proteins other than enzymes by entrapment in optically clear, porous glasses prepared by sol-gel techniques appears to be a promising approach to optical affinity sensing. However, little is known about the physical environment of the immobilized protein or the mechanism(s) of entrapment. Absorbance and fluorescence spectroscopies have been used to characterize the properties of bovine serum albumin (BSA) entrapped in wet sol-gel glass bulks [41]. The fluorescence behavior of dissolved and entrapped BSA in the presence of acid, a chemical denaturant, and a collisional quencher was examined. The results show that a large fraction of the BSA added to the sol is entrapped within the gelled glass in a native conformation. However, the reversible conformational transitions that BSA undergoes in solution are sterically restricted in the gel.

Copper–zinc superoxide dismutase (CuZnSOD), cytochrome c, and myoglobin retained their characteristic reactivities and spectroscopic properties when encapsulated in silica glass under the mild conditions of the sol-gel technique [40]. Chemical reactions of the immobilized proteins could be monitored by means of changes in their visible absorption spectra. Encapsulated CuZnSOD was demetallated and remetallated, encapsulated ferri-cytochrome c was reduced and reoxidized, and met-myoglobin (Mb) was reduced to deoxy-Mb and then reacted either with dioxygen to make oxy-Mb or with carbon monoxide to make carbonyl-Mb. Such studies give excellent prospects for the use of these novel materials in biosensors.

Also, the photosynthetic membrane protein of bacteriorhodopsin (bR) was encapsulated in an optically transparent and porous silica matrix using a modified sol-gel procedure [42]. The absorption spectra and the kinetics of the photocycle characteristic of the proton pumping function of bR were studied systematically throughout the different stages of the glass formation process. This new biomaterial was characterized by means of its optical absorption, circular dichroism (CD), and Raman spectra, its photocycle kinetics, the characteristic activation parameters of its photocycle, and its deionization and cation regeneration properties. Surprisingly, the bR structure, the local structure of the retinal chromophore, and the proton pumping function of bR were not affected by the encapsulation process. It was also found that the bR glass formed allowed transport of small ions such as Ca^{2+} into and out of the glass medium, and those ions were found to affect the properties of the protein just as they do in aqueous suspensions. When immobilized, the protein retains its light-sensitive properties [43].

In order to study the effects of biogel ageing on the dynamics of bovine and human serum, albumin was monitored by first labelling the protein with acrylodan (a polarity-dependent fluorescent probe) and then entrapping it into a sol-gel-type silica [44]. Effects of ageing and drying were studied by following the acrylodan steady-state and time-resolved emission, the decay of anisotropy, and the dipolar relaxation kinetics. The results indicate that there is a substantial amount of nanosecond and subnanosecond dipolar relaxation within the local environment surrounding cystein-34 in both proteins, even when they are fully encapsulated in a dry biogel. Time-resolved anisotropy experiments show that the acrylodan residue and the protein are able to undergo nanosecond motion within the biogel. The results suggests that the "pocket" hosting the acrylodan reporter group opens as the biogel dries.

Finally, it has been demonstrated that even whole cells may be immobilized in sol-gel [45]. Yeast cells (*Saccharomyces cerevisiae*) were incorporated into a sol-gel layer on a glass slide. The immobilized cells displayed uncompromised activity and kinetic behavior during degradation of sucrose. The morphological and other physical features of silica sol-gels did not exclude full viability and cell reproduction.

2.3 Probing the Structure of Sol-Gels

Sol-gels do undergo temporal changes in their physical and chemical structure, and this must be kept in mind when characterizing chemical sensors. The effects of ageing have been demonstrated in numerous, mainly spectroscopic studies. Luminescent molecules have been widely used as probes of the sol-gel process. The efficiency of the proton transfer in the excited state of 8-hydroxy-1,3,6-pyrenetrisulfonate to surrounding water molecules, for example, is a sensitive fluorimetric probe for following directly and in detail the kinetics of water consumption during the early stages of the TMOS sol-gel polymerization process [46]. Changes in the water/silane ratio and the pH of the sol were found to markedly affect the kinetics of water consumption. Fluorescent probes may also be used to obtain unique insights into local chemistry and structure during the sol-gel-xerogel transition [47].

In an alternative approach [48], two types of silica glass were doped with o-phenanthroline (o-phen) (S2) and without organic reagent (S1). Paramagnetic probes [Mn(II), VO(II), Cu(II)] were introduced into both systems by doping or by impregnation. The sol-gel-xerogel transition in these systems was investigated by electron paramagnetic resonance (EPR). In the S1 system, hydrated metal ions were shown to react weakly with silanol groups during each stage of silica polymerization. The effect of sol-to-gel transition is not observed in S1 spectra. Contrary to this, EPR spectra of VO(II) ions in S2 demonstrate significant changes during sol-gel-xerogel transition. Free radical spectra parameters (rotation correlation time, isotropic hyperfine interaction parameter and line width) reflect changes in viscosity and polarity in the vicinity of the spin label [49].

Static and dynamic fluorescence spectroscopy of rhodamine 6G (R6G) in a sol-gel matrix was used to study the effect of ageing time and hydrolysis pH on the local micro-viscosity [50]. These results demonstrate that the sol-gel formation cycle is composed of several distinct regions and that the average local microviscosity sensed by R6G changes by 2 orders of magnitude (!) throughout the cycle. However, the probe is never completely immobilized as in a vitrified solvent. Discrete microdomains are postulated and discussed in terms of a simple model. When R6G-doped thin sol-gel films were cast on glass microscope sildes and characterized with the help of steady-state and time-resolved fluorescence spectroscopy [51], the excited-state decay kinetics of the highly doped films show clear evidence for R6G aggregation.

The process of fluorescence resonance energy transfer (FRET) between rhodamine 6G (the donor) and malachite green (the acceptor) contained within a wet sol-gel matrix has been investigated as the time of ageing for the gel was increased [52]. The efficiency of the FRET process was seen to increase as the ageing time, and hence the pore diameter, was increased. The opposite to this would have been expected if a volume change had been the physical quantity reported on. The observations were interpreted in terms of a predominant 3-dimensional transfer process across the pore. Furthermore, the initially rough surface became smoother as the silica repositioned itself through the process of dissolution and redeposition. Therefore, the changing physical quantity reported on in these measurements is surface area. Using the data from the FRET measurements, it has been possible to calculate the surface area of the gel at various points in the ageing process. The surface area was seen to change from approximately 448 $m^2 g^{-1}$ in the unaged wet gel to 254 $m^2 g^{-1}$ in a gel that had undergone ageing for 7 h.

The luminescence of inorganic luminophors may be used to monitor structural changes in sol-gels prepared by firing [53]. Sol-gel membranes were doped with zinc oxide composed of nano-sized wurtzite crystals and supported on either quartz slides or silicon wafers. The change in visible luminescence intensity during firing of the membranes was examined from the standpoint of the change in surface properties of the ZnO crystals. Membranes fired at temperatures lower than 180 °C exhibited strong visible luminescence, but the intensity decreased markedly when the membranes were fired at temperatures above 180 °C. The structure changes dramatically during firing of the membrane. Most significant changes occur at around 180 °C.

A review [54] of the processing of organic-doped sol-gel glasses for optics and electrooptics also covers the utilization of the properties, the sensitivity and the photophysical behavior of the trapped organic molecules to study the structural changes that occur in a polymerizing system of silicon alkoxides in ceramer composites and in a reversed-phase silica. Similarly, fluorescent polarity probes have been used to study the micro-polarity and its temporal variation of inorganic and mixed organic/inorganic sol-gels [33].

The durability, high adsorption capacity based on large surface area, and the chemical selectivity based on controlled pore size as well as acid/base, ion exchange, or chelation chemistries make sol-gel coatings very suitable for acoustic wave devices [55]. Their porosity can be determined by N-adsorption isotherms and the chemical sensitivity and selectivity obtained with this class of coatings is illustrated using several examples including hydrous titanate ion exchange coating, zeolite/silicate microcomposite, and surface-modified silicate films.

2.4 Optical Sensing Techniques

Various methods of spectroscopic interrogation have been applied to monitor changes of the optical properties of sol-gel sensors [56]. In the simplest case, a glass support with a chemically reactive sol-gel coating is placed in a spectrometer and analyzed. There are, however, more sophisticated schemes used in optical sensing. These include the use of optical waveguides such as planar glasses, optical fibre systems and integrated optics, of disposable cuvettes, and of advanced spectroscopic techniques including evanescent wave spectroscopy. A final trend is towards miniaturization and near-field microscopy, both of which hold great potential with respect to sensing applications. These techniques will be briefly discussed in this section.

Sol-gel-derived photometric sensors particularly attracted the attention of analytical chemists because of the favorable optical properties of silica. In the vast majority of cases, the reagent was doped into a sol-gel matrix and deposited on planar glass supports. When exposed to a sample, the analyte interacts with the immobilized reagent and alters its optical properties. This is quantified by a photodetector using the typical arrangements shown before. Indeed, all major types of photometric sensors, including absorbance, have been successfully demonstrated in doped sol-gel matrices as will be shown below. However, the sol-gel may as well be deposited on more complex optical structures such as integrated optical waveguides, and interrogation is preferably performed then by evanescent wave spectroscopy in either absorption or fluorescence.

2.4.1 Fiber-Optic Sensors

This type of sensors has aroused particular interest because of several distinct advantages over conventional sensors [9]. Four modes of fiber-optic photometric sensors have been proposed. In the simplest, optical fibers are used only to guide the light to and from the doped sol-gel sensor. This approach was applied to sensing pH in both the transmission and reflection mode (Fig. 7A,B), and in fluorescence (Fig. 7B) [57, 58, 59]. The sol-gel may also be used to modify the optical fibre itself (Fig. 7C,D). An unclad section of a homemade, porous optical fibre was soaked with a starting sol-gel solution containing the pH indicators bromocresol purple or bromocresol green [60]. This configuration (Fig. 7C)

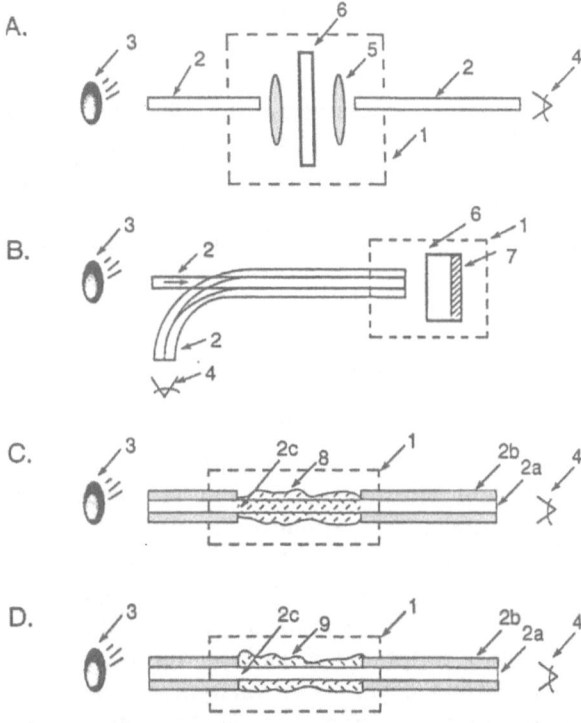

Fig. 7A–D. Configurations of reagent-doped sol-gel waveguides [56]. **A** conventional, direct mode sensor; **B** fluorescence- or reflection-mode sensor; **C** sol-gel modified porous optical fiber; **D** evanescent wave optical fibre sensor. In each case: (*1*) is the sample; (*2*) an optical fibre where (*2a*) is the core, (*2b*) the cladding, and (*2c*) the unclad section; (*3*) is a light source; (*4*) the photodetector; (*5*) an auxiliary coupler; (*6*) the sol-gel sensor; (*7*) an optional mirror; (*8*) doped sol-gel material, and (*9*) the doped sol-gel film

benefits from a long optical pathway, short analyte diffusion length, and simple optical coupling auxiliaries. These conditions resulted in a response time of < 10 s and no detectable leaching of the reagent.

A fourth approach (Fig. 7D) [61] is based on evanescent wave fluorescence excitation, involves pH-sensitive sensors made by dip-coating unclad, fused-silica optical fibres with a fluorescein-doped sol-gel film (Fig. 7D). Such evanescent wave sensors benefit from a relatively simple apparatus because optical coupling is not required. Additional advantages include long (though indirect) optical pathlength and relatively low dye bleaching, which is often responsible for loss of sensitivity in laser-based sensors. A similar arrangement based on the quenching of the fluorescence of doped ruthenium complexes by oxygen was used for oxygen sensing [62]. Reversible and fast (seconds) response as well as zero leaching were observed in both cases.

Rather than coating the fibre with a doped gel, a microsphere may be placed at its tip. Respectively doped porous silica microspheres can be prepared from liquid solution at near ambient temperatures, and their diameters controlled to between 5 μm and > 1 mm. The resulting microspheres can be attached to the distal end of an optical fibre in which the proximal end is attached to a spectrophotometer. Depending upon the organic species doped into the microsphere, a wide variety of sensing functions are possible [63]. The use of microsensors for measuring pH, temperature, and the solvent content of aqueous solutions has been demonstrated. For example, fluorescein-doped porous silica microspheres were immersed in aqueous solutions of various pH values. Fluorescence emission was measured after a few minutes of immersion, and a significant variation in fluorescent intensity, particularly for pH values between 1 and 7, was observed.

2.4.2 Integrated Optical Structures

An alternative to optical-fibre sensors is the planar integrated optical waveguide (IOW), a substrate-supported dielectric film, typically less than 3 μm thick, in which light can propagate via total internal reflection in only one or a few discrete modes [64]. Applications of IOW in optical chemical sensing have been partly limited by the difficulty of producing high-quality glass IOWs. The fabrication of IOWs by the sol-gel method from methyltriethoxysilane and titanium tetrabutoxide precursors was described, and the physical, chemical, and optical properties of the films during and after high-temperature annealing studied using a variety of analytical techniques. The results show that the catalyst used to accelerate the sol-gel reaction strongly influenced the optical quality of the IOW. Catalysis by hydrochloric acid produced waveguides with propagation losses close to $1 \, dB \, cm^{-1}$, whereas in the case of $SiCl_4$ catalysis, propagation losses were $< 0.2 \, dB \, cm^{-1}$, a value significantly less than any previously reported for sol-gel-derived IOWs. The use of $SiCl_4$ is thought to retard formation of a micro-heterogeneous network containing Si-rich and

Ti-rich domains, which is favored under acid catalysis and contributes to the higher observed losses. Atomic force microscopy images of the surfaces of a $SiCl_4$- and an HCl-catalyzed IOW are shown in Fig. 8.

Planar IOWs have several advantages over multimode fibre optics in that (a) they can be fabricated from a wider variety of materials; (b) the planar geometry is more compatible with established surface modification and deposition technologies; (c) the electric field distributions of the polarization-conserved-mode structure are spatially inhomogeneous and easily calculated; and

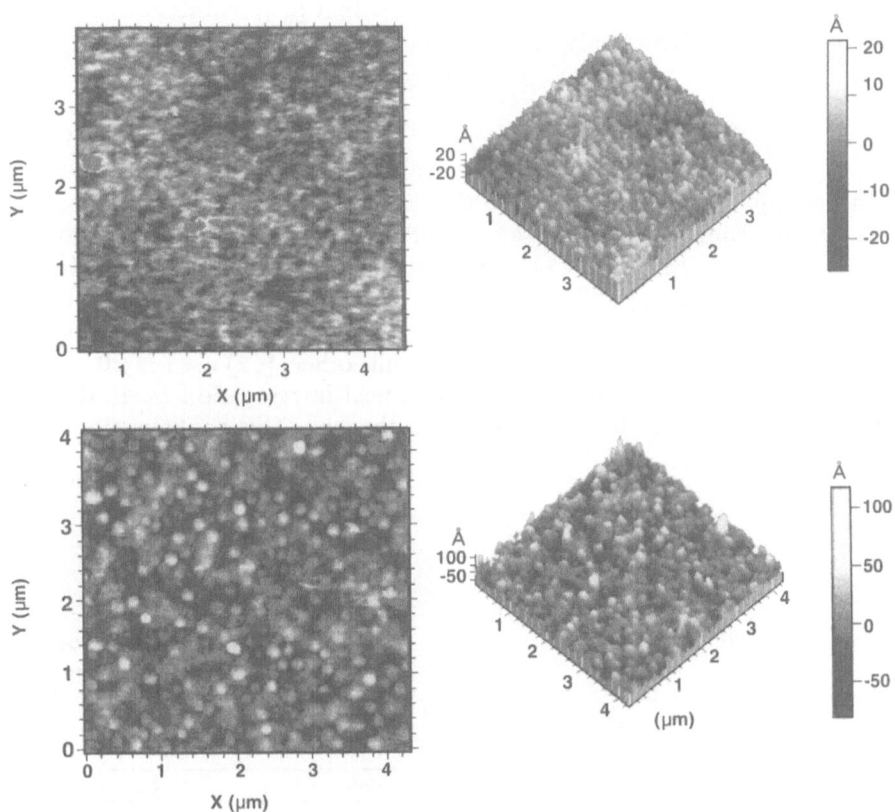

Fig. 8. Two- and three-dimensional atomic force microscopy images of the surfaces of integrated optical waveguides. *Top*: $SiCl_4$-catalyzed fabrication; *bottom*: HCl-catalyzed fabrication. The respective scale bar (in Å) for each set of images is displayed on the right. Two anomalously high spots (likely dust) were removed from the 2-D image of the $SiCl_4$-catalyzed IOW to show the underlying surface texture and to allow direct comparison of the surface roughness of the two samples. These spots are plotted as a uniform gray color [64]

(d) the much higher density of total reflections (up to several thousand per centimeter of beam propagation) yields a concomitant increase in the evanescent path length.

These advantages have been exploited in recent applications of planar IOW spectroscopy to research in thin-film structure, surface characterization, and chemical sensing. Several groups have used the IOW geometry to excite Raman scattering from polymer films [65, 66], Langmuir-Blodgett multilayers deposited on waveguide surfaces [67], and protein films [68, 69]. Planar IOW chemical and biochemical sensors based on ATR, fluorescence, and refractive index detection have been reported [10, 70–73].

Another interesting optical sensing platform was presented that consists of two, sol-gel-derived, submicrometer thick glass layers supported on an optically thick glass substrate [74]. The lower layer is a densified glass that functions as a planar integrated optical waveguide. The upper layer is an undensified glass of lower index doped with an optical indicator that is immobilized, yet remains sterically accessible to analytes that diffuse into the pore networks.

2.4.3 Other Waveguide Configurations

Sensing applications often require unusual sensor configurations; for example, planar and capillary tube waveguide configurations are useful in flow cell detectors. Lee & Saavedra [14] have devised a planar waveguide pH sensor made from a microscope slide coated with bromophenol blue-doped thin silica film (Fig. 9). In this configuration, the microscope slide operates as a waveguide and the doped film as a sensing cladding, while others [12] used capillaries with a reagent-doped inner coating (Fig. 5). Optical interrogation is via the evanescent wave of a laser beam coupled into the wall. This configuration can be

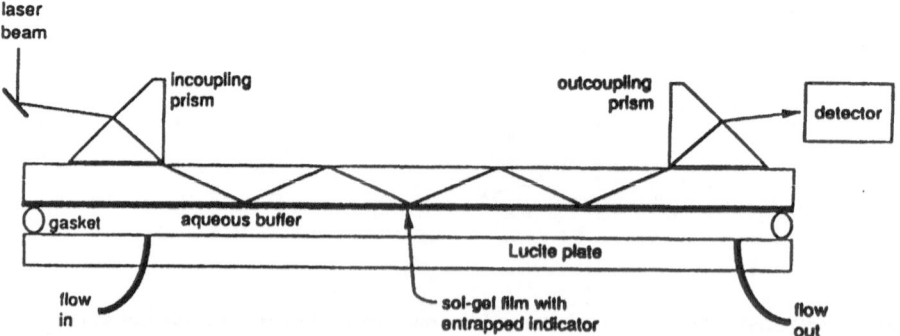

Fig. 9. Instrumental arrangement for ATR spectrometry of a sol-gel film coated on a glass slide. The 458-nm line of an argon ion laser was coupled into and out of the coated slide using a pair of fused silica prisms. The slide was mounted in a flow cell that allowed solutions to be introduced to and flushed from the coated side of the slide [14]

used for flow-through detectors and in disposable capillary samplers (similar to sol-gel-derived tube detectors) and has been described in numerous modifications [75].

2.4.4 Microsensors

Recently, optical techniques have been reported that enable the development of submicron-sized optical fibre sensors [76, 77]. The technology is based on nano-fabricated optical fibre tips and near-field photo-initiated polymerization. Multimode or single mode optical fibers are drawn into submicron optical fibre tips and then coated with aluminum to form submicron optical fibre light sources. Submicron pH sensors, as an example for small sensors, have been prepared by incorporating fluoresceinamine into an acrylamide copolymer covalently attached to a silanized fibre tip surface by photo-initiated polymerization. The sensors have demonstrated their spatial resolving abilities in measuring the pH of buffer solutions inside micron-size holes in a polycarbonate membrane. These submicron pH sensors have millisecond response times due to their extremely small sizes.

2.4.5 Near Field Optical Sensing

The reduced size of present day light sources along with the enhanced molecular-excitation cross section and the good spectral and time resolution have enabled the development of rugged, ultraslim, ultrasensitive, and ultrafast fibre-optic chemical sensors that require only attoliters of samples, zeptomoles (i.e., 10^{-21} moles) of analyte, and yet have milliseconds response times [78, 79]. In addition, the 100-nm size sensors are small enough to slip in and out of a cell's membrane without any damage or leakage. Such pH sensors have been used to investigate blood cells and, in particular, rat embryos, high quality, non-perturbative, in vivo measurements [76, 79] have contributed to new information on organo-genesis stages. These include studies on anaerobic-to-aerobic transformations, and on physical and chemical insults (including changes in pH, consumption of oxygen, or the addition of new drugs to the environment) during particular time windows.

In some analytical and, in particular, medical applications of sensors, miniaturization is indispensible. The feasibility of miniaturizing a sol-gel sensor down to the micron scale for near field optical pH sensing was demonstrated [31, 77]. For this purpose, a metal-coated micropipette was filled with a silica xerogel doped with the pH-sensitive fluorophore pyranine.

2.4.6 Evanescent Wave Sensing

When light propagates in an optical fibre or waveguide, a fraction of the radiation extends a short distance from the guiding region into the medium of lower refractive index which surrounds it. This evanescent field, which decays exponentially with distance from the waveguide interface, defines a short-range sensing volume within which the evanescent energy may interact with molecular species. Optical waveguide sensors for chemical and biological species based on such evanescent wave (EW) interactions have attracted considerable research interest [80]. Two distinct approaches may be adopted in these sensors. First, the evanescent wave can interact directly with the analyte if the interrogating wavelength coincides with an absorption band of the species. Such *direct spectroscopic* EW sensors are of particular interest in the infrared spectral region, where many species absorb strongly. Alternatively an intermediate reagent, which responds optically (e.g., by fluorescence or absorption change) to the analyte, may be attached to the waveguide. Often *reagent-mediated* EW sensors provide greater sensitivity than direct spectroscopic devices. MacCraith [81] elaborated theoretical considerations regarding the critical design parameters for EW sensors. Sol-gel coatings have been proposed as both direct spectroscopic and as reagent-mediated materials, if doped in the latter case. Their specific advantages and enhanced performances have been described [80, 81]. Although the examples given concentrate on optical fibres, the principles are equally applicable to planar waveguide devices.

3 Sensor Applications

The increasing popularity of sol-gels in sensor applications results from a number of interesting features including the following: (a) Sol-gels can manufactured at low temperatures and physiological pH's; (b) their micro-structure can be controlled to some extent; (c) they are optically transparent from the UV to the near-infrared; (d) the index of refraction can be varied to some extent; (e), they provide a certain permeation selectivity in that only small ions and molecules can enter their network, while large molecules such as proteins and whole cells remain outside the matrix (hence, interactions inside the matrix are usually limited to small molecules and ions); (f) they are chemically and mechanically stable; and (g) the fabrication procedure enables any desired geometric shape to be tailored. Typical configurations of doped sol-gel glass are, for example, thin films deposited on to glass plates or microscope slides, tubes, powders, or monolithic discs (see also ref. [82]), to say nothing of optical waveguides.

Sol-gels with immobilized reagents, such as organic or bioorganic molecules, can be found in a large number of variations. These combine the processing versatility of sol-gel technology, the photometric and chemical properties of silica and metal oxides, and the specificity and optical properties of their organic and bioorganic dopants. The diagnostic applications of this new type of material, their advantages, limitations and future prospects, also with respect to sensors, have been reviewed [17, 18, 23, 24, 35, 82, 83, 84], the recent short review by Lev et al. [56] being a most readable introduction to the immobilization of organic and biomolecules in sol-gels, and their use as sensors.

3.1 Studies of General Interest

3.1.1 Gas Diffusion

The microviscosity of the interior of sol-gel matrices has been studied by static and dynamic fluorescence spectroscopy using rhodamine 6G (R6G) as a molecular reporter [50]. Obviously, the sol-gel formation cycle is composed of several distinct cycles. Changes in structure and microviscosity both affect diffusion inside a matrix. Sol-gel coatings were deposited on tin-oxide films and the gas-sensing properties were studied [85]. It was found that hydrogen-sensitivity of the tin oxide film in the double-layer SiO_2/SnO_2 structure increased with the thickness of the sol-gel film. Significant enhancement in sensitivity was also achieved with the coating films having a large volume fraction of small pores.

Gases such as hydrogen sulfide are capable of penetrating a sol-gel network. This was demonstrated by entrapping lead and cadmium ions into a silica glass and reacting it with hydrogen sulfide which resulted in the formation of microcrystalline PbS and CdS, respectively [86, 87]. Similarly the penetration of oxygen into the interior of a sol-gel was studied [88] by doping it with pyrene. The fluorescence of pyrene is strongly quenched even when the fluorophor is located in the deep interior of the xerogel, the activation energy for penetration being 18 kJ mol^{-1}. Pyrene was also covalently immobilized onto micro-porous glass [89] and this material served as the sensing layer of a very fast-responding oxygen sensor. Again, quenching was found to be very efficient even though the fluorophore is partially located in the inner cavities of the glass support.

3.1.2 Effects of Temperature

Most kinds of sensors are strongly dependent on temperature. In the case of optical sensors, one of the reasons is the temperature coefficient of the refractive index of the material used. A set-up has been described [90] for varying the temperature of planar optical waveguides that are useful in sensing applications from 5 to 90 °C. The principle is to control the temperature of the cover medium (in this case water). The propagation constants of guided modes are measured via a grating coupler. Only that part of the waveguide in the immediate vicinity of the incoming external light beam is heated or cooled. The temperature coefficients of the thickness and refractive index of the waveguide material can be described by polynomial expressions.

Sol-gels have been used in electrodes, but much less frequently than in optodes. They have been used [91] to modify platinum and transparent indium oxide electrodes. The silicone dioxide layer reduces electron transfer to the electrode but does not completely prevent it. This was exemplified in a study on the electrochemical and photo-electrochemical properties of the $Ru(bipy)_3^{2+}$ cation.

3.1.3 Electrical Conductivity and Electrochemical Sensing

Sol-gel-derived ceramic carbon electrodes (CCE) made of organically modified silica and carbon filler are highly versatile electrochemical tools [2–94]. A comparison of several types of carbon powders reveals that higher carbon loading and large surface area electrodes can be attained by the incorporation of dense graphite powder [92]. When high surface area, small size carbon-balck powder is used, a homogeneous distribution of microelectrodes, separated by insulating modified silica is formed. This ensemble of microelectrodes increases the sensitivity of the CCE by more than two orders of magnitude as compared to glassy carbon electrodes and graphite CCEs. A comparison of the voltametric characteristics of sol-gel-derived indicator CCEs with other classes of graphite electrodes has been performed [93]. Similar to the carbon paste electrodes, the indicator CCEs can easily be renewed by mechanical polishing with good repeatability. Despite their rough surface the CCEs exhibit less background current compared with glassy carbon electrodes, though it is still much larger than that of the carbon paste electrodes.

An interesting application of sol-gels is their use in composite electrodes made of sol-gel-derived carbon-silica materials [95]. It has been shown that modified porous composite carbon-silica electrodes can exhibit hydrophobic or hydrophilic surface characteristics and can serve as an indicator (inert) electrode, as a potentiometric electrode, and in amperometric sensing and biosensing. The composite carbon ceramic electrodes are rigid, porous, easily modified chemically and have a renewable external surface. The electrodes offer higher stability than carbon paste electrodes, and they are more amenable to chemical modification than monolithic and (organic) composite carbon electrodes. Experimental examples demonstrating the scope of electroanalytical applications of the composite carbon-silica electrode and its modifications have been presented.

3.2 Gas and Vapor Sensors

This chapter is related to sensors for measurement of gases such as oxygen, carbon dioxide, methane, and vapors of liquid species such as water (i.e. humidity), and ethanol. Again, the sol-gel can act as a support for organic molecules or biomolecules, and at the same time as a perm-selective barrier to keep out potential interferents. In addition, certain dyes have higher luminescence quantum yields in sol-gels with uncompromised indicator properties than in water or organic solvents. A typical example is the strong enhancement of the $Ru(bipy)_3^{2+}$ complex when placed into a sol-gel [96].

Sensing gaseous oxygen is a subject of general interest, as it has numerous applications such as in catalytic converters, in biotechnology, medicine, biosensing, and environmental monitoring. Intrinsic evanescent wave fibre optic sensors for oxygen have been reported [62, 97]. The sensors are based on the quenching of fluorescence from a metal-organic ruthenium complex trapped in the nanometer-scale cage-like structure of a sol-gel-derived porous film on a declad section of a multimode optical fibre. The sensor exhibits excellent performance using 488-nm Ar ion laser or blue LED excitation and silicon photodiode detection, and establishes the viability of low-cost portable sensor devices based on the sol-gel process. Gas phase measurement data over the range 0–100% oxygen exhibit high signal-to-noise ratio (approximately 150), good repeatability and a response time of < 5 s. A typical response curve is shown in Fig. 10.

Sol-gel derived silica films have been optimized for gas-phase oxygen sensing and dissolved oxygen sensing [98]. The ratio R of water to precursor was found to be important for tuning the range of the detectable concentration. For R = 2, the response is highly non-linear and this film shows a high sensitivity to oxygen at low concentration. At R = 4, the response is more linear and this film is sensitive over a wider range of oxygen concentration. In aqueous environments, the quenching response is in general lower than in gas phase.

The design potential of the sol-gel process for sensor fabrication is demonstrated by the achievement of a substantial increase in sensitivity when the process parameters are adjusted to increase the pore volume. Similarly, the oxygen permeability of sol-gels may be tuned by adjusting the composition of the organosilane precursor, resulting in various organic–inorganic composite coatings of variable hydrophobic character and pore sizes [99]. The oxygen

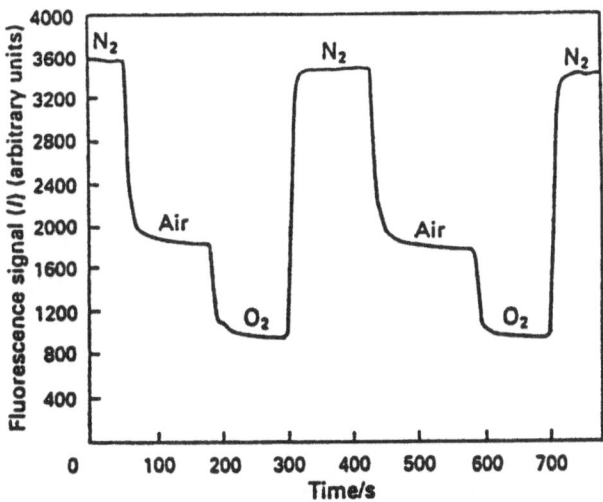

Fig. 10. Response of optical fibre oxygen sensor coated with a silica sol-gel film containing ruthenium (tris-diphenylphenanthrolin) on cycling from nitrogen to air to oxygen, and back [62]

permeability was measured using the selective quenching by oxygen of phosphorescent probes, such as platinum octaethylporphine. The application of such sol-gel coatings for optical fibre oxygen sensors was discussed as well.

A new class of sol-gel-derived electrochemical gas sensors (CCEs) employs a homogeneous dispersion of catalyst-modified carbon powder in porous, organically-modified silica [100]. Their hydrophobic surface rejects water, leaving only a very thin layer at the outermost surface in contact with the electrolyte and thus minimizing the effects of liquid side mass transfer. Heat-treated cobalt-tetramethoxymesoporphyrin-modified CCEs exemplify this new class of gas sensors. A schematic diagram of such as sensor is shown in Fig. 11.

Detection of several gases by anodic (sulfur dioxide) or cathodic (carbon dioxide and oxygen) reactions is demonstrated. All the calibration curves for oxygen exhibited zero intercept and linear response in the range from 0 to 0.21 atm oxygen. In the 0.21 to 1 atm range, the current density overpotential is best described by an electrochemical charge transfer step, preceded by a reversible Langmuir-type oxygen adsorption. The kinetics and mechanism of Co-porphyrin-catalyzed oxygen reduction by sol-gel-derived hydrophobic CCEs has been studied [101]. At low cathodic overpotential, oxygen reduction in acidic solution on heat-treated Co-tetramethoxymeso-porphyrin-supported graphite powder exhibits first-order kinetics with respect to adsorbed di-oxygen, which obeys the Langmuir adsorption isotherm. The activation energy of oxygen reduction equals $7.5 \pm 0.4 \text{ kcal mol}^{-1}$. Half-order oxygen dependence is

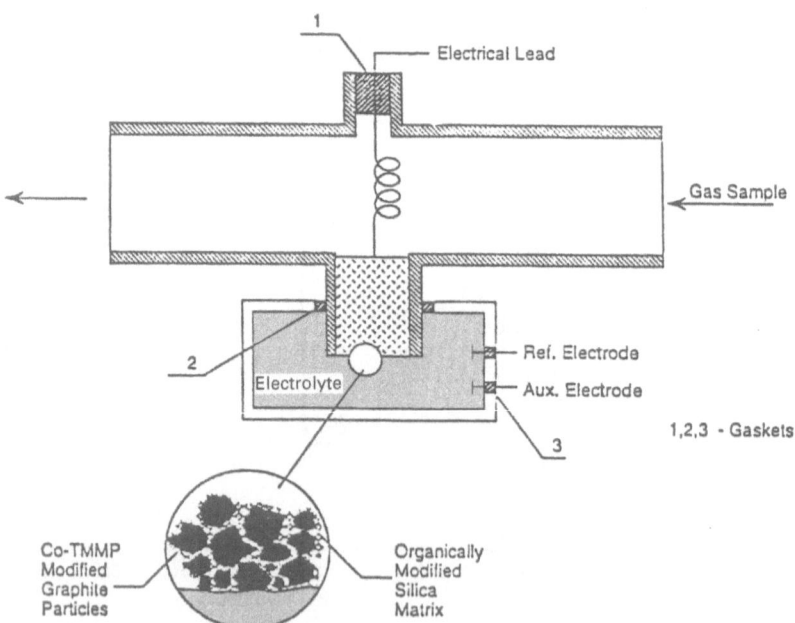

Fig. 11. Schematic of a composite carbon-ceramic electrochemical gas sensor [100]

observed at high overpotentials, and the reaction is limited by the rate of hydrogen peroxide conversion.

Similarly, palladium-modified CCEs were prepared by mixing a Pd salt and carbon powder with sol-gel precursors and subsequent reduction of palladium ions after xerogel formation [94]. The morphology and surface characteristics of the composite materials were examined by electron microscopy, X-ray microanalysis, nitrogen adsorption and water-wetting angle measurements. Oxygen reduction was used as a model reaction to probe the characteristics of the wetted section of the Pd/CCEs. Voltametric studies and polarization curves were employed to elucidate the size and structure of the active layer as a function of palladium content.

High-temperature sintering of porous pellets of metal oxides such as tin dioxide to form flammable gas sensors, frequently referred to as *Tagushi* sensors or *Figaro* sensors, remains one of the most straightforward and popular fabrication methods in use today, but is unsatisfactory in many ways, including lack of control over the structure and morphology of the pellets. The performance of Tagushi sensors strongly depends on the morphologies and environment of the metal oxide. Typically, such sensors respond to oxidizable, i.e., flammable gases and vapors including hydrogen, methane, and ethanol. In addition to the poor control of its structure, the chemical and physical role of catalytic dopants is often unclear and further sintering during operation can lead to long-term drift and unreliable behavior. The use of sol-gels as starting materials for better control of the composition and morphology is expected to result in distinctly improved sensor performance.

Thus, a sensor for ethanol vapors was obtained [102] by producing small chips of highly porous (> 50% v/v) semiconducting SnO_2/Pd by a sol-gel process combined with photochemical Pd-deposition. Heated to 300–400 °C in air, the chips exhibit an electrical *conductance* which is nearly linear with ethanol-concentration and shows only a low humidity-dependence. High uniformity of the material yields gas sensors with a narrow distribution of sensitivity.

A method has been developed [103] for making gas sensors of high specific surface sol-gels on tin oxide via the sol-gel method: this has resulted in the development of highly sensitive flammable gas detectors. To reduce the hydrolysis rate and therefore to obtain a well linked network $Sn(O-^tBu)_4$ was chosen as the precursor. Synthetic routes to obtain tin oxide, applying organic template with and without removal of the template, and the use of a template with binder, have been used. Response to 1% methane and to low ppm levels of toluene vapor in air could be shown. Comparative studies also showed that the sensor prepared by first using an organic template and then removing the template, followed by hydrolysis of the organometallic precursor resulted in improved response to toluene when compared with a commercially available tin dioxide sensor. The kinetics of the transient conductance change following a change in sensor temperature have been modelled using a barrier-layer at the approach to grain boundaries and been shown to fit an empirical model.

Because sol-gels are transparent to near-infrared light, they lend themselves to NIR-based spectroscopic sensing of gases that display useful NIR absorption. This has been exploited [81] for developing a sensor for methane in which the NIR overtone of methane gas is being measured in a fibre optic device where the core is coated with porous sol-gel.

To enhance the stability and selectivity of metal oxide conductivity sensors for organic gases in air, a new sensor concept has been realized [104]. A special chemical vapor deposition technique, based on the hydrolysis of tetraethylorthosilicate, was developed to produce selectively permeable SiO_2 coatings of a few nanometers thickness on top of SnO_2 detector films at 300 °C. Electrical conductivity measurements at both the uncovered and the coated SnO_2 were performed while an air flow, with pulsed additions of methane and propane at various concentrations, was passed over the sample. The SnO_2 membrane enhances the sensitivity to methane by a factor of eight, whereas the sensitivity to propane is not affected. The response time was measured to be about 50 s, with or without coating, at concentrations larger than 500 ppm.

Effects of membrane thickness on response time, selectivity, permeation, and recovery after exposure to hydrogen, methane and related gases, have been investigated [85], while sol-gels prepared from $Sn(OEt)_4$ have been used for sensing hydrogen and ethanol vapors [105]. Lanthanum(III) oxide and Pt may be added to tin oxide thin-film gas sensors based on the sol-gel process [106]. The films were sensitive to ethanol gas as demonstrated by a change in electrical conductivity. The sensitivity was markedly promoted by loading with La_2O_3/Pt and increasing film thickness. The response time increased with the amount of La_2O_3 and film thickness. The response time was in the range of 40–70 s at 300 °C and could be decreased to 35 s by loading with Pt. The doubly-loaded film sensors, $Pt–La_2O_3–SnO_2$, displayed enhanced ethanol-gas sensing properties.

There is now increasing interest in thin films of ferroelectric ceramics for application as piezoelectric sensors. Typical ferroelectric ceramics being considered are mixed oxides of lead, lanthanum, zirconium and titanium. The characteristics of ferroelectric ceramics and the preparation of thin films by sputter deposition and by the sol-gel process have been described [108]. High surface-area, porous, thin films of semi-conducting ceramic oxides have been applied gas sensors. The operation of such sensors, based on tin oxide, for the detection of reducing gases in air have been described and an example has been given of a thin film sensor produced by the sol-gel process. The use of a solid electrolyte electrochemical cell to investigate the reactions taking place on the catalytic surface of tin oxide has been described.

In another type of gas sensor, the electrical conductance of MgO-doped chromia, from pure Cr_2O_3 to the composition of spinel $MgCr_2O_4$ was shown to vary with the concentration of alcohol vapor [107, 108]. The solubility limit of MgO in Cr_2O_3 can be extended from the previously reported 1 mol% to 2.5 mol%, at a temperature as low as 650 °C, by the sol-gel process. The solubility limit is confirmed by the variation of the unit cell constant for doped chromia. The electrical conductivity increases with increasing MgO content inside the solid-solution range, and decreases outside that range, with the appearance of a conductance peak at the boundary. The alcohol sensitivity of the MgO-doped chromia sensor also varies with MgO content in a similar manner, but decreases inside the solubility limit with increasing MgO doping.

Integrated optical ammonia sensors based on evanescent field absorption were reported [10, 109]. The sensitive element is similar to Fig. 3 and consists of a multimode strip waveguide spin-coated with a sol-gel-immobilized indicator such as bromocresol purple. The strip waveguide is fabricated by field-assisted ion exchange in a glass. The dye is embedded in a porous silica matrix, fabricated using the sol-gel technique. Five ppm ammonia were detectable but only when the relative humidity was higher than 85%. No – or negligible – sensitivity is reported for SO_2, CO_2, NO, N_2O, and vapors of organic solvents such as acetone or ethanol. In another ammonia sensor [55], it was noted that organically modified glasses (ormosils), derivatized by a diethylenetriamine compound $(MeO)_3Si(CH_2)_3NH(CH_2)_2NH(CH_2)_2NH_2$, are capable of chelating copper ion. The resulting material can be used to sense atmospheric ammonia by virtue of the formation of an intensely blue copper/amine complex.

Detection of ammonia in water is more difficult because it requires that the sensor chemistry does not leach out of the sol-gel network. Preininger & Wolfbeis [34] have immobilized the rhodamine/bromophenolblue ion pair in a 50/50 mixture of TMOS and phenyl-trimethoxysilane at alkaline pH in order to obtain sensor layers which were capable of detecting 50 ppb ammonia in water by fluorescence. Response times are in the order of several minutes. In an alternative approach, the sol-gel layers were covered with an 8-μm film of teflon which is gas-permeable but ion-impermeable. Dye leaching was completely suppressed, while the response to ammonia in water solution was uncompromised, except for much longer response times which are in the order of 5 min for the forward and 10–15 min for the reverse response.

Optical detection of nitrogen monoxide (NO) has been investigated [26] using cobalt $\alpha,\beta,\gamma,\delta$-tetrakis(5-sulfothienyl) porphine, which is doped into a silica thin film by a sol-gel method. The absorption band at 420 nm is weakened significantly on contact with gaseous NO at 200 °C but is not sensitive to O_2 and CO. The band change is reversible and depends on the pressure of NO in the pressure range from 1.3 to 3.2×10^3 Pa (see Fig. 12).

The surface of sol-gels and glasses in general can be fully reversibly hydrated and dehydrated, and this can be used for sensing purposes. Obviously, the measured effect strongly depends on the specific surface. Films produced by the sol-gel method have been optimized with respect to pore size and pore distribution so as to obtain a material for thin-film humidity sensors [110]. The resulting capacitive humidity sensor is characterized by good reversibility, high sensitivity, good linearity, a response time of < 10 s, and low cross-sensitivity towards other gases such as SO_2, CO_2, and ammonia.

Sol-gel doped with polycrystalline Prussian Blue and spread as thick films can act as planar humidity sensors. Their impedance depends on relative

Absorbance

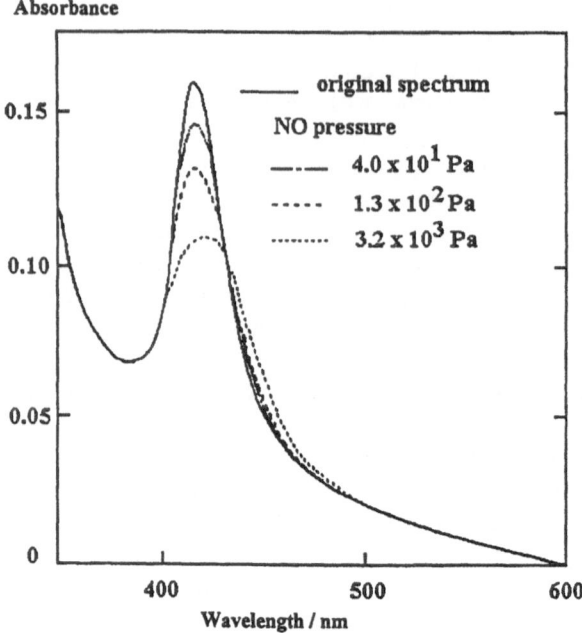

Fig. 12. Absorption spectra of Co [α,β,γ,δ-tetrakis(5-sulfothienyl)porphine] doped silica before and after exposure to NO at 200 °C [26]

humidity (RH) of the environment [111]. The equivalent circuit used allows simulation in the 10–100% RH range. The region of changes of sample resistance is about five orders, which is enough for high measuring sensitivity.

Properly cut quartz crystals have an eigenfrequency that depends on the mass of the quartz resonator. This effect has been exploited in another type of RH sensor, in which a porous silica coating was deposited on a quartz crystal by the sol-gel process [112]. Hydrochloric acid catalyst was used to accelerate hydrolysis and polymerization of a silica solution. Experimental results show the non-crystalline silica material to have nm-size pores. The resonant frequency of the coated quartz resonators decreases with increasing relative humidity. A maximum frequency shift of 3.72 kHz due to humidity is achieved.

Among other gases that have been sensed using sol-gel sensor interfaces, mention should be made of sulfur dioxide and carbon monoxide. Gas sensors based on interdigital capacitors, with organically modified silicates as a sensitive layer, have been developed [113] for the measurement of SO_2 concentrations between 0.5 vpm and some 6000 vpm. The gas-sensitive coating is produced by the sol-gel process and a spin-on technique. These sensors are distinguished by good reproducibility and reversibility, response times of < 20 s, low cross-sensitivity to other gases, and low dependence on temperature and humidity. The sensor concept, design, manufacture and measurements in a gaseous atmosphere as well as the cross-sensitivity to humidity are discussed.

Composite thin films, on exposure to carbon monoxide (CO) form a dicarbonyl complex with the rhodium complex $[Rh(TMPP)_2(CO)][BF_4]$, where TMPP stands for the ligand tris(2,4,6-trimethoxyphenyl)-phosphine, entrapped in sol-gel [114]. This is accompanied by dramatic spectroscopic and electrochemical changes which can be monitored by IR or electronic spectroscopy, or by cyclic voltametry. Titania and zirconia glass films with a pore size of 100–130 nm enable a facile diffusion of CO into the films. The layers with the rhodium complex are selective towards CO, and do not add oxygen, carbon dioxide, nitrogen and hydrogen under ordinary atmospheric conditions.

An electrolyte of the NASICON-type was prepared by the sol-gel method using three-component alkoxides as starting materials, sintered at temperatures of 900–1000 °C for 50 h and used for CO_2 sensing. Electromotoric force cells were constructed by fixing the NASICON disk to the end of a quartz tube. The devices with (Li, Ba)CO_3 as sensing electrode and air as reference electrode showed high sensitivity to 300–5000 ppm of CO_2 gas [115].

Vapors of acidic or basic compounds such as ammonia, amines or organic acids can be detected by embedding pH indicators in sol-gels [116]. The work is based on the introduction of malachite green into sol-gel glass, whereby coatings were applied to three kinds of glass: capillary glass tubes of 1 nm inner diameter, glass tubes of 4 mm inner diameter, and flat glass (70 × 15 × 2 mm). The spectral characteristics of the dye are dependent on the mode of preparation of the glass. In acidic glass, the yellow protonated dication is predominant, the green monocation exists in water and other media at pH 4 to 6, whereas the colorless carbinol base occurs at higher pH values. The detection limit for ammonia was found to be 1–2 ppm. Such sol-gel films may be deposited on the inner wall of a capillary which serves as a flow cell and undergoes a color change on exposure to ammonia [117].

A synthetic organic receptor, dansyl-tethered β-cyclodextrin, was doped into an inorganic sol-gel matrix and spin cast as a film on fused-silica plates [118]. The entrapped receptor exhibits changes in its fluorescence in the presence of vapors of borneol, obviously as a result of molecular recognition. Thus, this new material can be used as a reversible sensor. Various aspects of a sol-gel matrix as a host for recognition molecules were explored. This system demonstrates the possibility of using artificial receptors and sol-gel porous glass films for chemical sensing purposes.

Gases may also be monitored via biomolecules. This was demonstrated for the cases of oxygen, carbon monoxide (CO), and nitric oxide (NO) [18]. The high affinity of hemoglobin (Hb) and myoglobin (Mb) for oxygen, coupled with the changes in visible absorption spectra that occur when oxygen is bound, provides an opportunity to develop an oxygen sensor. Because myoglobin also binds CO, it can also used for CO sensing. Biorecognition of in-situ-generated NO was demonstrated using gel-entrapped manganese myoglobin. Myoglobin encapsulated in sol-gel glass exhibited the same chemical and spectroscopic properties in the gel as in solution upon reduction of metMb with dithionite to give deoxyMb [119]. Dissolved oxygen was determined in the concentration range from 2 to 8 ppm (i.e., air-saturated water).

3.3 pH Sensors

Sol-gels have been used in several cases to construct optical pH sensors ("*pH optodes*"). In all cases, a pH-sensitive dye was incorporated. The resulting sensors exploit two specific properties of sol-gels, namely the optical transparency and the permeability to ions while retaining the dye in the matrix. Also, the sol-gel process enables oxide glasses to be prepared at room temperature with little or no heating. By using this method, it is possible to encapsulate a wide variety of organic indicators in the inorganic matrix. A review [47] emphasizes the ability of encapsulated luminescent molecules to provide unique information on the local chemistry, and gives several examples of how organic-doped sol-gel materials are emerging as an important means of producing photonic materials for use in optical sensors.

pH sensors were among the first sol-gel-based sensors designed. In initial experiments, pH indicators were entrapped in the matrix, giving planar materials of various shapes. In later studies, pH-sensitive coatings were designed for an optical-fibre sensor by placing them on optical-fibre cores, or tips, and by coating integrated optical waveguides with pH-responsive sol-gels.

Plain sol-gel-based pH-sensitive layers have been prepared by numerous authors [57, 59, 120, 121]. A variety of color or fluorescence pH tests has been presented by these authors by incorporating pH indicators in sol-gel porous glasses by polymerization of tetramethoxysilane, occasionally in the presence of a surface active agent. The properties of the sensing materials [57] including spectral shifts, shifts in the pH-sensing range, cycle repeatability, leachability, rates of response and isosbestic points were described. It cannot be excluded, though, that the shifts in the sensing range observed in the sol-gel glasses result from the addition of the surfactant which is an anti-cracking agent.

Aside from a shift in the pK_a of a dye upon immobilization in a matrix the shape of the titration curve usually also undergoes considerable changes. This was demonstrated [59] in a report on a fluorescence-based optical sensor material suitable for measurement of pH in the 6–9 range using a new, LED-compatible fluorescent dye. Its base form shows strong absorption between 580 and 630 nm that matches the emission band of conventional yellow or orange light-emitting diodes. The dye was physically entrapped into a sol-gel matrix which can be used for plain sensing membranes, but also for optical-fibre tip coatings and in evanescent wave-type sensors. Fig. 13 shows how the shape of the titration plot of the same dye changes on going from water solution to a sol-gel and a cellulose matrix, respectively. The resulting sensor materials were studied with respect to dynamic pH ranges (6–8), response times (5–20 min), sensitivity toward ion strength, and stability.

The process parameters for preparing optochemical pH sensors by doping sol-gel glasses were optimized [122]. The operation range of the pH sensors, the response time and the stability was varied by modifying the composition of the starting mixture and optimizing the thermal aftertreatment, respectively. The response time of thin film sensors was less than 60 s and the sensitivity 0.09 pH

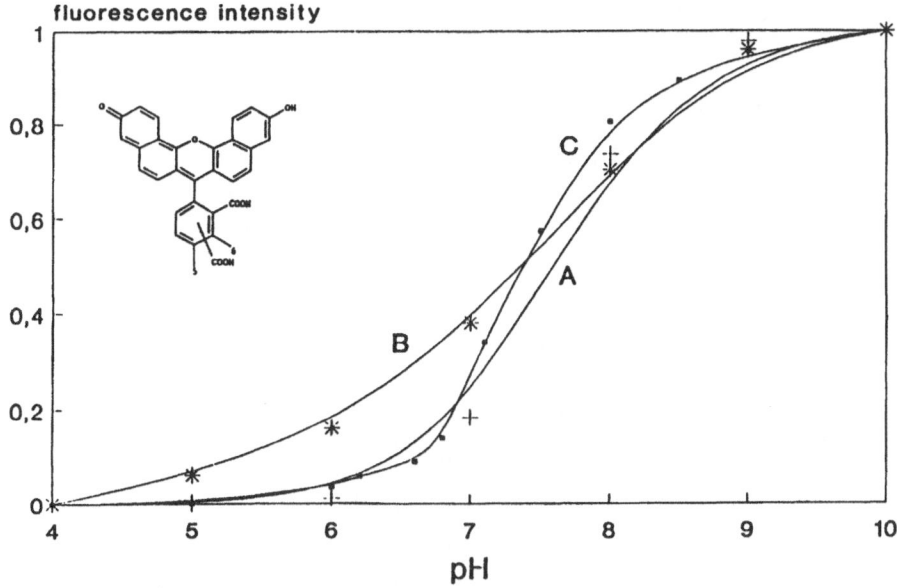

Fig. 13. pH titration plots of carboxy-naphthofluorescein in (*A*) a sol-gel; (*B*) on cellulose, (*C*) in free form in water [59]

units in the linear part of the calibration curve. Monolithic glass sensors showed an improved long-term stability and an increased response time. The reproducibility of thin film sensor preparation and a typical response over time is shown in Fig. 14.

A optical sensor for pH that is based on lifetime measurements rather than intensity measurements that makes use of an effect called resonance energy transfer has been described [123]. The pH-dependent increase in the absorbance of sol-gel-entrapped bromothymol blue acceptor results in an increased transfer of electronic energy to a fluorophore (Texas Red), and this reduces the fluorescence lifetime. The lifetimes are measured by the phase and modulation of the emission, relative to the modulated incident light, and are found to be insensitive to the total signal level and fluctuations in light intensity. The sensor provides stable readings for days and can be repeatedly autoclaved without loss of sensitivity to pH. Because the efficiency of energy transfer depends on the 6th power of the distance between absorber and acceptor, it is mandatory that the two dyes do not leach at all, in order to avoid drifts in the baseline signal.

Sol-gels are considered to be of particular use in connection with evanescent-wave optical-fibre or integrated optical-waveguide sensing. Bromophenol blue, a pH indicator, was entrapped in porous glass films prepared by the sol-gel method [14]. The colorimetric response of the films to pH was assessed visibly and by attenuated total reflectance spectrometry. The entrapped indicator was non-leachable and responded to the pH changes of the solution in

rel. intensity

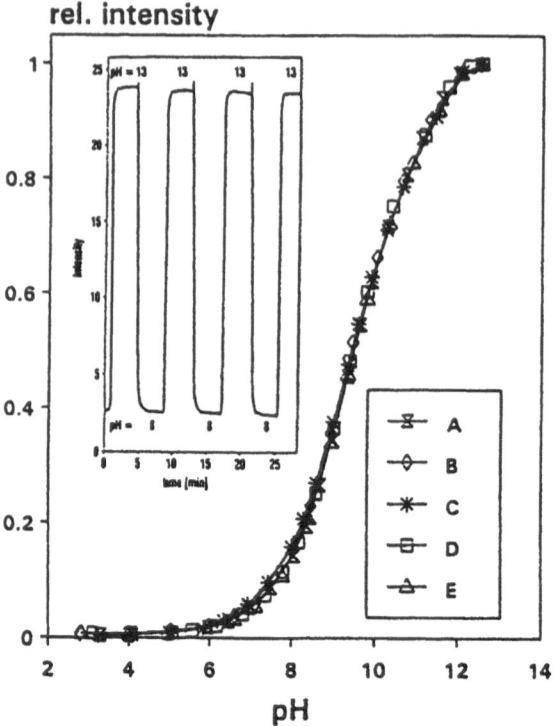

Fig. 14. Reproducibility of thin-film sensor preparation [122]. Calibration curves are shown for five different pH sensors. The pK_a value was 9.50 $\pm$ 0.03, the response time < 1 min

a reversible manner. The immobilization of indicator molecules in porous sol-gel films appears to be a particularly promising approach to the development of optical-waveguide sensors. A pH-dependent response over the entire interval of pH 3.5–9.5 was observed for bromocresol purple entrapped in a sol-gel-based, laminate planar IOW [74]. The broad range over which the IOW-ATR pH sensor responds clearly indicates that immobilization of bromocresol purple in the sol-gel layer produced a distribution of chemically inequivalent microenvironments that differ in response to the pH of the solution.

Sol-gel technology was applied in optical-fibre sensor technology [58, 124], in which the fluorophore FITC was immobilized in a gel coating on a fibre, and its characteristic optical spectra were analysed as a function of pH. A response to pH was obtained in the range pH 5 to pH 9. Numerous other examples of specific, sol-gel-based optical fibre pH optodes have been published. For example, the sol-gel process for low temperature fabrication of porous glass has been used to entrap chemically-sensitive dyes for fibre sensor applications [125]. Evanescent wave fluorescence excitation was used to examine the behavior of an unclad fibre which was dip-coated with a thin layer of a dyed sol-gel cladding within which a pH-sensitive dye was entrapped. The sensor covers a pH range of 3.5 to 6.5. An extended-range fibre-optic pH evanescent-wave sensor was obtained by entrapment of absorbance-based pH indicators in sol-gel [126]. When two indicators are being used, the pH range covered is from 3 to 9.

An improved optical-fibre sensor with extended lifetime for pH measurements has been reported [127]. The sensor is based on the change in absorbance of organic indicators, immobilized in a silica matrix that is coated as a thin film onto a porous-glass optical fibre by the sol-gel technique. This approach is said to result in highly stable, chemically durable and ruggedized optical-fibre pH sensors. Such sensors also have potential application in aqueous solutions at high temperatures and in samples containing organic solvents.

Conversely, minute sensors are needed in in-vivo applications, and this is a major challenge in the design of microsensors. Recently, the feasibility of miniaturizing a sol-gel sensor down to the micron scale for near-field optical pH sensing was demonstrated [31, 77]. A metal-coated micropipette was filled with a silica xerogel, doped with the pH-sensitive fluorophore pyranine. The pH range was from 5 to 8. The pH sensor was constructed by placing a HPTS-doped sol-gel matrix at a 10-μm tip of a metal-coated micropipette. Subsecond response time was achieved by further downscaling to 1 μm. Using an optical near-field microscopy configuration, these observations demonstrated the potential for constructing a variety of miniaturized sensors and chemical manipulators, by doping the sol-gel matrix with a suitable reagent.

A major problem in optical pH sensors based on sol-gel-entrapped dyes is leaching, and the only reliable way to overcome this seems to be covalent immobilization of the dye. The pH indicator methyl red was covalently linked to a sol-gel by first introducing a functional trimethoxysilane group into the dye and then copolymerizing it with TMOS during the sol-gel process [31]. While the dye was now present in immobilized form, its pK_a was shifted from 5.2 in solution to 3.0 in the glass. In an alternative approach to immobilize a dye covalently [32], the surface of the sol-gel network was derivatized with the reagent aminopropyl-triethoxysilane, resulting in a sol-gel with free amino groups on its surface, onto which the pH indicator fluorescein was covalently immobilized via its isothiocyanate. The resulting material was characterized in great detail and may serve to measure physiological pH values.

Another pH sensor with covalently immobilized dye was obtained by first covalently binding aminofluorescein (AF) to the sol-gel precursor 3-thiocyanatopropyltriethoxysilane (TCP-TS) [33]. A cocktail made up from AF, TCP-TC, ethanol, tetramethoxysilane, Triton X and 1 mM NaOH was allowed to gelate for two days at room temperature in closed glass bottles. Glass platelets were then coated with the gel, dried at 100 °C for five days, and stored in distilled water. The layers respond fully reversibly to pH. The pK_a was 6.43 but dropped to 6.38 after the layer had been stored in water for 5 days at room temperature. The response time was as fast as 50 s.

3.4 Ion Sensors

In the area of ion sensing, it is the unique capability of sol-gels to retain indicators, while allowing the penetration of small metal ions, that is exploited in sensors using this material. In order to detect metal ions, in particular heavy metal ions, metallo-chromic dyes are almost exclusively employed. They bind (chelate) metal ions and thereby undergo a color change. In such sensors, advantage is taken of the optical transparency of sol-gels along with their low costs. Numerous color tests have been designed by entrapping indicators, typical examples being sensors for iron(II), nickel(II), copper(II) and sulfate [121]. Another disposable sensor for determining iron(II) in aqueous solution has been described [11].

These sensors exemplify a new class of disposable field tests for field analysis of water pollutants. The latter consist of capillary glass tubes filled with porous sol-gel silica powder doped with o-phenanthroline. When a sample solution is passed through a tube detector, the iron ions are complexed by the immobilized o-phenanthroline and a stained section of the capillary develops (see Fig. 4). The metrological characteristics of these detectors, including precision and accuracy and chemical interferences by heavy metals and humic acids, has been discussed [11].

The method of immobilization of organic reagents in porous silica glass via the sol-gel procedure has been used [25] to produce quantitative photometric sensors. Silica glass detectors doped with 1-nitroso-2-naphthol for detecting cobalt ions and o-phenanthroline for determining divalent iron serve as test cases that demonstrate the advantages and the current limitations of the immobilization procedure. Similarly, porphyrin – which is known to bind the mercury ion – was entrapped in sol-gel [128]. When exposed to a sample for 20 min, as little as 2 ppb of mercury can be detected by fluorimetry. Unfortunately, the most stable films were those prepared under acidic conditions, but these exhibited the least sensitivity to mercury. Films prepared under alkaline conditions, in contrast, were more sensitive and less stable.

A novel and high-sensitivity photometric method to determine trace concentrations of water pollutants is based on the preconcentration of the analyte in the matrix [83, 129]. Organic reagents such as 1,10-phenanthroline were entrapped in sol-gel matrices without resorting to complicated covalent bonding procedures. When these doped sol-gel glass detectors are immersed in water, they concentrate almost an analyte such as iron(II) from the solution into the glass matrix. The glass detectors may then be analyzed by a conventional spectrophotometer. Thus, exceedingly low concentrations can be determined. For example, the detection limit demonstrated for ferrous ion in aqueous solution is < 100 ppt. The resulting chemically sensitive materials may be used in disposable sensors such as in tube detectors [11, 129], in "displacement chromatography sensors", and in monolithic sensors [25, 130]. The latter are comprised of a plate a few mm thick and molded by the usual sol-gel technique. The specific surface of monolithic sensors is as high as 300–1000 m^2 per gram, depending primarily on the methanol/water ratio and the type of catalysis employed (acid or base). The void fraction in the interior can be controlled in the range from 30 to 70%. Usually, a small amount of cetyl ammonium bromide is added in order to prevent fracture of the glass both during the gelation step and when using the sensor. Typically, 0.01 to 5% (w/w) of a photometric reagent are added.

In a comparison of immobilization methods and polymers for an optical ion sensor [131], the photometric reagent zincon was entrapped in various polymers and studied with respect to its capability for optically sensing copper(II) ions. Immobilization into a sol-gel was accomplished via a lipophilic ion pair. By spreading the dyed sol-gel onto a planar support, an irreversible test strip was obtained which responded to copper in the 1 to 100 μmol l^{-1} concentrations range [131]. A kinetic method, in which the increase in absorbance at the maximum of the copper complex at 620 nm is monitored over 2 min, was applied for data evaluation. On exposure to a sample, the indicator-loaded sol-gel layers showed a color transition from pink to blue. The effect of various parameters of the sol-gel protocol on the performance of the sensor was investigated. The highest sensitivity was obtained with layers based on alkaline catalysis, while the best stability in terms of dye leaching upon storage in buffer was provided by layers fabricated under acidic catalysis.

Xylenol orange was used as the colorimetric indicator for an integrated (i.e., noncontinuous) optical indicator for sensing lead ions [74]. The sensor measures absorption changes by attenuated total reflection and responds to lead concentrations ranging from 50 nM to 50 mM. Due to the small thickness of the layer (which is in the μm range), response times are very short, typically 4 min for 1 mM lead, which is significantly less than is necessary to equilibrate a monolithic sol-gel sensor.

Yoverdin, a fluorescent pigment biosynthesized by *Pseudomonas fluorescens*, was entrapped in a sol-gel glass and in this form used as a reactive solid phase for the determination of trace amounts of ferric ions [132]. No leaching or bleaching of the trapped reagent was observed. The fluorescent siderophore was more stable in sol-gel glass than on controlled-pore glass. Iron(III) was determined by either continuous flow or flow injection methods. Both methods provided fast response. Detection limits were 3 ng ml^{-1} and 20 ng ml^{-1}, respectively.

An ion-sensitive electrode (ISE) for sodium ion was obtained [133] by doping a sol-gel thin film with NASICON, a ceramic material acting as a fast ionic conductor for sodium ions. The capacitance-voltage characteristics showed the sensitivity to be a function of the Na$^+$ concentration in aqueous solution. NASICON was also dispersed into sol-gel [134]. Physicochemical characteristics such as structure, conductivity, and solubility were described. Studies on ISEs made with this material included ion-exchange kinetics, detection limit, and selectivity coefficients. The performance characteristics are compared to those of a commercial glass membrane, the detection limit is slightly poorer than for the commercial ISEs but selectivity is always better by up to 10 times. A considerable sintering-temperature influence is observed on the interfering-proton phenomenon is observed: at higher temperatures the effect becomes smaller.

A major drawback of practically all existing sensors for heavy metals (HMs) based on chromogenic dyes is their irreversibility, which is in constrast to the corresponding sensors for pH, oxygen, carbon dioxide, and ammonia. Recently, however, a successful attempt was made to reversibly detect and determine an HM ion. A cadmium sensor was prepared [135] by the sol-gel method, using the reagent 8-hydroxyquinoline-5-sulfonic acid (HQu). In aqueous solution, it forms a luminescent complex with cadmium, the excitation/emission maxima being at 390/530 nm. In order to achieve reversibility (which is not possible in aqueous solution), the reagent was incorporated into a glass film deposited on a glass support by dip coating from solutions of TEOS, which was hydrolyzed and polycondensed at room temperature. HQu is a weak acid and served both as a reagent for cadmium and as a catalyst. Glass plates were coated with a sol-gel, doped with HQu, and inserted into a solution of cadmium ions in the arrangement shown in Fig. 15. A fluorescent complex is formed between cadmium and HQu, and emitted light is collected at a right-angle geometry. The Cd-HQu complex obviously is easily decomposed by water, thereby making the reagent available for another reaction and, hence, making the system reversible. The reversible response of this system to cadmium is shown in Fig. 16. The limits of detection are 10 ppb.

Little work has been performed in the area of anion sensing. Quarternary ammonium compounds were entrapped in sol-gel glasses and reported to be useful for anion exchanger applications. Specifically, sol-gel glasses were doped with Aliquat 336, cetyltrimethylammonium bromide, and cetylpyridinium bromide, and found to exhibit selectively properties that were similar to those of strongly basic ion exchangers, the relative order being $SO_4^{2-} \gg NO_3^- > Br^- > Cl^- \gg OH^-$ [136]. The feasibility of fabricating a sol-gel-derived titanium carboxylate thin film for optical test strips was performed [137] using

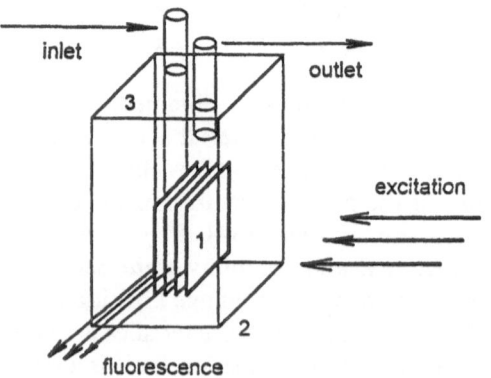

Fig. 15. Schematic diagram of an arrangement for testing the performance of glass plate sensors covered with a sol-gel film, incorporating the reagent (HQu). The fluorescence of the layer is concentrated at the edges (as in luminescent solar collectors) and this results in an increased sensitivity. (1) glass plates, (2) transparent container; (3) container with inlet and outlet tubings

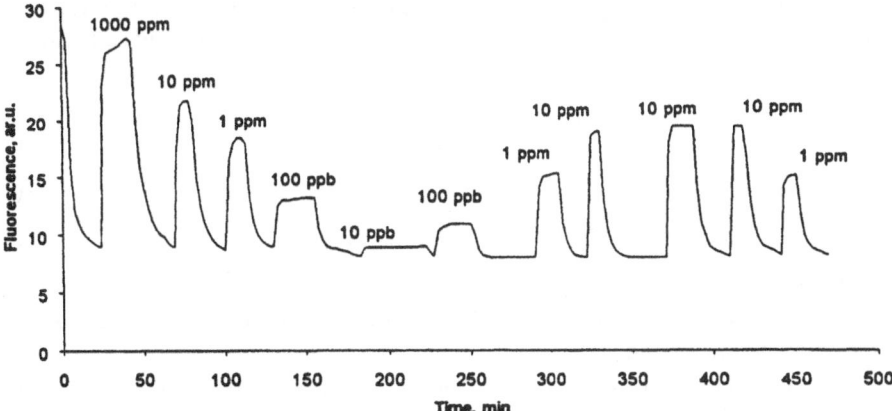

Fig. 16. Response of a cadmium sensor to different concentrations of cadmium ions in water. An increase in the signal is obtained on exposure to metal solutions, while descending signals are obtained after switching to pure water [135]

a model probe/analyte combination. An iron(III) porphyrin reagent was chosen as the probe molecule for detecting free cyanide-ion concentrations in aqueous solution. Cyanide ions were detected in the 40 to 25,000 ppm range. Finally, a sensor for the highly toxic chromate anion has been developed that is based on the oxidation of sol-gel entrapped colorless diphenylcarbazide by the chromate anion to give diphenylcarbazone which, upon complexation by the chromium(III) ion gives a purple coloration which can be exploited for monitoring the chromate ion [138].

3.5 Enzyme-Based Biosensors

Attachment of enzymes to, or into, an insoluble matrix is an essential step in the development of biocatalysts. Since sol-gels can be manufactured under mild experimental conditions, they are ideally suited for the incorporation of sensitive, even labile, biomolecules including enzymes and antibodies. In addition, its transparency, large hydrophilic surface, and good chemical and thermal stability make it an ideal material for optical sensor devices. The work up to 1993 has been reviewed [17, 18].

Transparent xerogels containing various enzymes including GOx were obtained by various authors [36, 39, 139, 140]. In essence, solutions of an enzyme were mixed with tetra-methoxy orthosilicate at room temperature followed by gelation and drying. Effective immobilization was usually obtained at initial pH values > 7, where there is a change in the gelation mechanism from predominant hydrolysis/condensation to predominant direct polymerization of silicate precursors.

Various examples of optical biosensors for glucose determination have been given in the past years, practically all being based on the following sequence of reactions:

$$\text{glucose} + \text{FAD} \xrightarrow{\text{GOx}} \text{gluconolactone} + \text{FADH}_2 \tag{4}$$

$$\text{FADH}_2 + \text{O}_2 \longrightarrow \text{FAD} + \text{H}_2\text{O}_2 \tag{5}$$

$$\text{gluconolactone} + \text{H}_2\text{O} \longrightarrow \text{gluconic acid} \tag{6}$$

The formation of FADH_2 (Eq. 4), the consumption of oxygen (Eq. 5), the formation of hydrogen peroxide (Eq. 5) or of protons (i.e. the decrease in pH, see Eq. 6) may be measured.

A simple solid-state optical biosensor for glucose was obtained [140] by immobilizing GOx in a sol-gel and monitoring the intrinsic absorption of the yellow enzyme. The initial rates of reduction of the FAD prosthetic group of the enzyme in the presence of various concentrations of glucose to give FADH_2 was monitored at 450 nm, which is the characteristic absorption band of oxidized FAD (Fig. 17). The analytical range of the sensor was 1–100 mM, and the measurement time was 2 min. The enzyme was shown to be considerably protected by the silicate matrix against leaching, thermal inactivation and even inactivation. The sensor was stable in daily use for 6 months. Similar results, however via fluorometry, with a sensor based on the measurement of the optical properties of the FAD coenzyme (Eq. 4) were reported for glucose [141] and lactate [142].

An indirect and amperometric glucose sensor was obtained by entrapment of glucose oxidase in doped vanadium pentoxide (V_2O_5) prepared from a colloidal suspension of V_2O_5 doped with tetravalent vanadium ion. The thin films

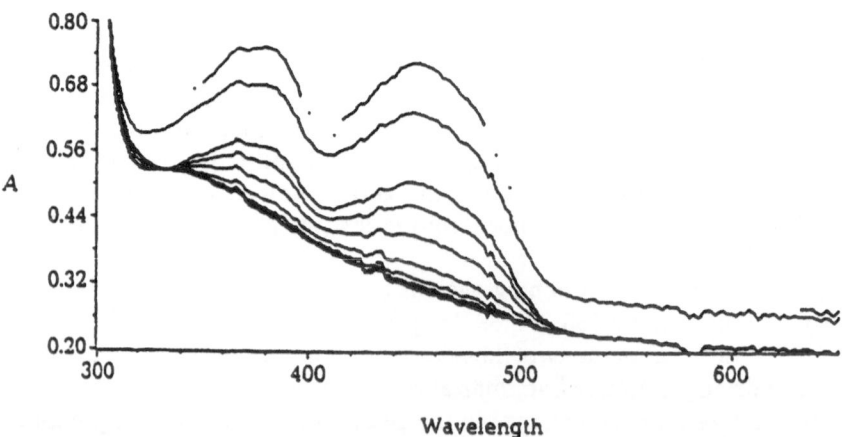

Fig. 17. Absorption spectra of glucose oxidase entrapped in sol-gel. Spectra were recorded at 1 min intervals in a sol-gel pellet immersed into 30 mM glucose. The upper line is the spectrum in pure water [140]

exhibited good electrochemical conductivity and adhered well to platinum and other conductive supports. On exposure to glucose and in the presence of excess oxygen, reactions (4) to (6) occur. Sensing is performed via the hydrogen peroxide produced, which can be electro-oxidized on the surface of the metal support. Typical cyclic voltamograms are shown in Fig. 18.

In another type of irreversible glucose probe, the enzyme glucose oxidase (GOx) was immobilized in a sol-gel along with horseradish peroxidase and the dye precursor, 4-aminoantipyrine, p-phenylenediamine, and p-hydroxybenzene sulfonate. The dye precursor, on reaction with hydrogen peroxide (Eq. 5), forms a quinone-imine dye with an absorption maximum at 510 nM. When the biogel monoliths was exposed to glucose solution, GOx caused the formation of hydrogen peroxide, and this, in turn, triggered the peroxidase-catalyzed formation of the dye. By monitoring the absorption of the quinoneimine dye, glucose levels could be determined in the range from 0.56 to 4.44 $mol\,l^{-1}$ [143]. An important aspect of this system is the demonstration that multi-enzymatic reactions can be performed in a gel matrix. GOx also was immobilized in and on a thin sol-gel film using physisorption, micro-encapsulation, and a new sol-gel: GOx: sol-gel sandwich configuration, and both electrochemical (amperometric) and photometric detection schemes were used to study the response profile and to quantify glucose [50].

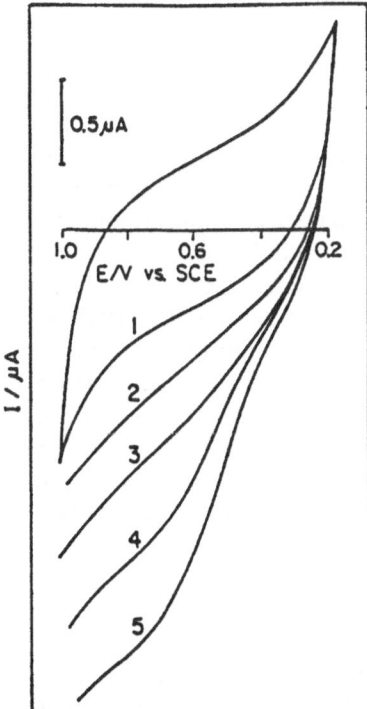

Fig. 18. Cyclic voltamograms of a Pt wire coated with a thin film of a glucose oxidase-vanadium pentoxide matrix at various concentrations of glucose. (*1*) Blank solution (0.1 M phosphate buffer of pH 5.6); (*2*) 1.2 mM; (*3*) 3.5 mM; (*4*) 5.5 mM; (*5*) 8.3 mM glucose solutions. Scan rate 10 mV s⁻¹ [139]

Another type of amperometric glucose sensor was obtained by covering a working electrode with a enzymatically active material obtained by polymerization of tetraethyl silicate in the presence of GOx [144]. The resulting gel showed very strong enzymatic activity. On raising the ageing temperature, the yield increased but the activity decreased. The sensor served as a detector in a flow injection device, i.e., a system where a sample containing glucose is injected into a carrier stream of buffer. When passing over the oxygen electrode covered with GOx-entrapped silica gel, oxygen is consumed as a result of the enzymatic oxidation of glucose (see Eq. 4), and this results in the measured response of the oxygen electrode to glucose, which occurs in physiological concentrations.

A method was reported [145] for electrochemically probing the activity of GOx embedded in various silica gels which, in turn, were immobilized on the surface of an electrode. In the presence of a ferrocene mediator and under controlled reaction conditions, the enzyme-catalyzed oxidation of glucose has been observed via the re-oxidation current of the mediator. The immobilized enzyme sometimes retains its fully activity, and the catalysis can be analyzed as if it were a homogeneous process, due to the open structure of the gel regarding the diffusion of small molecules.

Recently, several types of amperometric biosensors made of immobilized GOx in chemically-modified sol-gel-derived ceramic carbon matrices were presented and compared [146, 147]. The electrodes can be prepared either in the form of quick supported films (useful for disposable electrodes) or as bulk-modified, disk-shape electrodes, which may be used a renewable surface electrodes. The shelflife stability of these electrodes is highly dependent on the preparation conditions. If prepared with no methanol at moderate pH, the electrodes are stable for at least two months at ambient conditions. The dynamic range of the glucose ceramic carbon electrodes depends on the structure and thickness of the active layer and typically exceeds the required range for biomedical applications. However, the batch-to-batch and inter-electrode reproducibility is poor.

Biodetection of oxalate was accomplished via oxalate oxidase and horse raddish peroxidase (HRP), both entrapped in a sol-gel [18]. By analogy to Eq. (5), the oxidation of oxalate results in the formation of hydrogen peroxide which, in an HRP-assisted reaction forms a blue dye on reaction with 3-methyl-2-benzothiazoline hydrazone in the presence of 3-dimethylaminobenzoic acid. The indamine dye has an absorption maximum at 590 nm. A linear response in the concentration range from 20 to 800 μmoll^{-1} oxalate and a response time of 5 min was obtained with the biogel sensor which, however, is irreversible.

3.6 Immunosensors and Affinity Sensors

The major issue in designing sol-gel based immunosensors is the retainment of affinity constants and selectivity patterns, along with the need for virtually unretarded access of the analyte to the immunoprotein. As can be seen below, the results are partially controversial. Integrated optical difference or polarimetric interferometers have been applied for sensing purposes [148]. SiO_2/TiO_2 waveguides were prepared by a sol-gel process on silicon wafers, with a pure SiO_2 buffer layers acting as substrates; depending on the firing temperature, the 140–200 nm thick waveguiding films are microporous or compact, the latter showing superior stability in sensor applications. The device was shown to be (a) a refractometer for liquid samples, (b) a sensor for monitoring protein adsorption, and (c) an affinity sensor.

The affinity reaction between avidin adsorbed on the waveguide surface and biotinylated bovine serum albumin was investigated in some detail, and another type of sensor, known as an input grating coupler, with planar monomode waveguides as sensors was later designed [149]. It can be used for real-time monitoring of protein adsorption, and monitoring of affinity- and immunoreactions. The adsorbed immunoglobulins (h-IgG) acted as the receptors on the waveguide surface for the corresponding ligand in the sample. The formation of complexes between avidin and biotinylated protein A (prot A) and biotinylated bovine serum albumin (BSA), between h-IgG and anti-h-IgG, and between h-IgG and prot A were investigated. From the data obtained, not only the thickness, but also the refractive index and the surface coverage of an adsorbed or bound single isotropic adlayer were determined. The $SiO_2–TiO_2$ waveguides are microporous and, in aqueous buffer solutions, show a slow but persistent increase of the measured effective refractive indices with time. This drift effect is greatly diminished in compact $SiO_2–TiO_2$ waveguides, annealed at a temperature of 900 °C instead 500 °C.

Recently, the encapsulation of antibodies in sol-gel glass matrices was successfully accomplished. Polyclonal antifluorescein retains an affinity for fluorescein when embedded in sol-gel, the affinity constant K(f) for the antibody-hapten complex being on the order of 10^7 l mol^{-1} [150]. The encapsulation process decreases K(f) by about 2 orders of magnitude compared to the native system in buffer solution. The intact antibody affinity can be maintained using simple storage protocols and the effects of ageing and drying on the K(f) are reported. Similarly, an antinitroaromatics immunoglobulin was entrapped in a sol-gel matrix, giving a final protein concentration of about 1 mg g^{-1} [31]. Sol-gel glasses containing *anti*-2,4-dinitrobenzene immunoglobulin retained 66% of 2,4-dinitrophenyl hydrazine after it had passed through a column filled with the glass. A final report deals with the activity of an immunoglobulin which remained unchanged when immobilized by reaction with free amino groups or thiol groups inside a sol-gel that was obtained by copolymerizing aminopropyltriethoxysilane with TEOS [151].

Acknowledgements. This work was supported by the Austrian Science Foundation (FWF) within Project P09869, and by the Austrian National Bank (Project 5412). This support is gratefully acknowledged. We are also very grateful to the Austrian Friends of the Hebrew University for their financial support and to M. Zevin for providing some of the experimental data.

4 References

1. Brinker CJ, Scherer GW (1990) Sol-Gel Science. Academic Press, New York
2. Hench LL, West JK (1990) Chem Rev 90: 33
3. Buckley AM, Greenblatt M (1994) J Chem Educ 71: 599
4. Levy D, Esquivias L (1995) Adv Mat 7: 120
5. Sanchez C, Livage J, Henry M, Babonnea F (1988) J Non-Cryst Solids 100: 65
6. Sorek Y, Reisfeld R, Tenne R (1994) Chem Phys Lett 227: 235
7. Henry M, Jolivet JP, Livage J (1992) in: Chemistry, Spectroscopy and Applications of Sol-Gel Glasses, Reisfeld R, Jorgensen CK (eds), Structure and Bonding 77: 153
8. Göpel W, Schierbaum KD (1991) in Sensors: A Comprehensive Survey, Göpel W, Hesse J and Zemel JN (eds) VCH Publ, Weinheim, vol 2, chap 9.
9. Wolfbeis OS, ed (1991) Fiber Optic Chemical Sensors and Biosensors, CRC Press, Boca Raton, Fl.
10. Brandenburg A, Edelhäuser R, Werner T, He H, Wolfbeis OS, Ammonia Detection via Integrated Optical Evanescent Wave Sensors (1996) Mikrochim Acta 112: xxx (in press)
11. Kuselman I, Lev O (1993) Originally-Doped Sol-Gel Based Tube Detectors-Determination of Iron(II) in Aqueous-Solutions, Talanta 40: 749
12. Weigl BH, Wolfbeis OS (1994) Capillary Optical Sensors, Anal Chem 66: 3323
13. Weigl BH, Draxler S, Kieslinger D, Lehmann H, Wolfbeis OS, Lippitsch ME, Proc. SPIE 2508: in press
14. Lee JE, Saavedra SS (1994) Evanescent Sensing in Doped Sol-Gel Glass-Films, Anal Chim Acta 285: 265
15. Lübbers DW, Opitz (1975) The pCO_2/pO_2-Optode: A new probe for measurement of pCO_2 and pO_2 of gases and liquids, Z. Naturforsch. Part C, 30C: 532
16. Avnir D, Ottolenghi M, Braun S, Zusman R (1994a) Doped Sol-Gel Glasses for Obtaining Chemical Interactions, US Pat 5,300,564
17. Avnir D, Braun S, Lev O, Levy D, Ottolenghi M (1994) in: Sol-Gel Optics – Processing and Applications, LC Klein, ed, Kluwer, USA, 1994, chap 23, p 539
18. Dave BC, Dunn B, Valentine JS Zink JI (1994) Sol-Gel Encapsulation Methods for Biosensors, Anal Chem 66: 1120A
19. Novak BM (1993) Hybrid Nanocomposite Materials – Between Inorganic Glasses and Organic Polymers, Adv Mater 5: 422
20. Komareni S (1992) Nanocomposites, J Mat Chem 2: 1219
21. McDonagh C, Sheridan F, Butler T, MacCraith BD (1995) J Sol-Gel Sci Tech (submitted)
22. McDonagh C, Sheridan F, Butler T, MacCraith BD (1995b) J Non-Cryst Solids (accepted)
23. Reisfeld R, Jorgensen CK (1992) Optical-Properties of Colorants or Luminescent Species in Sol-Gel Glasses, Structure & Bonding 77: 207
24. Avnir D, Braun S, Ottolenghi M (1992a) Encapsulation of Organic-Molecules and Enzymes in Sol-Gel Glasses – A Review of Novel Photoactive, Optical, Sensing, and Bioactive Materials, ACS Symp Ser 499: 384
25. Iosefzon-Kuyavskaya B, Gigozin I, Ottolenghi M, Avnir D, Lev O (1992) Spectrophotometric Detection of Heavy-Metals by Doped Sol-Gel Glass Detectors, J Non-Crystall Solids 147: 808
26. Eguchi K, Hashiguchi T, Sumiyoshi K, Arai H (1990) Optical Detection of Nitrogen Monoxide by Metal Porphine Dispersed in an Amorphous Silica Matrix, Sensors and Actuators B1: 154
27. Kamitani K, Uo M, Inoue H, Makishima AJ (1993) J Sol-Gel Sci Technol 1: 85
28. Fournier T, Tran-Thi TH, Herlet N, Sanchez C (1993) Chem Phys Lett 208: 101
29. Zaytsev VN, Trophymchuk AK (1984) Ukrainian Chem J 50: 1126
30. Ikoma S, Kawakita K, Yokoi H (1990) Characterization of Polyamine Copper(II) Complex-Doped Alumina Gels Prepared by the Sol-Gel Technique, J Non-Cryst Solids 122: 183
31. Aharonson N, Altstein M, Avidan G, Avnir D, Bronshtein A, Lewis A, Lieberman K, Ottolenghi M, Polevaya Y, Rottman C, Samuel J, Shalom S, Strinkovski A, Turniansky A (1994) in Better Ceramics Through Chemistry, VI, Sanchez C, McCartney ML, Brinker CJ, Cheetham A (eds) Materials Res Soc Symp Proc 346: 519; publ. by the Mat. Res. Soc., Pittsburgh
32. Badini GE, Grattan KTV, Teseun ACC (1995) 120: 1025
33. Lobnik A, Oehme I, Wolfbeis OS (1995) unpublished results

34. Preininger C, Wolfbeis OS (1995), unpulished; PhD dissertation of CP (1996).
35. Avnir D, Braun S, Lev O and Ottolenghi M (1992b) Chemically active organically doped sol-gel materials: Enzymatic sensors, chemical sensors and photoactive materials, Proc SPIE 1758: 456
36. Braun S, Rappoport S, Zusman R, Avnir D, Ottolenghi M (1990) Biochemically Active Sol-Gel Glasses – The Trapping of Enzymes, Mater Lett 10: 1–5
37. Avnir D, Braun S, Lev O, Ottolenghi M (1994) Chem Mat 6: 1605
38. Shtelzer S, Rappoport S, Avnir D, Ottolenghi M, Braun S (1992) Properties of Trypsin and of Acid Phosphatase Immobilized in Sol-Gel Matrixes, Biotechnol Appl Biochem 15: 227
39. Braun S, Shtelzer S, Rappoport S, Avnir D, Ottolenghi M (1992) Biocatalysis by Sol-Gel Entrapped Enzymes, J Non-Cryst Solids 147: 739–743
40. Ellerby LM, Nishida CR, Nishida F, Yamanaka SA, Dunn B, Valentine JS, Zink JI (1992) Encapsulation of Proteins in Transparent Porous Silicate-Glasses Prepared by the Sol-Gel Method, Science 255: 1113
41. Edmiston PL, Wambolt CL, Smith MK, Saavedra SS (1994) Spectroscopic Characterization of Albumin and Myoglobin Entrapped in Bulk Sol-Gel Glasses, J Coll Interface Sci 163: 395
42. Wu SG, Ellerby LM, Cohan JS, Dunn B, Elsayed MA, Valentine JS, Zink JI (1993) Bacteriorhodopsin Encapsulated in Transparent Sol-Gel Glass-A New Biomaterial, Chem Mater 5: 115
43. Weetall HH, Robertson B, Cullin D, Brown J, Walch M (1993) Bacteriorhodopsin Immobilized in Sol-Gels Glass, Biochim Biophys Acta 1142: 211
44. Jordan JD, Dunbar RA, Bright FV (1995) Anal Chem 67: 2436
45. Inama L, Dire S, Carturan G, Cavazza A (1993) J. Biotechnol. 30: 197
46. Kaufman VR, Avnir D, Pines-Rojanski D, Huppert D (1988) Water Consumption During the Early Stages of the Sol-Gel Tetramethylorthosilicate Polymerization as Probed by Excited State Proton Transfer, J Non-Crystall Solids 99: 379
47. Dunn B, Zink JI (1991) Optical-Properties of Sol-Gel Glasses Doped with Organic Molecules, J Mat Chem 1: 903
48. Shames A, Lev O, Iosefzon-Kuyavskaya B (1993) Study of Sol-Gel Glass-Formation and Properties by Paramagnetic Probes, J Non-Crystall Solids 163: 105
49. Shames A, Lev O, Berkovich Y, Iosefzon-Kuyavskaya B (1994) J Sol-Gel Sci Tech 2: 255
50. Narang U, Wang R, Prasad PN, Bright FV (1994) Effects of Aging on the Dynamics of Rhodamine-6G in Tetramethyl Orthosilicate-Derived Sol-Gels, J Phys Chem 98: 17
51. Narang U, Bright FV, Prasad PN (1993a) Characterization of Rhodamine 6G-Doped Thin Sol-Gel Films, Appl Spectrosc 47: 229
52. Black I, Birch DJS, Ward D, Leach MJ (1996) J Fluoresc xxx: yyy (submitted)
53. Sakohara S, Tickanen LD, Anderson MA (1992) Luminescence Properties of Thin Zinc-Oxide Membranes, Prepared by the Sol-Gel Technique – Change in Visible Luminescence 58. Grattan KTV, Badini GE, Palmer AW, Tseung ACC (1991) Use of Sol-Gel Techniques for Fiberoptic Applications, Sensors & Actuators A26: 483
54. Levy D (1992a) Glasses, Ceramics, Powders, Membranes and Composites; 10: Sol-Gel Glasses for Optics and Electrooptics, J Non-Crystall Solids 147: 508
55. Frye GC, Brinker CJ, Ricco AJ, Martin SJ, Hilliard J, Doughly DH (1990) Sol-gel coatings on acoustic wave devices: thin film characterization and chemical sensor development, Mater Res Soc Symp Proc 180: 583
56. Lev O, Tsionsky M, Rabinovich L, Glezer V, Sampath S, Pandratov I, Gun J (1995) Organically Modified Sol-Gel Sensors, Anal Chem 67: 22A
57. Rottman C, Ottolenghi M, Zusman R, Lev O, Smith M, Gong G, Kagan ML, Avnir D (1992) Doped Sol-Gel Glasses as pH sensors, Mat Lett 13: 293
58. Grattan KTV, Badini GE, Palmer AW, Tseung ACC (1991) Use of Sol-Gel Techniques for Fiberoptic Sensor Applications, Sensors & Actuators A26: 483
59. Wolfbeis OS, Rodriguez NV, Werner T (1992) LED-Compatible Fluorosensor for Measurement of Near-Neural pH Values, Mikrochim Acta 108: 133
60. Shariari MR, Ding JY (1994) p 279 in Sol-Gel Optics: Processing and Applications, Klein LC ed., Kluwer, Boston
61. MacCraith BD, McDonagh C, O'Keeffe G, Butler T, O'Kelly B, McGlip JF (1994b) Sol-Gel Sci Technol 2: 661
62. MacCraith BD, McDonagh CM, O'Keefe G, Keyes ET, Vos JG, O'Kelly B, McGlip JF (1993b) Fibre Optic Sensor Based on Fluorescence Quenching of Evanescent-Wave Excited Ruthenium Complexes in Sol-Gel Derived Porous Coatings, Analyst 118: 385

63. Pope E.J.A 1994, Proc SPIE 2288, p. 410
64. Yang L, Saavedra SS, Armstrong NR, Hayes J (1994) Fabrication and Characterization of Low-Loss, Sol-Gel Planar Wave-Guides, Anal Chem 66: 1254
65. Miller DR, Han OH, Bohn PW (1987) Appl Spectrosc 41: 249
66. Zimba CG, Hallmark VM, Turrell S, Swalen JD, Rabolt JF (1990) J Phys Chem 94: 939
67. Rabe JP, Swalen JD, Rabolt JF (1987) J Chem Phys 86: 1601
68. Bolton BA, Scherer JR (1989) J Phys Chem 93: 7635
69. Walker DS, Hellings HW, Saavedra SS, Reichert WM (1993) J Phys Chem 97: 10217
70. DeGrandpre MD, Burgess LW, White PL, Goldman DS (1990) Anal Chem 62: 2012
71. Choquette SJ, Locasio-Brown L, Durst RA (1992) Anal Chem 64: 55
72. Tiefenthaler K, Lukosz W (1985) Thin Solid Films 126: 205
73. Nellen PM, Tiefenthaler K, Lukosz W (1988) Sensors Actuators 15: 285
74. Yang L, Saavedra SS (1995) Anal Chem 67: 1307
75. Wolfbeis OS (1995) Trends Anal Chem (in press)
76. Tan W, Shi ZY, Smith S, Birnbaum D, Kopelman R (1992b) Anal Chem 64: 2985
77. Samuel J, Strinkovski A, Shalom S, Lieberman K, Ottolenghi M, Avnir D, Lewis A (1994b) A Metallized Micropipette Tip Filled with Doped Xerogel for Near Field Optical pH Sensing, Mater Lett 21: 431
78. Kopelman R, Tan W (1993) Near-Field Optics: Imaging Single Molecules, Science 262: 1382
79. Tan W, Shi ZY, Smith S, Birnbaum D, Kopelman R (1992a), Science 258: 778
80. Lieberman RA (1991) in Fiber Optic Chemical Sensors and Biosensors, Wolfbeis OS (ed) CRC Press, Boca Raton vol 1, chap 5
81. MacCraith BD (1993) Enhanced Evanescent Wave Sensors Based on Sol Gel-Derived Porous-Glass Coatings, Sensors & Actuators B11: 29
82. Lev O (1992a) Diagnostic Applications of Organically Doped Sol-Gel Porous Glass, Analysis 20: 543
83. Lev O, Kuyavskaya BI, Sakharov Y, Rottman C, Kuselman A, Avnir D, Ottolenghi M (1993) Photometric sensors based on sol-gel porous glass doped with organic reagents; Proc SPIE 1716: 357
84. MacCraith BD, McDonagh C, McEvoy AK, Butler T, O'Keefe G, Murphy V (1995) submitted to J Sol-Gel Sci Tech
85. Feng CD, Shimizu Y, Egashira M (1994) Effect of Gas-Diffusion Process on Sensing Properties of SnO_2 Thin-Film Sensors in a SiO_2/SnO_2 Layer-Built Structure Fabricated by Sol-Gel Process, J Electrochem Soc 141: 220
86. Nogami M, Nagasaka K, Kotani K (1990) Microcrystalline PbS Doped Silica Glasses Prepared by the Sol-Gel Process, J Non-Crystall Solids 126: 87
87. Nogami M, Nagasaka K, Takana M (1990) Cadmium Sulfide Microcrystal-Doped Silica Glass Prepared by the Sol-Gel Process, J Non-Crystall Solids 122: 101
88. Samuel J, Polevaya Y, Ottolenghi M, Avnir D (1994) Chem Mater 6: 1457
89. Wolfbeis OS, Offenbacher H, Kroneis HK, Marsoner HM (1984) A Fast Responding Fluorescence Sensor for Oxygen, Mikrochim Acta (Vienna) I: 153
90. Saini S, Kurrat R, Prenosil JE, Ramsden JJ (1994) Temperature-Dependence of Pyrolyzed Sol-Gel Planar Waveguide Parameters, J Phys D: Appl Phys 27: 1134
91. Dvorak O, De Armond MK (1993) Electrode Modification by the Sol-Gel Method, J Phys Chem 97: 2646
92. Gun G, Tsionsky M, Lev O (1994) Mater Res Soc Symp Proc 346: 1011
93. Gun G, Tsionsky M, Lev O (1994) Anal Chim Acta 294: 261
94. Gun J, Tsionsky M, Rabinovich L, Golan Y, Rubinstein I, Lev O (1995) J Electroanal Chem (in press)
95. Tsionsky M, Gun G, Glezer V, Lev O (1994) Sol-Gel-Derived Ceramic-Carbon Composite Electrodes – Introduction and Scope of Applications, Anal Chem 66: 1747
96. Reisfeld R, Brusilovsky D, Eyal M, Jorgensen CK (1989) Chimia 43: 385
97. MacCraith BD, O'Keeffe G, McDonagh C, McEvoy AK (1994) LED-Based Fibre Optic Oxygen Sensor Using Sol-Gel Coating, Electron Lett 30: 888
98. McEvoy AK, McDonagh C, MacCraith BD (1995) J Sol-Gel Sci Tech, (submitted)
99. Liu HY, Switalski SC, Coltrain BK, Merkel PB (1992) Oxygen Permeability of Sol-Gel Coatings, Appl Spectrosc 46: 1266
100. Tsionsky M, Lev O (1995) Anal Chem 67: 2409
101. Tsionsky M, Lev O (1995) J Electrochem Soc 142: 2154

102. Forster M, Eberle J, Strassler S, Pfister G (1990) Sol-Gel Derived Highly Porous SnO_2 for Semiconductor Gas Sensors, Inst Phys Conf Ser 111: 479
103. Wilson A, Wright JD, Murphy JJ, Stroud MAM, Thorpe SC (1994) Sol-Gel Materials for Gas-Sensing Applications, Sensors & Actuators B19: 506
104. Althainz P, Dahlke A, Goschnick J, Ache HJ (1994) Low-Temperature Deposition of Glass Membranes for Gas Sensors, Thin Solid Films 241: 344
105. Zhang JC, Shen JQ, Shan ZJ, Zhou SX, Zhu JL (1994) Radiation Effects in Tin Dioxide as Gas Sensors, Nucl Instrum Meth Phys Res Sect B 91: 663
106. Park SS, Zheng HX, Mackenzie JD (1993) Ethanol Gas-Sensing Properties of SnO_2-Based Thin-Film Sensors Prepared by the Sol-Gel Process, Mater Lett 17: 346
107. Atkinson A, Moseley PT (1993) Thin-Film Electroceramics, Appl Surface Sci 65/66: 212
108. Chen ZX, Colbow K (1992) MgO-Doped Cr_2O_3 – Solubility Limit and the Effect of Doping on the Resistivity and Ethanol Sensitivity, Sensors & Actuators B9: 49
109. Klein R, Voges E (1993) Integrated-optic ammonia sensor, Sensors & Actuators B 11: 221
110. Lin J, Heurich M, Obermeier E (1993) Manufacture and Examination of Various Spin-on Glass-Films with Respect to Their Humidity-Sensitive Properties, Sensors & Actuators, B13: 104
111. Vaivars G, Pitkevics J, Lusis A (1993) Sol-Gel Produced Humidity Sensor, Sensors & Actuators, B13: 111
112. Sun HT, Cheng ZT, Yao X, Wlodarski W (1993) Humidity Sensor Using Sol-Gel-Derived Silica Coating on Quartz-Crystal, Sensors & Actuators, B13: 107
113. Lin J, Moller S, Obermeier E (1991) Thin-Film Gas Sensors with Organically Modified Silicates for the Measurement of SO2, Sensors & Actuators B5: 219
114. Dulebohn JI, Haefner SC, Berglund KA, Dunbar KR (1992) Reversible Carbon-Monoxide Addition to Sol-Gel Derived Composite Films Containing a Cationic Rhodium(I) Complex: Toward the Development of a New Class of Molecule-Based CO Sensors, Chem Mater 4: 506
115. Lee DD, Choi SD, Lee KW (1995) Sensors & Actuators B24/25: 607
116. Chernyak V, Reisfeld R (1993) Photochemical Sensor Based on Malachite Green in Glass Films, Sensors and Materials 4: 195
117. Chernyak V, Reisfeld R, Gvishi R, Venezky D (1990) Oxazine-170 in Sol-Gel Glass and PMMA Films as a Reversible Optical Waveguide Sensor for Ammonia and Acids, Sensors & Materials 2: 117
118. Narang U, Dunbar RA, Bright FV, Prasad PN (1993) Chemical Sensor-Based on an Artificial Receptor Element Trapped in a Porous Sol-Gel Glass Matrix, Appl Spectrosc 47: 1700
119. Chung KE, Lan EH, Davidson MS, Dunn BS, Valentine JS, Zink JI (1995) Anal Chem 67: 1505
120. Knobbe ET, Dunn B, Gold M (1988) Organic molecules entrapped in a silica host for use as biosensor probe materials, Proc SPIE 906: 39
121. Zusman R, Rottman C, Ottolenghi M, Avnir D (1990) Doped Sol-Gel Glasses as Chemical Sensors J Non-Cryst Solids 122: 107
122. Kraus SC, Czolk R, Reichert J, Ache HJ (1993) Optimization of the Sol-Gel Process for the Development of Optochemical Sensors & Actuators B15: 199
123. Bambot SB, Sipior J, Lakowicz JR, Rao G (1994) Sensors & Actuators B22: 181
124. Badini GE, Grattan KTV, Palmer AW, Tseung ACC (1989), Development of pH-Sensitive Substrates for Optical Sensor Applications, In: Arditty HJ, Dakin JP, Kersten RTh (eds) Optical Fibre Sensors, Springer Proceed in Phys, vol 44, Springer, Berlin-Heidelberg
125. MacCraith BD, Ruddy V, Potter C, Kelly B, McGilp JF (1991) Optical Wave-Guide Sensor Using Evanescent Wave Excitation of Fluorescent Dye in Sol-Gel Glass, Electron Lett 27: 1247
126. Butler TM, MacCraith BD, McDonagh CM (1995) Proc SPIE 2508: xxx (in press)
127. Ding JY, Shahriari MR, Sigel GH (1991) Fibre Optic pH Sensors Prepared by Sol-Gel Immobilization Technique, Electron Lett 27: 1560
128. Plaschke M, Czolk R, Ache HJ (1995) Anal Chim Acta 304: 107
129. Kuselman I, Iosefson-Kuyavskaya B, Lev O (1992) Disposable Tube Detectors for Water Analysis, Anal Chim Acta 256: 65
130. Lev O, Kuyavskaya BI, Gigozin I, Ottolenghi M, Avnir D (1992) A High-Sensitivity Photometric-Method Based on Doped Sol-Gel Glass Detectors Determination of Sub-ppb Divalent Iron, Fresenius J Anal Chem 343: 370
131. Oehme I (1995) PhD Dissertation, Karl-Franzens University, Graz, Austria
132. Barrero JM, Camara C, Perez-Conde MC, San Jose C, Fernandez L (1995) Analyst 120: 431

133. Fabry P, Huang YL, Caneiro A, Patrat G (1992) Dip-Coating Process for Preparation of Ion-Sensitive Thin-Films, Sensors & Actuators, B6: 299
134. Caneiro A, Fabry P, Khireddine H, Siebert E (1991) Performance-Characteristics of Sodium Super Ionic Conductor Prepared by the Sol-Gel Route for Sodium-Ion Sensors, Anal Chem 63: 2550
135. R. Reisfeld, unpublished results
136. Levy D, Gigozin I, Zamir I, Kuyavskaya BI, Ottolenghi M, Avnir D, Lev O (1992) Immobilization of Quarternary Ammonium Anion Exchangers in Sol-Gel Glasses, Separat Sci Tech 27: 589
137. Dunuwila DD, Torgerson BA, Chang CK, Berglund KA (1994) Sol-Gel Derived Titanium Carboxylate Thin Films for Optical Detection of Analytes, Anal Chem 66: 2739
138. Zevin M, Reisfeld R, Oehme I, Wolfbeis OS (1994) unpublished results
139. Glezer V, Lev O (1993) Sol-Gel Vanadium Pentaoxide Glucose Biosensor, J Am Chem Soc 115: 2533
140. Shtelzer S, Braun S (1994) An Optical Biosensor Based upon Glucose-Oxidase Immobilized in Sol-Gel Silicate Matrix, Biotechnol Appl Biochem 19: 293
141. Trettnak W, Wolfbeis OS (1989) Anal Chim Acta 221: 195
142. Trettnak W, Wolfbeis OS (1989) Fresenius Z Anal Chem 334: 427
143. Yamanaka SA, Nishida F, Ellerby LM, Nishida CR, Dunn B, Valentine JS, Zink JI (1992) Enzymatic-Activity of Glucose-Oxidase Encapsulated in Transparent Glass by the Sol-Gel Method, Chem Mater 4: 495
144. Tatsu Y, Yamashita K, Yamaguchi M, Yamamura S, Yamamoto H, Yoshikawa S (1992) Entrapment of Glucose-Oxidase in Silica-Gel by the Sol-Gel Method and Its Application to Glucose Sensor, Chem Lett 8: 1615
145. Audebert P, Demaille C, Sanchez C (1993) Electrochemical Probing of the Activity of Glucose-Oxidase Embedded Sol-Gel Matrices, Chem Mater 5: 911
146. Pankratov I, Lev O (1995) J Electroanal Chem 393: 35
147. Sampath S, Pankratov I, Gun J, Lev O (1995) J Sol-Gel Sci Tech (in press)
148. Stamm C, Lukosz W (1993) Integrated Optical-Difference Interferometer as Refractometer and Chemical Sensor, Sensors & Actuators, B11: 177
149. Nellen PM, Lukosz W (1993) Integrated Optical Input Grating Couplers as Direct Affinity Sensors, Biosensors & Bioelectron 8: 129
150. Wang R, Narang U, Prased PN, Bright FV (1993) Affinity of Antifluorescein Antibodies Encapsulated within a Transparent Sol-Gel Glass, Anal Chem 65: 2671
151. Chaput F, Boilot JP, Riehl D and Levy Y (1994) Proc. SPIE 2288, 286

New Materials for Nonlinear Optics

Renata Reisfeld*

Inorganic Chemistry Department, The Hebrew University of Jerusalem, 91904 Jerusalem, Israel

The field of nonlinear optics has been characterized by a remarkable increase in research activities during the past decade. We summarize the specific research into nonlinear optical materials prepared by the sol-gel technology. The chapter describes the preparation of glasses doped by semiconductor nanoparticles such as CdS, CuCl, CuBr, Bi_2S_3 and PbI_2, metal nanoparticles and organic dyes. The third order susceptibilities of the various materials are tabulated. Glasses doped by organic dyes, their preparation, and properties of second and third order nonlinearity are discussed. The potential of these materials for industrial and technical applications is stressed.

*Enrique Berman Professor of Solar Energy

Structure and Bonding, Vol. 85
© Springer-Verlag Berlin Heidelberg 1996

1 Introduction

Nonlinear optical materials will play with a key role in future photonics. Examples of nonlinear optical phenomena that are potentially useful include the ability to alter the frequency (or color) of light and to amplify one source of light with another, switch it, or alter its transmission characteristics through the medium, depending on its intensity [1–3]. In order to implement nonlinear optical materials for the performance of photonics functions, there is a need for materials of high optical quality with large and stable optical nonlinearities [4, 5]. Progress during the last decade in the design of organic systems allows the preparation of new materials with promising lasing and nonlinear optical properties [6]. A novel way of constructing nonlinear materials is by the sol-gel process [7]. The sol-gel process provides a method to make single or multicomponent oxide glasses with optically active species at low temperature [8, 9]. The process involves hydrolysis followed by the polycondensation of inorganic esters of silicic acid. Due to the low temperature processing conditions, this method allows the introduction of organic species into a glass during gelation, or by impregnation after drying due to the high porosity of the final glass.

Future applications could be in the field of electrooptic and optical switching. Because photons travel three order of magnitude faster than electrons in media, they have obvious advantages for signal transmission [10]. The combination of electronics and photonics that has been used in optical signal processing and particularly in optical switches can be dramatically improved if the slowest member of the chain (that is, the electric signal used to induce refractive index changes necessary for optical switching) is eliminated. Recent theoretical and experimental results indicate a dramatic change in the linear and nonlinear optical (NLO) response as a function of decreasing the size of semiconductor atomic clusters [11] to the point of quantum confinement. The use of the NLO response of these near "molecular" clusters provides a means to optically switch using photons alone at correspondingly faster response times.

An excellent example is the recent announcement by IBM Yorktown Heights of the development of experimental computer chips for transmitting and receiving data over fiber optic lines at speeds of one billion bits per second. Each chip has more than 8000 transistors with characteristic features as small as 1 μm. One "receiver" chip holds 50 times more optical and electronic components than ever previously assembled on a chip for optoelectronic data transmission. Quantum-well confinement technology is used to create the light pulses for data transmission [5].

Electrons in recently developed devices can be confined to panes, lines or quantum dots as mathematical points, corresponding to two, one and zero dimensions. In addition to the exciting basic science, quantum dots promise properties that could be harnessed for a range of electronic and optical applications. Arrays of densely packed dots could form a substrate for computers of unprecedented power. Dots could also constitute materials capable of absorbing and emitting light at wavelengths predetermined by their designers, or could even serve as the basis for semiconductor lasers that are more efficiently and precisely tuned than any now in existence.

Electrons confined in a plane with negligible thickness have no freedom of motion in the third dimension. Those confined in quantum wire are free in only one dimension, and those confined in a quantum dot are not free in any dimension. For common semiconductors, the length scale for a free conduction electron is about 100 ångstrom (= 10 nanometer) [12].

Nanoparticles of metals, semiconductors and dye aggregates have atomic arrangements in aggregates of the size ranging between molecular clusters and infinite solid state arrays of atoms. Their properties are determined by the extent of confinement of highly delocalized valence electrons. When introduced into glasses their spectroscopic properties can be measured and the raising of the discrete energy levels compared to the continuous band energies determined. These energy states of the quantum dots are positioned between the discrete energy levels of the atoms or bands of molecules and the broad band of the condensed phase. Glasses doped by semiconductor, metal and dye particles are known to have nonlinear properties. Glass films can be prepared by gold, silver and platinum metal aggregates and CdS, $CuCl$, $CuBr$, and PbI_2 and Bi_2S_3, etc. semiconductor quantum dots. The relation between the energies of the electronic states and size of the particles influences the nonlinear properties. Dye aggregates having characteristic electronic spectra also possess nonlinear properties, as do glass films obtained from oxides of large refractive index (see the chapter by Sakka in this volume).

The ability to generate nanostructure quantum-confined materials atom by atom in the context of recent scientific developments in nonlinear phenomena [6] is both an intriguing and potentially rewarding challenge. Changes associated with quantum confinement, resulting in restriction of the electronic wave function to smaller and smaller regions of space, will have profound effects on the nonlinear displacement of charge with optical electromagnetic fields.

In the nanoparticles there is strong spatial delocalization of valence electrons, and therefore a small crystallite must grow fairly large in order to achieve the limiting bulk electronic structure. The intermediate-sized clusters can have unique properties, characteristic of neither the molecular nor the solid state limits. Examples include metal clusters (as predicted by free electron shell theory) semiconductor crystallites which exhibit the bulk unit cell but have only partial band structure development and dye aggregates of appropriate size. Nanoclusters have optical spectra that can be tuned in wavelength simply by varying the crystallite size. They represent, in three-dimensions, an analogy to the quantum well semiconductor heterostructures that show a one-dimensional quantum size effect.

2 Quantum Structures and Atom Assemblies

The basic ideas of quantum confinement or quantum wells were introduced in the early 1970s. In these nanostructures, the dimensions of the wave function of the electron-hole pair (exciton) in the lowest excited state of the nanocluster are comparable to the physical size of the particle. This quantum confinement of the exciton means that the continuum band of energies becomes more molecular in character, with narrow ranges of energy and line structure in the optical spectra. For the chemist, less delocalization means less energy stabilization; a reflection of this is that the absorption band for direct transitions of nanosized semiconductor clusters is shifted to higher energies than in the extended bulk parent materials. The "photo-charging" of the semiconductor particle, that is, photo-inducing high electron concentrations into a narrow conduction band with a small effective density of states, also changes the band gap and the resulting absorption edge. In addition to NLO applications, the resulting phenomena can be used to carry out electron-transfer processes that are otherwise inaccessible to the bulk semiconductor.

Most important for NLO is that strong nonlinear responses are observed near exciton resonances, a feature not available in comparable materials that have no quantum confinement [13, 14]. In quantum structures, the number of states that are available in the relatively small energy band can be very limited, so that adding or removing only a few carriers can substantially change the absorption edge and thus the optical coefficients (such as the index of refraction). Conversely, perturbation of the quantum cluster by light can modify the optical coefficients and the absorption edge. Small changes in the dielectric properties of the confining structure or the surface states of that cluster can also introduce large changes in optical properties. Quantum confinement can therefore be used to change the position and nature of the resonance absorption transitions, as well as to make the NLO response particularly susceptible to modification by optical and low-frequency electric fields. An additional and equally important property of quantum-confined materials is that ultra-fast response times become accessible because of the increased importance of surface states.

The bulk NLO response depends on the number of polarizable electrons per unit volume so that a high-density collection of quantum nanoclusters is required, both for achieving a basic understanding of the quantum-structure phenomena and for device development. In designing such devices, one must necessarily consider not only the atomic construction of the individual quantum wires, boxes, sheets, and dots, but also the assembly of these into a macroscopic unit. Glasses and polymers have the advantage in that they can readily be made into thin films for device application, but the disadvantage of having particle size distributions, which in present commercial samples typically vary by more than 10%. The consequence is that the exciton resonances are broadly distributed in energy so that wavelength-selective NLO phenomena are difficult to achieve.

The first hints that zero-dimensional quantum confinement was possible came in the early 1980s, when A.I. Ekimov [15] and his colleagues at the Ioffe Physical-Technical Institute in St. Petersburg noticed unusual optical spectra from samples of glass containing the semiconductors cadmium sulfide or cadmium selenide. The samples had been subjected to high temperatures; Ekimov suggested tentatively that the heating had caused nanocrystallites of the semiconductor to precipitate in the glass and that quantum confinement of electrons in these crystallites caused the unusual optical behavior.

How small must a nanocrystallite be for quantum confinement to be visible? In a vacuum the effects of confinement would begin to appear when the electron was trapped in a volume about 10 Å across. This size implies an electron wavelength of 20 Å and therefore an energy of about one fortieth of an electron volt.

Semiconductor physics comes to the aid of the nanotechnologist in the basic understanding of spectral behaviour. The wavelength of an electron depends on its energy and its mass. For a given wavelength, the smaller the mass, the larger the energy and the easier it is to observe the energy shift that confinement causes. The electrostatic potentials of the atoms in the crystalline lattice superimpose to provide a medium in which electron waves propagate with less inertia than they do in free space. The "effective mass" of the electron is thus less than its actual mass. In gallium arsenide the effective mass is about 7 percent of what it would be in a vacuum, and in silicon it is 14 percent. Quantum confinement (QC) in semiconductors occurs in volumes roughly 100 Å across (Fig. 1).

In the QC regime hybrid molecular-solid state, the bulk unit cell is present, yet the spectra show three-dimensional confinement and are in principle discrete. If the QC radius a_0, $R > a_0$ (where R is the bulk exciton Bohr radius), for

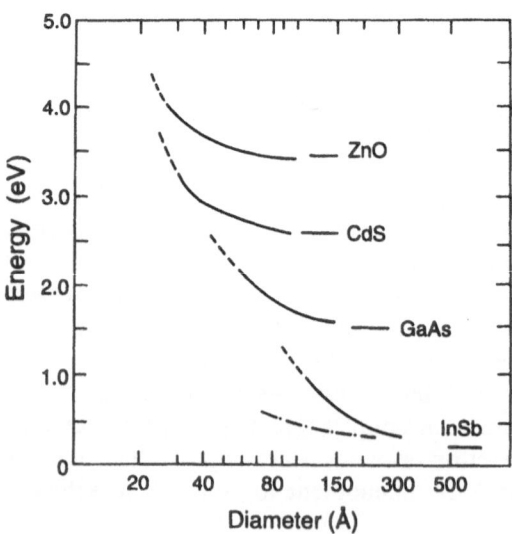

Fig. 1. Energy of the lowest excited electronic state as a function of crystallite diameter as calculated via wavefunction (13) from Ref. 12. Short horizontal solid lines are the bulk band gap energies of the materials indicated. The dot-dashed line for InSb incorporates the surface carrier charge density, as described in the text

example $R = 10$ Å in II–VI crystallites, the electron and hole are principally individually confined with minor electron-hole correlation and no hydrogenic exciton formation. This is the strong confinement limit. As an example let us take CdSe nanocrystallites with a radius of 16 Å; the Bohr radius of the exciton in CdSe is 30 Å [16].

In the weak confinement limit, when a_0 is 2–4 times R, a hydrogenic exciton forms whose center of mass motion is quantized and exhibits discrete states within the crystallite [17]. CuCl crystallites provide the only well-characterized examples in this regime: the a_0 value here is 7 Å. In principle QCs show only discrete transitions that are size tunable. Their optical properties are essentially molecular. Much of the QC nonlinear optical interest comes from the fact that the absorption coefficient at one wavelength might be much larger than in superlattices, and so the QC nonlinear effects per unit mass could be much larger.

The quantum-structure size that is desirable for optimum NLO response depends on the exciton diameter, the effective separation of hole and electron in the first excited state of the semiconductor. For large band gap semiconductors such as zinc sulfide (ZnS, $E_g = 3.6$ eV) or cadmium sulfide (CdS, $E_g = 2.58$ eV), quantum confinement will not take place until the particle size is ≤ 15 Å or ≈ 35 Å, respectively, although for the small band gap indium antimonide (InSb, $E_g = 0.23$ eV) the exciton will be delocalized over $\approx 10^6$ atoms with a diameter of ≈ 1400 Å. These numbers are "particle-in-the-box" estimates for quantum boxes or wells and different considerations apply to quantum wires or sheets.

The optical absorption threshold for nanocrystallites of this size shifts to higher energies – away from the red end of the spectrum – as the crystallite becomes smaller. This effect appears most elegantly in cadmium selenide clusters; the progression from deep red to orange to yellow as the diameter of the cluster declines can be clearly seen by the naked eye. As seen above, the near band gap electronic absorption is predicted to be discrete and very sensitive to physical size.

Quantum confinement effects in semiconductor micro crystallites have been discussed by Efros and Efros [18] and the first quantitative effort using an idealized model for evaluating the electronic levels of the quantum dots was performed by Brus [19]. He used wave-mechanical treatment to calculate the lowest eigenstate of an exciton confined in a spherical box, having either infinitely high walls or a finite potential barrier. The calculations gave the correct trend of the blue shift of the lowest exciton level with the decrease of the size of the semiconductor particle. An improvement of the model was proposed by Fojtik et al. [20]. In their model, the electron having the reduced mass $\mu = (1/m_e + 1/m_h)^{-1}$ (where m_e and m_h are the effective masses of electron and hole, 0.19 and 0.8 for bulk CdS, respectively) moves around a massless hole placed at the center of the particle. The allowed energy levels are found by solving a semiclassical phase integral, as in the models of Brus [19].

Generally, the blue shift of the excition absorption peaks is discussed in two cases: one is valid for the excitons having an effective Bohr radius R much larger than the crystal size, a_0, $a_0/R < 2$, and the other is for the exciton having an effective Bohr radius much smaller than the crystal size, $a_0/R > 4$. These two regimes are named the 'electron-hole confinement' and the 'exciton confinement', for which energies are approximately given by, respectively,

$$E = E_g + (\hbar^2\pi^2)/(2\mu a_0^2),$$ (1)

$$E = E_g + (\hbar^2\pi^2)/(2M a_0^2)$$ (2)

where μ and M is the reduced mass ($\mu = m_e * m_h$) and the translational mass ($M = m_e + m_h$), respectively, of the exciton [20].

In general, the lower excited state energies can be calculated within the effective mass approximation using a different mass for the conduction and valence bands. In addition, the Coulomb interaction can be included to give an approximate expression for the lowest state energy [21]

$$E^* = E_g + \frac{\hbar^2\pi^2}{2a_0^2}\left[\frac{1}{m_e} + \frac{1}{m_h}\right] - \frac{1.8e^2}{\varepsilon a_0}$$ (3)

The expected behavior for different materials as calculated by Brus from formula (13) given in Ref. 19 is shown in Fig. 1. Small band gap III–V materials display their bulk energy gaps only at large diameters because of their very small electron effective mass. II–VI materials typically have larger masses and smaller dielectric coefficients. They display their bulk gaps at smaller diameters, and show one diameter at which significantly non-zero effective mass and Coulomb terms accidentally cancel.

In both materials, transitions at much higher energies (typically in the far ultraviolet) involve localized discrete states near the center of the Brillouin zone [22]. These states, with energy insensitive to size, are present in the smallest crystallite. In a sense, they are σ to σ^* transitions within one unit cell.

3 Sol-Gel Glasses as Transparent Media for Nanoparticles

As discussed earlier an ideal medium for incorporation of nanoparticles are glasses prepared by the sol-gel method [4, 23–27].

The sol-gel technique offers a low temperature method for synthesizing amorphous materials which are either totally inorganic in nature or composed of inorganics and organics. The process is based on the hydrolysis and condensation reactions of organometallic compounds in alcoholic solutions. The most widely-investigated system involves silica-based glasses which are prepared by polymerization of a silicon alkoxide, $Si(OR)_4$ [28–30]. The process is based on inorganic polymerization reactions, starts from molecular precursors, and a macromolecular oxide network is obtained via hydroxylation condensation reactions which can be controlled by the chemical design of molecular precursors. The viscosity properties of sols can be adjusted allowing easy deposition of transparent coatings onto glass, ceramic, or polymeric substrates [31, 32]. Sol-gel chemistry is performed in solution at lower temperatures than in conventional chemical methods. Homogeneous doping by mixing components at a molecular level, synthesis of metastable or amorphous phases allowing larger concentrations of chromophores, and synthesis of mixed organic-inorganic materials can then be performed.

The ability to synthesize inorganic polymers using sol-gel processing with little or no heating makes it possible to dope these gels with a variety of organic and organometallic molecules [33–36], especially dyes that possess the required luminescent properties. The emission properties of the molecules have been used to optically probe sol-gel chemistry and structure.

In the case of organosilicate glasses (organically modified ceramics or ormocers), the silicate network may be modified by organic substituents such as alkyl (e.g. methyl) groups or other functional groups (e.g. 3-glycidoxypropyl) which may form organic copolymers that penetrate the silicate structure. In principle, Si may be substituted by Al, B, Ti, or Zr to yield ceramics of variable mechanical properties (see also the chapter on sensors for more details).

A schematic representation of formation of oxide glass by the sol-gel method is:

Hydrolysis:

$$M(OR)_4 + 4H_2O \rightarrow M(OH)_4 + 4ROH \tag{4}$$

Condensation:

$$M(OH)_4 + M(OH)_4 \rightarrow (OH)_3 - M - O - M - (OH)_3 + H_2O \tag{5}$$

where M can be Si, Ti and Zr, and R is an alkyl group. An example of Ormocer–Ormosil formation is polycondensation of polydimethyl siloxane (PDMS form) with silicon alkoxide. PDMS is

$$
\begin{array}{cccc}
\text{O} & \text{Me} & \text{Me} & \text{O} \\
| & | & | & | \\
\text{O--Si--O--} & \text{Si--O--} & \text{Si--O--} & \text{Si--O} \\
| & | & | & | \\
\text{O} & \text{Me} & \text{Me} & \text{O}
\end{array}
\qquad \text{PDMS} \tag{6}
$$

and

Polycondensation reaction of PDMS and TEOS:

$$
2Si(OH)_4 + HO-\left[\begin{array}{cc} \text{Me} & \text{Me} \\ | & | \\ \text{Si--O--Si--} \\ | & | \\ \text{Me} & \text{Me} \end{array}\right]_n -OH \longrightarrow \tag{7}
$$

$$
\begin{array}{c}
\text{OH} \\ | \\ \text{OH--Si--O--} \\ | \\ \text{OH}
\end{array}
\left[\begin{array}{cc} \text{Me} & \text{Me} \\ | & | \\ \text{Si--O--Si--} \\ | & | \\ \text{Me} & \text{Me} \end{array}\right]_n
\begin{array}{c}
\text{OH} \\ | \\ -\text{O--Si--OH} - 2H_2O \\ | \\ \text{OH}
\end{array}
\tag{8}
$$

Polycondensation reaction of metal alkoxide and the copolymer
(phenylmethylsiloxane):

$$2M(OH)_4 + H-[copolymer]-OH \rightarrow -\overset{\displaystyle OH}{\underset{\displaystyle O}{\overset{|}{\underset{|}{M}}}}-O-[copolymer]-O-\overset{\displaystyle O}{\underset{\displaystyle O}{\overset{|}{\underset{|}{M}}}} + H_2O \quad (9)$$

In the investigation of glassy systems, several analytical methods [37] have been used among which are neutron and light scattering. However, at present there is no consistent picture about the microstructure of these materials. However, new developments are in progress [38, 39, 40].

Most published work reports the use of sol-gel-derived SiO_2 glass as the matrix material. Takada et al. [41] decided to use a $Na_2-O-B_2O_3-SiO_2$-gel-derived glass at a temperature of 580 °C. CdS crystallites ranging from 40–60 Å were prepared with concentrations up to 8 wt%. The glass nanocomposites have excellent optical qualities and high values of $\chi^{(3)}$[4]. When better control of particle size distribution is achieved, the $\chi^{(3)}$ values are expected to increase further, thus rendering these gel-derived nanocomposites as acceptable candidates for a variety of photonic applications. The Ormosils have also been used as matrices for the CdS semiconductor quantum dot materials in thin films [4, 26].

Particle-size tailoring can be carried out with other systems like semiconducting quantum dots, based on semiconducting oxides or sulfides (ZnO, CdS, PbS), or metals that also become semiconducting. Recently we have shown that the particle size can be controlled by first forming a complex with a bulky ligand and subsequently adding thiourea to the glass precursors.

3.1 CdS-Doped Films

The most studied of all semiconductor particles is cadmium sulfide. The incorporation of small semiconductor particles into thin films with wave guiding properties has potential applications in optoelectronics and nonlinear optics. The nonlinear properties of CdS incorporated in thin films prepared by the sol-gel method and in composite organic–inorganic polymers was recently described by us [4].

The preparation of the glasses doped with CdS was carried out according to the general procedures for preparing glass films, as described in references [42–45]. The hickness of the films is a function of the speed of withdrawal of substrate glass from the precursor solution, pH, concentration of alkoxide in the solution and orientation of the sample with respect to the direction of withdrawal. Control of these conditions is necessary in order to guarantee reproducibility of the coatings.

The films are prepared from precursor solution having the typical molar ratios: tetramethoxysilane (TMOS) or tetraethoxysilane (TEOS): water: alcohol

= 1:4(6):11, where the number in parentheses refers to TEOS. The molar ratio of Cd varied from 0.0077 to 0.195 (0.12) moles of Cd to 1 mol of TMOS (TEOS). ORMOCER (organically modified ceramics) films are prepared according to the recipes given in Ref. 43 and modified in our group for incorporating dyes and colorants [46, 47]. The fresh films are then exposed to gaseous hydrogen sulfide for a few minutes, dried and heated to 250 °C. Alternatively, first a complex of Cd with bis-1,1,1,5,5,5-hexafluorpentane-2,4-dionate(hexafluoroacetylacetonate) is formed [6]. The complex is then dissolved in the composite glass precursor [48] and thiourea is added. The resulting sulfide complex of Cd in the glass forms CdS particles of 20–30 Å in size. The thickness of the films is determined by an interference method.

The average sizes of the CdS particles in the films are measured by two independent methods. The first is transmission electron microscopy (TEM) of thin films containing CdS deposited directly on copper grids. The second method is based on X-ray diffraction on powders, where Scherrer's equation is used to find the correlation between the width (in the units of 2θ) of the diffraction peaks and the average size of the particles. Concentration of CdS particles in the films is of the order of 10^{21} molec./cm^3. However, the number of molecules in a slab of $1 cm^2 \times 10^{-5} cm$ (which is the order of magnitude of thickness of the films) is about 10^{16} molecules. Such a small number of molecules requires very long scan times of 100–300 seconds/0.05–0.02°.

The X-ray diffraction spectra reveal the existence of both hexagonal (wurtzite) and cubic (zinc blende) forms of CdS. The TEM reveals an average size for CdS particles of 20–50 Å although some particles as large as 200 Å can be seen. The inhomogeneous distribution of particle sizes in the glass is expected and is analogous to the inhomogeneous spectral broadening, where there is statistical distribution of crystallographic sites having slightly different energies. However, the dominant behavior (the maximum intensities of absorption and emission) in the spectroscopically active sites, the inflection point and nonlinear optical properties are due to the fact that most of the particles have similar sizes.

For the analysis of our results, we have used the approach of Schmidt and Weller [49]. Here the exciton is allowed to move inside the sphere defined by the radius of the quantum dot, i.e. both electron and hole are allowed to move, whereby the potential term is defined as the Coulomb interaction between the two particles V(r) and is proportional to $e/\varepsilon|r_1-r_2|$, where r_1 and r_2 are the coordinates of the electron and the hole respectively. We start from the general Hamiltonian [23],

$$H = -(\hbar/2m_e)\nabla_1^2 - (\hbar^2/2m_h)\nabla_2^2 - e/\varepsilon|r_1 - r_2| \qquad (10)$$

where m_e and m_h are the reduced masses of the electron hole r_1 and r_2 are their corresponding coordinates, the first two terms are the kinetic energies of the electron and the hole respectively, and the third term represents the Coulomb interaction between the two particles, screened by factor ε, the high frequency dielectric constant of bulk CdS.

The eigenvalues of the system represent the excited energy levels of the system counted from the lower edge of the conduction band. In the infinitely extended material, the exciton exhibits hydrogen-like energy levels, proportional to $-(1/2n^2)$ with $n = 1, 2, 3, 4$. As long as the crystal of CdS is much larger than the excitonic radius, these levels are bounded slightly below the conduction band edge.

When the electron and the hole are constrained in a small volume of radius a_0 comparable to the radius of the exciton (about 19.3 Å for CdS), all degeneracies are removed and the excitonic levels become blue-shifted with respect to the conduction band edge.

Such a situation can be treated by the method used for a two-electron atom. The Hamiltonian can be scaled by $r \rightarrow r/a_0$ to get the form.

$$H = H^0 + R(-1/r_{12})$$ (11)

where a_0, the radius of the quantum dot is treated as a perturbation term. The eigenvalues are then approximated by expansion around $R = a_0$;

$$E = 1/R^2(E^{(0)} + RE^{(1)} + (R^2)E^{(2)}...)$$ (12)

where $E^{(0)}$ is the sum of kinetic energies of the electron and the hole and $E^{(1)}$ is the corresponding expectation value of $-1/r_{12}$.

Eigenfunctions are then a combination of hydrogen-like functions: 1s, 2s, 3s for angular momentum $L = 0$, and 2p, 3p... for angular momentum $L = 1$. If R is smaller than the excitonic radius, the lowest state eigenfunction is just a product of the lowest-state functions of the electron and the hole. For larger values of R, however, the function of state should be expanded in terms of the complete orthogonal set of functions.

$$\Psi_a = \Sigma c_{a,b} \Phi_a(r_1)\Phi_b(r_2)$$ (13)

Predicted energy levels are calculated from Table 1 of Ref. 23 and Eq. (12) in the first-order approximation and drawn in Fig. 2. These are compared with the energies calculated from the excitation spectra in Fig. 3. Thin glass films doped with cadmium sulfide emit perceptible fluorescence only at 77 K. Fig. 3 presents the excitation and emission spectra of two TEOS films, curve 1 with a concentration of 0.051 mol of Cd per mol of TEOS and curve 2 with a concentration of 0.127 mol of Cd per mol of TEOS. Curve 3 presents the emission and excitation spectra of cadmium sulfide (0.117 mol per mol) in GLYMO film. The excitation spectra are well structured and the wavelengths of the maxima can be determined. The calculated levels fit fairly well the experimental results of the CdS particles introduced into glass films.

The much smaller nanocrystalline semiconductors CdS, ZnS, PbS, CdSe, CdTe, CuCl and CuBr particles with diameters ranging from 2 to 10 nm were doped in the glasses by the sol-gel process of Nogami [50]. The optical absorption edge associated with those crystals exhibits a large blue shift compared to that of the bulk crystal, and the energy shift was reciprocally propor-

Table 1. Determined parameters used for calculating $\chi^{(3)}$ and $\chi_m^{(3)}$. The values in the brackets are the maximum experimental errors [73]

Ligand	Absorption $\alpha\,(\text{m}^{-1})$	Refractive index n_d	Peak-max.	FWHM (nm)	ε_2 [15]	p (vol. ‰)	$\chi^{(3)}$ (esu)	$\chi_m^{(3)}$ (esu)
Heating temperature = 300 °C								
APTS	1.64×10^5	1.51	535	76	3.04	1.39	9.6×10^{-9} $(\pm 1.4 \times 10^{-9})$	3.0×10^{-7} $(\pm 4.8 \times 10^{-8})$
DIAMO	3.08×10^5	1.55	532	90	3.60	2.80	6.3×10^{-8}	1.4×10^{-6}
TRIAMO	3.29×10^5	1.53	536	86	3.44	3.02	2.9×10^{-8} $(\pm 9.4 \times 10^{-9})$	5.6×10^{-7} $(\pm 2.2 \times 10^{-7})$
Thiosilane	1.90×10^5	1.53	522	104	4.16	2.06	$-$ $(\pm 4.3 \times 10^{-9})$	$-$ $(\pm 8.9 \times 10^{-8})$
Heating temperature = 500 °C								
APTS	4.93×10^5	1.46	535	84	3.36	5.86	2.5×10^{-8} $(\pm 3.8 \times 10^{-9})$	3.4×10^{-7} $(\pm 5.4 \times 10^{-8})$
DIAMO	4.92×10^5	1.48	532	77	3.08	4.45	6.3×10^{-8} $(\pm 9.5 \times 10^{-9})$	7.0×10^{-7} $(\pm 1.1 \times 10^{-7})$
TRIAMO	5.51×10^5	1.47	535	80	3.20	5.55	6.0×10^{-8} $(\pm 9.0 \times 10^{-9})$	6.9×10^{-7} $(\pm 1.1 \times 10^{-7})$
Thiosilane	4.31×10^5	1.49	524	92	3.68	4.47	$-$	$-$
Ref. 3: Au: DIAMO 1:1, UV-curing, + thermal treatment (200 °C)								
Diamo	3.81×10^5	1.52	539	60	2.4	2.3	2.8×10^{-7} $(\pm 4.2 \times 10^{-8})$	2.0×10^{-6} $(\pm 3.1 \times 10^{-7})$

Notes: APTS: 3 amino propyl triethoxy silane. DIAMO: N-2 amino ethyl 3 amino propyl trimethoxy silane. TRIAMO: Trimethoxy silyl propyl diethylene triamine

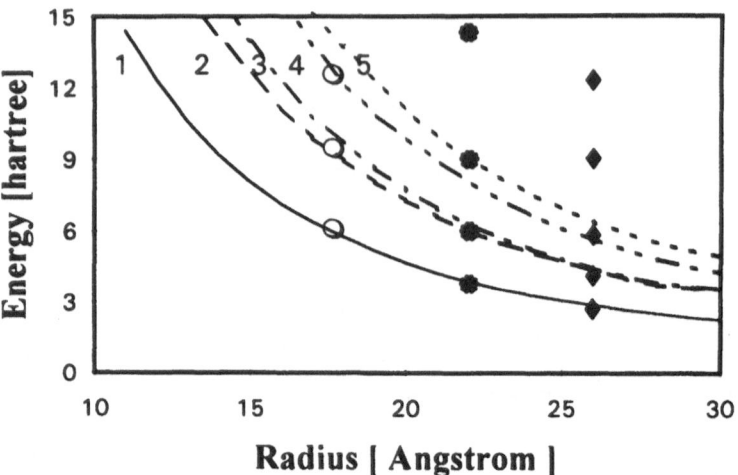

Fig. 2. Five lowest quantum levels of quantum dots of CdS obtained by first-order approximation of equation (XI) L = 0. *Curve 1*: 1s1s. *Curve 2*: 1s2s; *Curve 3*: 2p2p. *Curve 4*: 1s3s. *Curve 5*: 2p3p. These are compared with energy levels obtained from the excitation spectra of thin films. *Circles*: low concentration of cadmium sulfide 0.051 mol/mol. *Start*: high concentration of cadmium sulfide 0.117 mol/mol. *Diamonds*: GLYMCO 0.127 mol/mol. Values of parameters for calculation of eigenvalues are given in Table 3

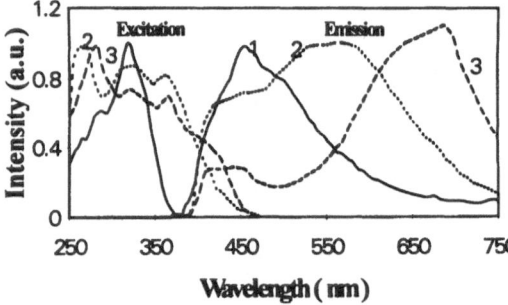

Fig. 3. Emission and excitation spectra of thin films containing CdS measured at 77 K. *Curve 1*: TEOS film containing initially cadmium nitrate 0.051 mol/mol TEOS. *Curve 2*: TEOS film containing initially cadmium nitrate 0.117 mol/mol TEOS. *Curve 3*: GLYMO film containing initially cadmium nitrate 0.127 mol/mol of GLYMO

tional in the square of the crystal size according to

$$a = a_0 + (\omega Im\,\chi^{(3)}/\varepsilon_0 n_0 c)E^2 \tag{14}$$

where n_0 and a_0 are the linear refractive index and absorption coefficient respectively, $Re\,\chi^{(3)}$ and $Im\,\chi^{(3)}$ are respectively the real and imaginary parts of $\chi^{(3)}$, ω is the frequency, and c is the velocity of light. $\chi^{(3)}$ was estimated to be 1.5×10^{-10} esu for CdS-doped glass and 1.1×10^{-8} esu for CuCl-doped glass, respectively. Large $\chi^{(3)}$ value gives rise to a large changes in n and a, which result in the development of nonlinear optical devices.

Improved semiconductor quantum dot materials by the sol-gel method are discussed by Chia-Yen Li et al. [51]. Pore-free sodium borosilicate (NBS) glasses and organically-modified silicates (ORMOSILs) were used as matrices for the CdS nanocrystallites. Results from both degenerate four-wave mixing and pump-probe techniques indicated large third-order nonlinear responses of the order of 10^{-6} esu from CdS-doped NBS glasses Table 4. By using potassium ion exchange, the first CdS-doped channel waveguides were fabricated in NBD glasses. Propagation of femtosecond laser pulses through the waveguide was investigated. A pulse breakup effect that may be the result of soliton formation was observed in these waveguides. Polydimethylsiloxane (PDMS) was used as the organic component to modify the silica gel. CdS-dope ORMOSILs derived from the PDMS-TEOS system exhibits improved mechanical properties and film-forming abilities compared to purely inorganic gels. Both CdS-doped PDMS-TEOS ORMOSILs and NBS glasses were much more resistant to photodarkening than the glasses made by melting. The use of the bifunctional ligand 3-aminopropyltriethoxysilane (APTES) significantly reduced the average particle size and narrowed the size distribution of CdS quantum in silica gels and densified NBS glasses.

Silica-titania optical waveguides doped with CdS and PbS prepared by the sol-gel method were also reported in Ref. 52. Thiourea was used as an internal source of sulfur. Cadmium and lead were introduced in the form of acetates. Using tetraethoxysilane (TEOS) and $Ti(OBu)_4$ as precursors for silica and titania, films with typical thicknesses of 0.15–0.2 µm were obtained. In order to obtain thicker layers TEOS was substituted with methyltritoxysilane (MTES).

These films were obtained by dipping, using two types of solutions where the silica precursor was TEOS and MTES respectively, Cadmium and lead sulfides were obtained by an "in-situ-reaction method" from the corresponding acetates and thiourea. X-ray diffraction performed both on powders and films demon-

Table 2. Nonlinear properties of glass [72]

Particle glass preparation method	$\chi^{(3)}$ esu	$\alpha\,(cm^{-1})$	$\tau\,(s)$
SiO_2	1×10^{-14}	10^{-5}	10^{-14}
PbO–SiO_2 (SF59)	7.5×10^{-14}	6×10^{-3}	
PbO–Bi_2O_3–GaO_3	4.2×10^{-13}	–	–
Chalcogenide (As_2S_3)	1.7×10^{-12}		–
$CdSSeK_2O$–ZnO–SiO_2 base filter Corning CS3-68	1.3×10^{-8}	~ 3	–
ibid Shott OG30	1×10^{-8}	–	50
CdSe Na_2O–ZnO–$B_2O_3SiO_2$ base glass melt-heat treatment	–	~ 50	0.3–3
CdS Silica Sol-gel	1.5×10^{-10}	–	–
CuCl Na_2O–B_2O_3–SiO_2 base glass melt-heat treatment	3×10^{-6}	~ 600	–
CdSe CdO–ZnO–P_2O_5 base glass melt-heat treatment	5×10^{-7}	~ 1200	–
CdTe silica Laser evaporation	6×10^{-7}	8000	–
CdS Na_2O–B_2O_3–SiO_2 base glass. Sol-gel	6.3×10^{-7}	–	–
Au-doped glass	5×10^{-11}	30	2

Table 3. Values of parameters for calculating eigenvalues

L	Configuration	$E^{(0)}$	$E^{(1)}$
0	1s1s	4.935	1.786
0	1s2s	7.776	1.780
0	2p2p	10.093	1.889
0	1s3s	12.511	1.777
0	2p3p	13.883	1.851
1	1s2p	5.926	1.620
1	2p1s	9.104	1.620
1	1s3p	9.715	1.724

strated the formation of the sulfides. CdS nano-particles were present both in powders and in films. PbS particles in the films are in the nano-size range, while in bulk they are almost 10 times bigger, suggesting that the gelation rate may have an important influence on crystal growth. Nonlinear characterization has been performed only on few CdS-doped samples heat-treated at 300 °C in nitrogen flux. The nonlinear refractive index n_2 was measured by the nonlinear m-line technique. High negative nonlinearities (n_2 in the order of 3×10^{-8} cm^2/kW were found between 530 and 560 nm. The nonlinear effect was found to remain almost unchanged after several measurements and after a period of one month.

3.2 Copper Chloride

Copper halides incorporated into films prepared by the sol-gel method have been recently prepared [53]. Copper chloride, a semiconducting material containing nanometer-sized particles, is a matter of increasing scientific interest. The reason is similar to that of the quantum dots of CdS, i.e. large optical nonlinearities. However, the small Bohr radius of these particles (about 6.8 Å as compared with the 19.7 Å for CdS) leads to different size ranges for the optical nonlinearities. As in the case of CdS, the most desirable form of nonlinear semiconductor materials would be thin films, which are easy for miniaturization and can serve also as an nonlinear waveguides. This poses a challenge for the chemist of sol-gel glassess, who wishes to prepare thin glass films incorporating quantum dots of CuCl that possess the desired optical properties.

Preparation of thin films containing mainly monovalent copper presents an experimental difficulty since copper chloride tends to be easily oxidized to its divalent state. The best samples that retain monovalent copper chloride in thin films are prepared by suspending Cu_2O in pure acetonitrile and adding the suspension to the standard sol-gel precursor solution, to which a stoichiometric amount of HCl has been added, to obtain a clear solution. The precursor solution of the glass film consists of TMOS, water and methanol in the molar ratios 1:9:5 acidified by 2.10^{-4} mol l^{-1} of HCl. The particle size is controlled

Table 4. Third-order susceptibility measured by DFWM for CdS quantum-dot borosilicate glasses [51]

CdS (wt.%)	$\lambda = 450$ nm		$\lambda = 460$ nm	
	Abs.Coeff. $\alpha(cm^{-1})$	$\chi^{(3)}$ (esu)	Abs.Coeff. $\alpha(cm^{-1})$	$\chi^{(3)}$ (esu)
8	1833	$\sim 4.5 \times 10^{-7}$	1715	$\sim 6.3 \times 10^{-7}$
3.8	915	$\sim 3.1 \times 10^{-7}$	849	$\sim 4.1 \times 10^{-7}$
1.4	493	$\sim 2.9 \times 10^{-7}$	417	$\sim 3.3 \times 10^{-7}$
0.5	217	$\sim 2.9 \times 10^{-7}$	184	$\sim 1.3 \times 10^{-7}$

by different concentrations of Cu_2O in the precursor solution. If kept in the dark in a vacuum, the films so prepared are stable over a period of six weeks.

It should be noted that films prepared from CuCl exhibit only cationic Cu^+ rather than monocrystallites of CuCl, as revealed from absorption spectrum when the starting material was Cu_2O as described above. The absorption due to Z_3 excitons allows the calculation of the size of the nanoparticles to be calculated, as was done for CuCl microcrystals by Masumoto et al. [54] and by Nogami et al. in silica glass [55]. When the crystal size is decreased from 77 to 25 Å, the energy of the Z_3 exciton is shifted from 3.234 eV to 3.275 eV, while the bulk value is 3.218 eV [56]. Fig. 4 presents the absorption spectra of CuCl microcrystallites prepared using the following initial concentrations of Cu_2O: 0.061, 0.0091 and 0.122 mole per mole TMOS. The increase of the crystallites with concentration is evident from the spectral shift to the longer wavelengths.

The absorption spectra of each glass has two bands which result from the excitons associated with two spin-orbit valence subbands. The double group symmetry types (formed in an excited configuration similar to $3d^9 4s$) are Γ_7 and

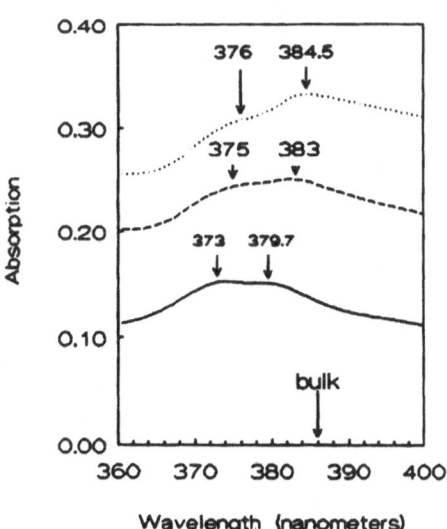

Fig. 4. Absorption spectra of CuCl particles in glass films measured at room temperature. *Solid line*: 0.061 mol. *Broken line*: 0.091 mol. *Dotted line*: 0.122 mol per mol of TMOS of Cu_2O in the starting mixture. Positions of the excitonic bands, $Z_{1,2}$ (short wave length) and Z_3 (long wavelength) are shown by arrows. Red spectral shift with increasing concentration is evident

Γ_8, respectively. The Z_3 exciton line originates from two-fold degenerate band Γ_7 and the $Z_{1,2}$ line originates from the four-fold Γ_8 degenerate valence band. The cubic symmetry of the CuCl quantum dots in thin glass films is obtained from X-ray diffraction [27].

From the shift of the position of the Z_3 exciton in microcrystallites of CuCl (Fig. 4) we can see that the particle size of CuCl in films can be determined by the starting concentration of Cu_2O in the precursor solution. Figure 5 presents the excitation and emission spectrum of the CuCl particles in a glass film at 77 K. The exciton spectrum consists of two peaks which are assigned to the excitonic levels $Z_{1,2}$ (at high energy) and Z_3 (at lower energy). Although the emission takes place from the Z_3 levels, it can be excited also via the $Z_{1,2}$ from which the energy relaxes with the assistance of lattice phonons to the luminiscent level Z_3.

The energies of the exciton Z_3 shift as a function of temperature and diameter of the microcrystallites of CuCl. The general trend reveals that an increase in temperature shifts the excitonic state to a higher energy (probably as a result of thermalization); a decrease in particle size raises the electronic states as a result of an increasing quantum effect. By comparing our results with those obtained by other groups in crystals and bulk glasses, we can see that the particle size of CuCl in films can be determined by the starting concentration of Cu_2O in the precursor solution.

3.3 Copper Bromide

CuBr doped films were obtained in the same way as films doped by CuCl. The best CuBr-doped film, in which copper retained valency I, were obtained from TMOS, water and methanol in the molar ratio $1:6:11$ and 2.10^{-4} mol l^{-1} HCl

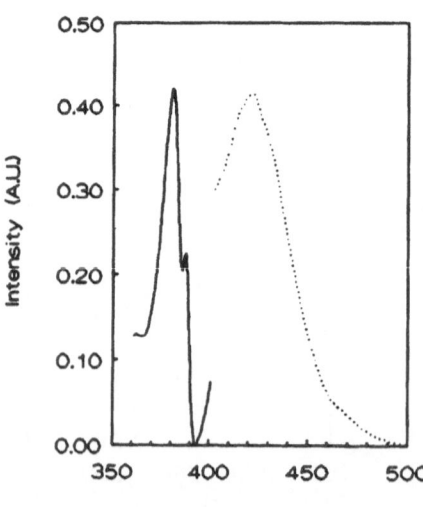

Fig. 5. Excitation and emission of CuCl in a thin glass film. The initial concentration of Cu_2O was 0.021 mol per mol TMOS at 77 K

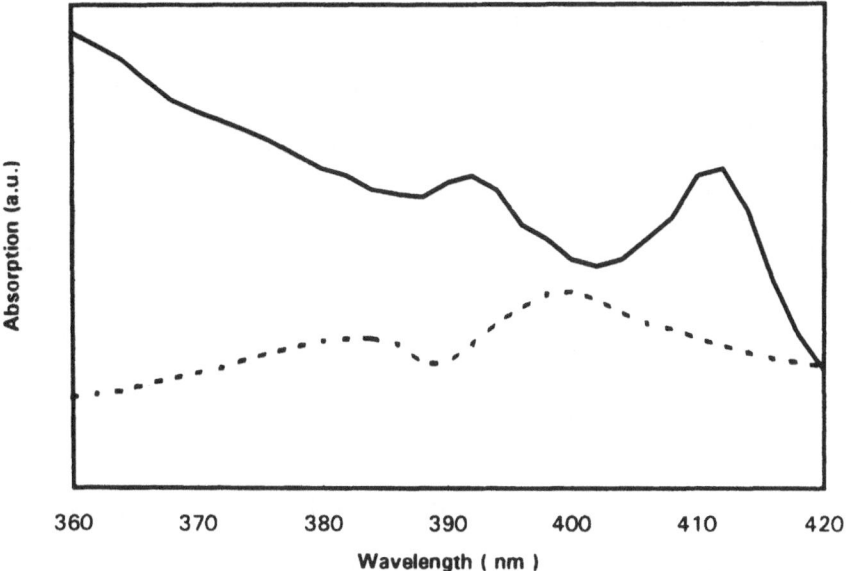

Fig. 6. CuBr films. Effect of Montmorillonite. Absorption spectrum of CuBr in thin films. The initial concentration of $Na_2SO_3/CuBr_2/HBr$ = 0.014/0.0046/ 0.0015 mol/mol TMOS (*full line*); the *broken line* is for the alternative with Montmorillonite

aqueous solution. Na_2SO_3, $CuBr_2$ and HBr were added after one hour of hydrolysis. Their corresponding ratios were 0.014/0.0046/0.0015 in mol/mol TMOS units.

As an alternative, we have added montmorillonite dispersed in TMOS to the standard solution of TMOS/water/methanol. The absorption spectra of the films are presented in Fig. 6. From X-ray diffraction, the average calculated diameter for the CuBr-doped sol-gel films with Montmorillonite is 239 Å, whereas in the CuBr-sol-gel films prepared under the same conditions, it is between 482 and 519 Å. A shift of the absorption bands in the films with Montmorillonite is observed at higher energies, indicating smaller microcrystallites.

The relative positions of the $Z_{1,2}$ and Z_3 excitons where found from the CuCl–CuBr solid solutions.

The electron and hole exchange interaction is found to play an important role in the relative intensity of $Z_{1,2}$ and Z_3 exciton bands in solid solutions of CuCl–CuBr. The relative positions of the $Z_{1,2}$ and Z_3 exciton bands in CuCl is opposite of that in CuBr.

We have not yet performed any measurements on nonlinear properties of copper halide quantum dots. CuCl was measured recently by other groups [55] who obtained the value of $\chi^{(3)}$ as 3×10^{-6} esu.

Further experiments on the introduction of copper halide nanocrystallites into sol-gel, including montmorillonite, and zeolite cages are in progress.

3.4 Lead and Bismuth Compounds

Semiconductor quantum dots of PbI_2 were recently reported by Lifshitz et al. [57, 58]. When embedded in sol-gel films, the lead iodide particle shows a blue shift in the optical spectra when the particle diameter size decreases from 100 to 25 Å.

Microcrystallites of Bi_2S_3 in sol-gel films also show quantum effects [59] and exhibit a shift to a higher energy of absorption maxima with a decrease of particle size.

4 Optical Nonlinearities of Small Metal Particles

Metal clusters forming quantum dots have been also prepared by chemical reduction of silver salts in silica films prepared by the sol gel method [25] or in ormosil nanocomposites [60].

The linear optical properties of small metallic particles are interesting and have attracted the attention of a large number of physicists. Since a small particle is necessarily surrounded by a dielectric, one unavoidably deals with a composite material made of small metal particles embedded in a dielectric that is usually water or glass. The interface between the metal and the dielectric leads, through the surface-plasma resonance, to the dielectric anomaly that is the cause of the colors in metal colloids [61, 62]. It is observed that the width of the absorption band of such a colloid, which is roughly proportional to the imaginary part of the susceptibility, increases when the size of the particles decreases [63]. This effect was accounted for classically in terms of a limited mean free path [63, 64] or quantum mechanically in terms of the quantum size effect [65, 66].

The optical properties of the noble metals are mostly due to the conduction electrons that behave almost as free electrons [67]. At frequencies slightly larger than the surface-plasma resonance, there is also a contribution from the d electrons. The nonlinearity is attributed to the conduction electrons. But, in a bulk metal, for a gas of free electrons, the electric dipole $\chi^{(3)}$ vanishes. On the other hand, when the electrons' motion is confined to a small particle, the situation is similar to that of a molecule [68], from which comes a nonzero dipolar $\chi^{(3)}$ that must be size-dependent.

4.1 The Optical Kerr Effect in Small Metal Particles and Metal Colloids

The linear optical properties of small metallic particles and metal colloids or metal-doped glasses have been extensively studied. Their beautiful colours have been known from decades to be due to the surface plasma resonance [1–3] occurring at the frequency ω, for which, approximately:

$$\varepsilon'_m(\varepsilon_s) + 2\varepsilon_d = 0$$

where $\varepsilon_m(\varepsilon) = \varepsilon'_m + i\varepsilon''_m$ (complex and frequency-dependent) is the dielectric constant of the metallic particle and ε_d (real) is that of the surrounding dielectric. The broadening of the absorption band corresponding to this resonance when the size of the small metal spheres is reduced was observed somewhat more recently [69] and assigned to the confinement of the electrons. It was accounted for classically in terms of a limited mean free path [64] or quantum mechanically in terms of the quantum size effect [65, 69]. In fact, these small metal spheres and their suspensions are attractive mainly because of these two phenomena: the surface plasma resonance and the quantum size effect.

Experimental observation of the nonlinear optical properties of silver and gold in metal colloids by optical phase conjugation showed the strongly resonantly-enhanced nonlinearity near the surface plasma resonance frequency [69]. Optical phase conjugation in the degenerate four-wave mixing configuration is a third-order process described by the Kerr susceptibility $\chi^{(3)}(\omega, -\omega, \omega)$. The dominant nonlinear response from the metal spheres is almost on the 28 ps time scale. The resonant enhancement is due to a local field effect which was very simply derived from an extension of the Maxwell-Garnett theory. The optical Kerr susceptibility of metals was found to be large, about 2 to 3 orders of magnitude larger than the third harmonic susceptibility of the same metals [70].

From theoretical considerations, it is deduced that two contributions play a major role: the hot-electron and the interband contributions. They both involve d-electrons, in the vicinity of the L point of the Brillouin zone in the first case and in the vicinity of the X point for the second one. They are both mainly imaginary but have different signs and different anisotropies.

Optical phase conjugation experiments, done by varying several parameters: size of the spheres, temperature, polarization of the incident beams as well as phase measurements and saturation measurements, allow us to understand the physical origin of the optical Kerr effect in small gold spheres. The dominant contribution is that of the hot electron but the interband contribution also plays an important role. It is actually the only one that contributes when the probe beam is cross-polarized.

In the case of silver, for which the experimental data are scarce, the hot-electron contribution is expected to be weaker since there is not the same resonant contribution from the L point of the Brillouin zone as there is for gold. This could explain why the measured value of $\chi^{(3)}_m$ was smaller for silver than for gold [71]. However, for Ag particles, the absorption peak frequency (surface plasma resonance) also coincides with direct interband transitions so that, even in this case, the interband contribution dominates the interband one.

4.2 Glass Doped with Small Metal Particles

Composite glass [72] doped with small metal particles (10 nm diameter) is also
a very attractive optical nonlinear material for optical switching and computing
because of its ultrafast nonlinear response [24]. Most of the available literature
on the silver and gold colloids report that the nonlinear response is shorter than
the laser pulse duration of several ps [24–27], which appears to be sufficient for
use as nonlinear materials [72a].

The large third-order susceptibility $\chi^{(3)}$ of these glasses is attributed to local
field enhancement near the surface plasmon resonance of the metal particles by
Hache et al. [69]. The local electric field E_i inside the particles, which usually
differs from the applied field E_0, is given by the well-known formula [28].

$$E_i = 3\varepsilon_d(\omega)E_0/\{\varepsilon_m(\omega) + 2\varepsilon_d(\omega)\} = f_i(\omega)E_0, \tag{15}$$

where $\varepsilon_m = \varepsilon'_m + i\varepsilon''_m$ is the dielectric constant of the metal particles, ε_d is the
dielectric constant of the matrix material and $f_i(\omega)$ is defined as the local field
factor. The nonlinear susceptibility $\chi^{(3)}$ of such a composite can be written as

$$\chi^{(3)} - pl\,f_i(\omega)l^2f_i^2(\omega)\chi^2, \tag{16}$$

where p is volume fraction of metal particles in the composite materials and
$\chi_m^{(3)}$ is the optical nonlinear susceptibility of a metal particle itself. Since the local
field factor $f_i(\omega)$ enters into the expression for $\chi^{(3)}$ to the fourth power, a large
enhancement is expected in the nonlinear susceptibility near the plasmon
resonance frequency ω_s for which

$$\varepsilon'_m(\omega_s) + 2\varepsilon_d = 0. \tag{17}$$

Another essential contribution to the nonlinear susceptibility is $\chi_m^{(3)}$. There
are three mechanisms that can contribute to the nonlinearity intraband
transition, interband transition and the hot electron effect. Based on the fact that
there is no substantial size dependence of $\chi^{(3)}$, the nonlinear response of metal
particles is thought to be due to nonequilibrium electron heating (hot electron)
[27]. The pump pulse energizes the conduction electrons, resulting in a high
electronic temperature while the lattice remains cool. This electron heating leads
to Fermi smearing, which affects the transition probability of the d-band
electrons to the conduction band energies near the Fermi level, leading to
changes in the reflectivity at the surface of the metal. The hot electron gas cools
to the metal lattice through electron-phonon scattering in $2 \sim 3$ ps. The value of
$\chi^{(3)}$ reported in Ref. 72 is $\sim 10^{-6}$ esu for gold particles and 2.4×10^{-9} esu for
silver particles. Despite these encouraging results, the value of $\chi^{(3)}$ for the
composite glasses are on the order of 10^{-12} esu, because of the very small
volume fraction p on the order of 10^{-6}.

Noble metal nanoparticle in SiO_2 and TiO_2 sol-gel films were prepared by
heating in air or in a reducing atmosphere [72a]. In the Au particle-dispersed
SiO_2, coating films obtained from solutions of $HAuCl_4 \cdot 4H_2O$ and $Si(OC_2H_5)_4$,

the size and shape of the Au particles were markedly affected by the sol ageing time, the amount of H_2O for $Si(OC_2H_5)_4$ hydrolysis and the exposure of the gel film to an amine vapor prior to heat-treatment. Dielectric matrices with high refractive indices such as TiO_2 were effective in shifting the surface plasma resonance peak of Au particles to longer wavelengths. Pt/TiO_2, Pd/TiO_2 and Ag/TiO_2 nanocomposite films have been prepared and shown to exhibit optical absorption in the visible range. This suggests a close relationship between the structure and nature of the gel matrix and the final microstructure of the composite films. In the dielectric matrix of TiO_2 with high refractive index, the surface plasma resonance peak of Au particles was at longer wavelengths than in the SiO_2 matrix, as predicted by Mie theory [61]. Pt/TiO_2, Pd/TiO_2 and Ag/TiO_2 nanocomposite films exhibited optical absorption in the visible range due to the high dielectric constant of the matrix.

An organic–inorganic synthesis route to Au-colloid containing, transparent SiO_2 coatings has been developed by Menning et al. [73] using four different types of functionalized silanes as stabilizing ligands for the Au. By varying the kind and the concentration of the stabilizing silane in the sol, the onset temperature for the colloid formation varies between 100 and 300 °C, and the final colloid radii can be controlled in a range between 3 and 30 nm after densification of the composite coatings on glass at 500 °C. The third order polarizability $\chi_m^{(3)}$ in the metal particles is one order of magnitude higher than in glass composites and exhibits a strong dependence on the ligand.

The experimental procedure for preparation of the gold particles is described in Ref. 73 as follows:

For the synthesis of gold-containing solutions, 200 mg H $[AuCl_4]$ $3H_2O$ was dissolved in 5 ml of ethanol and a functional silane of the group 3-amino propyl triethoxy silane (APTS), N-2-amino ethyl)-3-amino proply trimethoxy silane (DIAMO) < trimethoxy silyl propyl diethylene triamine (TRIAMO) or 3-mercapto propyl trimethoxy silane (THIO) was added in an equimolar ratio. The coating sols were synthesized from glycidoxy propyl triethoxy silane (GPTS) and tetra ethoxy orthoxy orthosilecate (TEOS) in a molar ratio of 80:20 by refluxing a mixture of 160 g GPTS, 40 g TEOS and 27 g 0.1 M HCl in 240 ml of ethanol for 24 h at 80 °C. After that, the gold-containing solution and surplus amounts of the appropriate functionalized silane were added and stirred for 30 min at room temperature. Microscopy slides were dip coated with this sol using withdrawal speeds between 3 and 5 mm/s. After drying for 15 min. at 80 °C the coatings were densified at temperatures between 100 °C and 500 °C for 1 h using a heating rate of 1 K/min. The formation of gold colloids was monitored by UV-VIS spectroscopy and transmission electron microscopy (TEM). The coating thickness and the refractive index were determined by spectral ellipsometry. The nonlinear optical experiments were performed by forward degenerate four wave mixing.

The $\chi^{(3)}$ values for the gold colloid containing coatings were calculated from the measured values of the diffraction efficiency η is the intensity relation of the transmitted first and zeroth order of the diffraction pattern, which appears by self-diffraction of a laser-induced grating.

$$\chi^{(3)} = \frac{8n^2c^2\varepsilon_0\alpha\sqrt{\eta}}{3\omega I_p(1-T)\sqrt{T}} \tag{18}$$

n = refractive index   c = velocity of light

α = linear absorption   ω = frequency of I_p

I_p = pump intensity   T = transmiccion of I_p

These values are valid for the composite. To evaluate the third order nonlinear susceptibility in the gold particles, $\chi^{(3)}$, Eqs. (19) and (20) are used:

$$\chi^{(3)} = pf_1^2|f_1|^2\chi_m^{(3)} \tag{19}$$

$$f_1 - f_1(\omega) = \frac{3\varepsilon_d}{\varepsilon_m(\omega) + 2\varepsilon_d} \tag{20}$$

where p is the concentration of the Au particles, f_1 is the local field factor, ε_m is the dielectric function of the Au, and ε_d is the dielectric constant of the matrix. All samples, except the one with thiosilane ligands, are measured at the maximum of the Au-colloid surface plasmon resonance. Because of wavelength restrictions due to the laser dye, the samples with thiosilane ligands are measured on the low energy side of the resonance, 2 nm away from the maximum, and it is assumed that this slight deviation has no significant influence on the NLO properties obtained. To get a complete characterization of the composite, all the material parameters needed in Eqs. (18)–(20) (namely particle concentration, the dielectric constants of the Au and the matrices at the excitation wavelength) were measured and used for the calculation.

$$\frac{w}{w_\omega} = \frac{\varepsilon''m}{\varepsilon''^\omega_m} \tag{21}$$

Table 5. Nonlinear characteristics of metal-doped glasses made by new fabrication process [72]

Metal	Process	Nonlinear coefficient $\chi^{(3)}$(esu)	Response time τ(s)
Au	[MH]	$\sim 10^{-2}$	28
Cu	[MH]	1.3×10^{-8}	5
Ag	[MH]	9×10^{-6}	
Au	[SPT]	1.3×10^{-6}	
Au	[SPT]	$\sim 10^{-7}$	
Au	[IMP]	1.2×10^{-7}	
Cu	[IMP]		5
Au	[Sg]	7.7×10^{-9}	
Au	[IMP]	1.7×10^{-10}	< 35
Cu	[IMP]	$\sim 10^{-8}$	

Notes: [MH] : melt-quench and heat treatment; [SPT]: sputtering; [IMP]: ion implantation; [Sg]: sol-gel process

where w = FWHM of the plasmon resonance with $w_\varpi \cong 60$ nm and $\varepsilon_m''^\varpi = 2.4$ for the values in the large particle limit. The volume fraction p of the Au colloid constant is extracted from the absorption spectra by Eqn. 22.

$$\alpha = p\frac{\omega}{n_d c}|f_1|^2 \varepsilon_m'', \quad \text{where } n_d = \text{refractive index of the matrix.} \qquad (22)$$

Recent investigations into a new fabrication technology offer high nonlinear susceptibilities of composite glasses doped with metal particles. Table 5 summarizes the nonlinear characteristics of various composite glasses made by these new glass fabrication processes.

4.3 Summary

Metal nanoparticles embedded in a dielectric medium exhibit optical properties different from metallic bulk materials. Gold and silver colloids have ruby and yellow colors respectively, due to surface plasma resonance. Recently nonlinear optical properties of gold and silver metal colloids were measured, and the potential applications for nonlinear optical devices were demonstrated.

A possible strategy to obtain a suitable nonlinear optical material might be to try to incorporate a larger volume fraction of metallic particles into the glass matrix. The conventional melting method, however, permits only a small fraction of metal colloids to be incorporated in the glass, for instance, about 10^{-4} for gold [75]. The sol-gel technique may be used alternatively as a simple process for preparing nanocomposite materials containing metallic particles. The low temperature approach permits us to incorporate a high volume fraction of metal particles, and to control the dimensions and shapes of the particles.

The sol-gel method was recently used to prepare coating films doped with gold, copper, platinum and palladium [81] metal nanoparticles. In particular, the sol-gel method was used with different approaches to obtain bulk and coating glassy materials containing silver nanoparticles. D. Brusilovsky et al. [76] obtained sol-gel thin films (0.2–0.4 µm) containing silver particles by chemical reduction. R. Reisfeld et al. [77, 78] prepared thin films containing silver particles, co-doped with organic dyes. M. Menning et al. [79] obtained silver colloids in a borosilicate system, by photoreducing silver ions, that were stabilized through complex formation with a bifunctional amine compounds. The volume fraction of silver particles was 10^{-5}. A Hinsch et al. [80] produced colloidal silver particles in thin SiO_2 sol-gel glass layers. Silver in the form of $AgNO_3$ or $AgBF_4$ salts was added and thermally or photochemically reduced. J.Y. Tseng et al. [60] doped ormocer in bulk and as coating films with Ag particles. A 1% volume fraction of silver particles (2 nm average size) was obtained in bulk material, heat-treated in argon.

Starting from a mixture of MTES and TEOS, it is possible to obtain uncracked and completely inorganic films 0.5–2 µm in thickness by a single dip-coating process. The content of MTES was also found to affect the critical thickness and porosity of coatings. This method was used by Kozuka et al. [109] to prepare gold-doped nanocomposite coatings, up to 0.7 μm thick. Different amounts of MTES changed the shape and size of the metal particles. The gold particles became more spherical and smaller when the content of MTES was increased. The volume fraction of gold particles was only limited by the solubility of the gold salt precursor in the solution and a volume fraction of several percent of the metal in the final material was reachable. A thicker coating obtainable with a single dip-coating process may be preferred for optical applications, for instance, to couple optical fibers with optoelectronic devices. To obtain a thick coating by the sol-gel method, a multideposition process is necessary, but during each dip-coating the surface may be damaged and the elimination of residual organic materials thus becomes more difficult.

Nonlinear optical properties of composite glass materials doped with nanometer-sized microcrystallites were reviewed. The nonlinearities of the glass doped with semiconductor microcrystallites mainly originate from quantum-confinement and band-filling effects and, as a result of the progress in new glass-fabrication technologies, are now on the order of 10^{-6} esu. The photo-annealing effects are now being intensely studied with the aim of modifying the response time. The nonlinear susceptibilities of the glasses doped with small metal particles originate from the local field effect and nonequilibrium electron heating effect, and are on the order of 10^{-7} esu. The nonlinear response is shorter than $2 \sim 3$ ps.

According to Flytzanis [110], the third-order susceptibility, $\chi^{(3)}$, of metallic nanoparticles consists of three contributions:

$$\chi^{(3)} = \chi^{(3)}_{intra} + \chi^{(3)}_{inter} + \chi^{(3)}_{he}$$

where $\chi^{(3)}_{intra}$ and $\chi^{(3)}_{inter}$ are respectively the contributions from intraband and interband resonant transitions and $\chi^{(3)}_{he}$ is the contribution from the electrons. It was shown 14 that the main contribution to $\chi^{(3)}$ is coming from the hot electron term, $\chi^{(3)}_{he}$. The only size-dependent contribution to $\chi^{(3)}$ is the intraband one but this is also the weakest; thus $\chi^{(3)}$ has effectively no size-dependence. Typical values of the third order susceptibility for gold and silver metal nanoparticles embedded in glass are 1.5×0^{-8} and 4×0^{-10} (esu/cm^3), respectively. The smaller measured value of $\chi^{(3)}$ for silver is probably due to the weakest contribution of the hot electrons.

A material with suitable third-order nonlinear properties may therefore be obtained by incorporating a large volume fraction of metal nanoparticles in the matrix.

The previous requirements imply that is necessary to have good control over the mechanisms of formation and growth of the metal particles.

Complying with these requirements, silica and silica-titania coating films doped with silver and gold nanoparticles were prepared by Innocenzi [81].

Acid-catalyzed solutions of methyltriethoxysilane (MTES) mixed with tetraethoxysilane (TEOS) or titanbutilate [titanium tetra(butyloxide)] were used to deposit silica and silica-titania coatings by dipping and spinning.

Inorganic coatings free of cracks were obtained with a thicknesses up to 1.5 μm, after densification at 500 °C. The MTES-derived coatings were used as matrices for metallic nanoparticles. Gold, silver and platinum nanoparticles were obtained in the coatings by thermal reduction. Gold particles were obtained at a temperature of 200 °C, while temperatures around 800 °C were necessary for silver and platinum. The effect of the composition of the matrix, the heat treatment temperature and the atmosphere on the formation of the particles was studied. The metallic particles were characterized by transmission electron microscopy, x-ray diffraction and UV-vis absorption spectroscopy. The content of MTES and the heat treatment temperature were found to affect the dimension and the shape of the metal particles. The distribution in particle size may be tailored by choosing the heat treatment temperature and the amount of MTES.

XRD spectra after thermal reduction of the films containing gold and silver particles revealed that the crystalline sizes, calculated from the half height widths of the diffraction peaks, were 17 nm for silver (800 °C) and 12 nm for gold (500 °C). The gold particles were formed from about 200 °C upwards, whereas the silver ions were reduced at temperatures of not less than about 800 °C.

The presence of MTES was found to affect the thickness of the coatings as well as the formation and growth of the metal particles. Homogeneous distributions of gold particles in the film were obtained using higher concentrations of MTES.

5 Nonlinear Properties of Organic Molecules in Glasses Prepared by the Sol-Gel Method

There is a great need for non-linear optical materials that can be used with low intensity light sources for applications such as phase-conjugation, image processing and optical switching. Organic materials in polymers for nonlinear optics prepared by the sol-gel method have been extensively studied in recent years by introducing these molecules into bulk glasses and into thin films [4, 25–27].

The origin of the large non-linear susceptibility of organic molecules, which have good singlet-triplet transfer at room temperature and are incorporated in a glass matrix, is as follows. Optical excitation from the ground state of the molecule S_0 to first excited state S_1 is followed by an intersystem crossing transfer to the lowest lying triplet state T_1. Provided that the energy difference between the first excited singlet S_1 and the triplet state T_1 is of appropriate value, delayed fluorescence from the excited single state occurs after a back intersystem crossing that returns the population to the excited singlet state occurs after a back intersystem crossing that returns the population to the excited singlet state from the vibrational excited states of the lowest-lying triplet manifold. Many of the relaxation mechanisms that can quench the triplet state do not exist when the dye is held rigidly in a solid matrix, and hence the lowest-lying triplet state has a very long lifetime, resulting in a significant population trapped in that state even by weak optical pumping. The saturation intensity I_s is therefore quite small, less than $100 \, \text{mW/cm}^2$, and since the non-linear susceptibility $\chi_{\text{eff}}^{(3)}$ varies inversely with I_s [82, 83], it becomes extremely large, about 0.1 esu, as compared with 10^{-12} for the liquid CS_2. Dynamic gratings formed by interaction of three beams in four-wave mixing experiments in such materials have long lifetimes.

The optical and nonlinear properties of dyes can be improved. When they are incorporated into inorganic glasses. These materials may be applicable for technological uses. The well-known technology of dyes in polymer matrices is not sufficient since both plastics and dyes undergo decomposition under prolonged irradiation and under harsh external conditions. On the other hand, one cannot use the current glass technology for incorporation of dyes because the dyes decompose at the high temperature needed for the process.

For this purpose, we incorporated organic dyes into glasses prepared by the sol-gel method. We studied fluorescein and its derivatives [84, 85] in sol-gel glasses, in composite glasses and in boric acid glasses, and determined triplet lifetimes and saturation intensities. Addition of a heavy ion, namely mercury in the form of mercury acetate, cause an increase in the $S_1 \rightarrow T_n$ transition rate followed by a decrease in saturation intensity I_s. Acridine-orange and acridine-yellow in composite glass and heavy glass have also been studied [87–89]. At low CW powers (typically up to $10 \, \text{W/cm}^2$) the materials behave as saturable absorbers with typical triplet lifetimes of milliseconds to seconds and saturation intensities of the order of milliwatt/cm^2. Methyl-orange belongs to the class of materials which possess nonlinear optical properties as a result of a change in conformation. The time scales and nonlinear optical properties of methyl-orange in the millisecond regime have been studied recently [88–91].

Pump-and-probe experiments performed on these materials have shown nonlinear properties also on a nanosecond scale. In these experiments, we irradiated the nonlinear materials by spectrally-narrow nanosecond laser pulses, while monitoring the changes of transmission of the samples using a much weaker spectrally broad probe beam. We found in these experiments that acridine-orange in composite glass shows induced transmission, i.e. bleaching at laser powers below $6.3 \, \text{MW/cm}^2$ and induced absorption at high powers [88]. Acridine-orange in heavy glass shows induced transmission at a laser power of $1.7 \, \text{MW/cm}^2$ and induced absorption at higher powers [89]. Both phenomena occur within the lifetime of the exciting pulse. Their recovery time (i.e. relaxation to the state before the excitation), however, is different: the induced absorption recovers on a time scale varying from minutes to hours, depending on the initial laser power, whereas the induced transmission conforms to the general scheme of saturable absorbers of this group of materials with recovery times of the order of tens of milliseconds.

In similar experiments done on methyl-orange, we observed induced absorption at both low and high excitation powers and a similar energy dependence of the induced absorption as in acridine [88]. However, in contrast to the acridine, the relaxation of the absorption after switch-off of the laser is completed within less than a minute. At a continuous-wave (CW), low power excitation, methyl-orange undergoes induced bleaching due to different absorption coefficients of its *cis* and *trans* forms [91, 92].

5.1 Nonlinear Behaviour of Acridine Orange on a Nanosecond Time Scale

The nonlinear behaviour of acridine orange on a nanosecond scale due to singlet transitions at high intensity MW excitation in the case of heavy glass, boric acid and in sol-gel glass was reported recently [90].

A typical procedure for boric acid and sol-gel glass is as follows: The acridine-doped heavy glass is prepared by the mixing and heating of 52.2 mol% of SnF_2, 10.5 mol% of SnO, 5.1 mol% of PbF_2 and 32.1 mol% of $[1/2P_2O_5]$. The viscous liquid is cooled to 275 °C and 1 mg of dye per each 10 grams of glass were added.

The glass was slowly cooled and then polished to 0.16 cm thickness. The final concentration of acridine orange in the sample is 3×10^{-3} M/L and the extinction coefficient is 6×10^4, which makes the number density of acridine molecules in this sample equal to about $1.8 \times 10^{18}/cm^3$.

The composite glass doped by acridine orange was prepared by impregnation of sol-gel porous glass by methylmethacrylate containing acridine. Concentration of the dye in the glass is also about 10^{-3} mol l^{-1}. Boric acid samples were prepared by admixing the dye into ortoboric acid, heating to 250 °C and sandwiching the viscous solution between two preheated microscopic slides. The concentration of the dye in such boric acid glass is about 10^{-3} mol l^{-1} (which is difficult to determine exactly, because boric acid sublimates in the course of melting).

The spectral data of these materials were compared with the data measured in alcoholic solution of acridine orange in a 0.1 cm glass cuvette having an optical density similar to that of the samples.

Power-dependent measurements of lifetimes were performed using a 10 ns. coumarin 485 nm dye laser pumped by an excimer laser in a single shot mode. The signals were captured by a streak camera and the luminescent signal was deconvolved from the shape of the laser pulse.

5.1.1 Power-Dependent Luminescence and Lifetimes

The lifetime of acridine orange in boric acid, which is power independent at pumping powers between 1.6 and 25 MW cm^{-2}, is 2.5 ns. This means that in that glass no nonlinear process takes place. For heavy glass, the lifetime changes from 2.1 ns at 1.6 MW cm^{-2} to less than 0.1 ms at 25 MW cm^{-2}.

The power dependence can be rationalized as a competition between a process that increases the population of excited monomers and a process that decreases this population. The most likely processes are: photodissociation of dimers which increases the population of monomers [94] and various processes of photoquenching, the most important of which seem to be excited state absorption or two-photon absorption. Apparently the photoquenching has a higher power threshold than the photodissociation of the dimers, which occurs between 1.6 and 6.3 MW cm^{-2} for heavy glass and is above 25 MW cm^{-2} in boric acid glass. It should be noted that in a pump-and-probe experiment done on the same sample in the power range of 1.5 MW cm^{-2} to 30 MW cm^{-2} we observed the onset of long-lived induced absorption at an energy level that was slightly higher than the lower limit [87–89]. This is due to the photodissociation of dimers to monomers, which, after switching-off of the laser radiation, recombine by diffusion back to dimers.

The behaviour of the dyes incorporated in sol-gel glasses on a pico- and femto-second scale requires further studies, which are now in progress.

5.1.2 Steady-State Measurements

The representative absorption and emission spectra of AO in a heavy glass matrix are presented in the upper half of Fig. 7. Compared to an alcohol solution of the same or slightly higher concentrations both spectra are shifted in the heavy glass matrix to longer wavelengths in accordance with previously reported results [93]. Furthermore, the absorption spectrum of AO in glass is broadened and was found to exhibit a substantial deformation when the sample is excited near 515 and 485 nm. Along with the main emission-band peaking at 560 nm, a pronounced band ranging from 600 to 700 nm is clearly observable. Sol-gel glasses containing AO exhibit the same peculiarities, but in these cases a short wavelength wing of the dye absorption strongly overlaps with the absorption edge of the matrix. For this reason, the samples of heavy glass containing AO will be analyzed more thoroughly below, whereas the data on AO in a sol-gel matrix will be used only to reveal some specific details connected with the influence of the matrix on the fluorescence kinetics.

The excitation spectra were found to be strongly dependent on emission wavelength (Fig. 8). The emission spectra recorded at excitation wavelengths (λ_{ex}) between 500 and 530 nm represent the extreme cases while the emission recorded for $\lambda_{ex} = 360$–530 nm has a spectral shape intermediate between these two spectra. The emission in the range $\lambda_{em} = 600$–650 nm was found to be most sensitive to the value of λ_{ex}. The excitation spectrum for $\lambda_{em} \geq 700$ nm reproduced the typical absorption spectrum for AO monomers, with a regular set of subbands due to transitions from the ground singlet state S_0 to the vibrational sublevels of the S_1-state except that they are shifted to longer wavelengths.

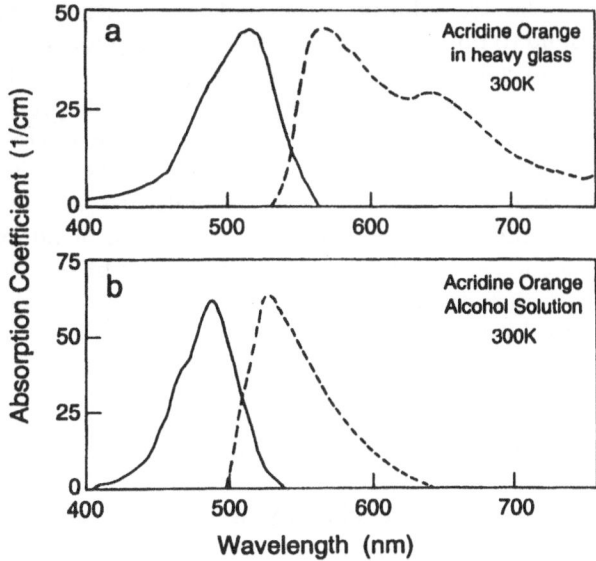

Fig. 7a,b. Absorption and emission spectra of AO in glass and in ethanol solution

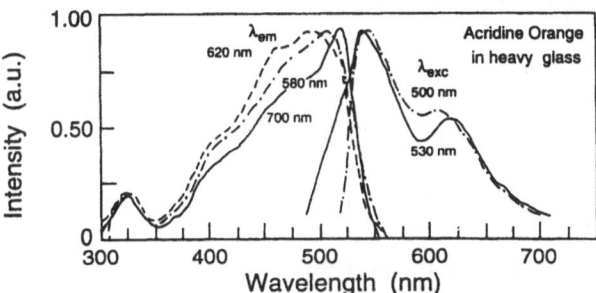

Fig. 8. Excitation spectra (*left*) and emission spectra (*right*) of acridine orange in a heavy glass matrix. Figures give the excitation wavelengths for the emission spectra and the registration wavelengths for excitation spectra

Table 6. Decay parameters of AO in glasses and in ethanol

Sample	Emission wavelength, (nm)	τ_1 (ns)	τ_2 (ns)	τ_3 (ns)
AO in heavy glass	560	1.79 ± 0.2	6.66 ± 0.6	24.6 ± 3
	630	2.03 ± 0.2	9.18 ± 1	20 ± 2
AO in sol-gel glass	560	2.36 ± 0.2	7.64 ± 0.7	22 ± 2
AO in ethanol 10^{-4} M/L	525	3.85 ± 0.3		

The dye solution in ethanol at a concentration of $C \leq 10^{-3}$ moll^{-1} was found to possess typical monomer absorption, excitation and emission spectra, i.e. no emission band in the range $\lambda_{em} \geq 620$ nm can be resolved, the absorption and emission spectra being of the same shape. This was also found for the excitation spectrum of AO in glass at $\lambda_{em} = 700$ (Fig. 2). For $C > 10^{-3}$ moll^{-1} the absorption spectra starts to change when C increases, indicating the presence of a small proportion of molecules in the form of dimers. However, a pronounced concentrational quenching of fluorescence does not allow the emission spectra for $C \geq 10^{-3}$ moll^{-1} to be recorded accurately. The intensity is in this case about two orders of the magnitude lower than that of glasses doped with AO.

5.1.3 Time-Resolved Measurements

A representative set of decay curves [93] is plotted in Fig. 9 where plots (a) and (b) provide a comparison of the decay curves of the main band that peaks at $\lambda = 560$ nm in sol-gel and heavy glass matrices, respectively, whereas plots (b) and (c) allow the kinetics of the two bands in the same, namely the heavy glass, matrix to be compared. Because the duration of the exciting pulse has a half width of about 3 ns close to the radiative lifetime of the excited singlet state of AO molecules, the registered decay signal I(t) must be considered as a convolution integral

$$I(t) = \int_0^t P(t')F(t-t')dt' \tag{23}$$

where P(t) is the pump pulse and F(t) the intrinsic fluorescence decay function. On the whole, the results of a nonlinear fitting procedure for the decay curves using the multiexponential expansion

$$F(t) = \sum_{i=1}^n A_i \exp(-t/\tau_i), \qquad i = 1,2,3 \tag{24}$$

lead to the following conclusions:

(i) no decay of AO in glass can be satisfactorily fitted by either a single or a double exponential function;

(ii) on average the decay in a sol-gel matrix is substantially slower than that in a heavy glass matrix;

(iii) the additional band at 630 nm possesses a slower decay rate than that of the main band at 560.

For the alcohol solution at low dye concentration (10^{-4} moll^{-1}) we obtained a monoexponential behaviour with $\tau = 3.85 \pm 0.05$ ns whereas a strong concentrational quenching in the solution at dye concentrations close to that in glasses does not allow the decay to be traced within more than one decade. The

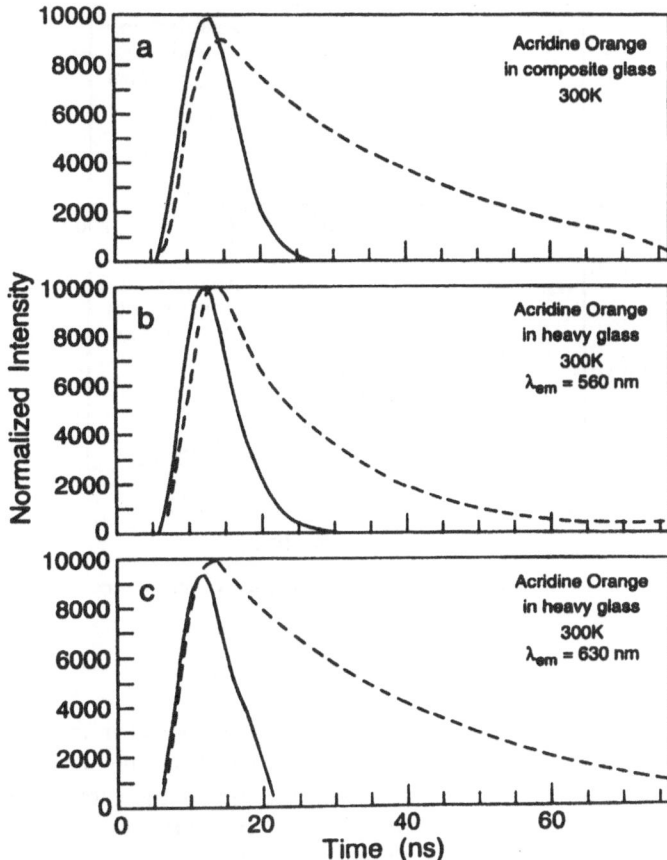

Fig. 9a–c. Fluorescence kinetics of acridine orange in composite (**a**) and heavy (**b, c**) glass matrices. Dots – emission signal, dashed line – excitation pulse, λ_{ex} = 337 nm, λ_{em} = 560 (**a, b**) and 630 (**c**) nm

single exponential-fit in this case gives $\tau = 1.6 \pm 0.1$ ns which should be considered only as a rough estimate because of the lack of experimental data.

The difference in the decay rates for different bands registered for AO in a glass matrix must lead to a spectral shift with time, as is confirmed in Fig. 10. At the initial stage, the emission spectrum looks very similar to the steady-state emission of the alcohol solution (compare the proper curves in Figs. 7–10), whereas the emission shifts for longer times to longer wavelengths. Additionally the spectrum becomes broader. The decrease in intensity did not allow us to follow the decay for a longer period.

5.1.4 Polarization-Resolved Measurements of Acridine Orange in Sol-Gel Glasses

To obtain additional information on the origin of the emission bands we used a polarization-resolved technique. The two emission bands were found to show

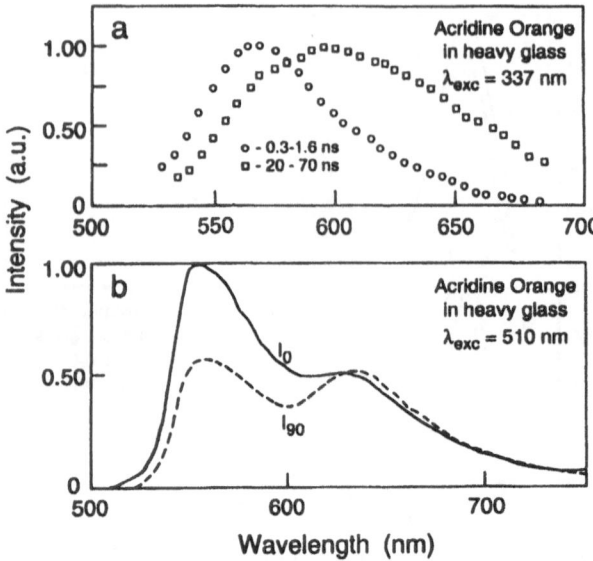

Fig. 10. Normalized time-resolved emission spectra of acridine orange in heavy glass. Registration interval 0.3–1.6 ns (*circles*), 20–70 ns (*squares*)

drastically different behaviors. When the detector analyzer is polarized perpendicularly to the excitation beam polarization, the main band (560 nm) considerably decreases in intensity, whereas under the same conditions the additional band (near 650 nm) shows either the same or even an enhanced intensity (Fig. 11). Accordingly, the polarization degree P defined as

$$P = \frac{I_\pi}{I_\pi +} - \frac{I_\perp}{I_\perp} \tag{25}$$

was found to be positive for the main band and close to zero or even negative for the additional band. These features were observed for all excitation wavelengths ranging from 350 to 530 nm. A negative degree of polarization is clearly observed for $\lambda_{ex} = 500...530$ nm. In Fig. 11 data for $\lambda_{ex} > 510$ nm are not presented, simply because the close position of excitation and emission wavelengths does not allow a reliable evaluation of P for the short-wave wing of the emission spectrum in this case.

To summarize the experimental date described in Sects. 3.1–3.3 we wish to outline the following peculiarities of fluorescence of AO-doped glasses:

(i) the steady-state emission spectrum contains an additional long-wavelength band;
(ii) the kinetics is not described by a single exponential;
(iii) the kinetics is different for different matrices and for different bands within the same matrix;

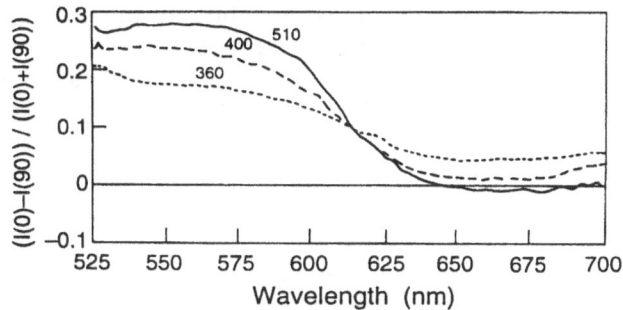

Fig. 11. *Upper panel*: emission spectra recorded in parallel and orthogonal analyzer geometry in relation to excitation beam polarization. *Lower panel*: polarization degree of fluorescence under excitation at different wavelengths

(iv) the main and the additional emission bands have drastically different degrees of polarization.

The origin of the additional long-wave emission band seem to be the most interesting point of the discussion. In what follows we shall show that all the data connected with this band can be consistently understood with the aid of the hypothesis of coexisting monomers and dimers in the glass matrix. The influence of aggregation on the absorption and steady-state emission spectra of AO has been extensively examined [93]. To avoid a collisional dissociation of the aggregates in liquid solution, the measurements were done at $-173\,°C$ using a water/ethanol mixture as solvent. The results of these studies can be summarized as follows; from them the characteristic peaks in absorption and emission spectra are plotted. At low concentration $(10^{-5}\,mol\,l^{-1})$ each spectrum consists of a set of vibronic sublevels which are not resolved in our case because of the high (room) temperature. The emission spectrum of AO in heavy glass, recorded during the first nanoseconds (Fig. 10) reproduces the typical shape of the monomer fluorescence. At a high concentration $(10^{-3}\,mol\,l^{-1})$, when most of the molecules are coupled in dimers, both emission and absorption spectra are changed [93].

The influence of the aggregation of AO molecules on fluorescence kinetics has been thoroughly analyzed [93]. The dimerization is accompanied by a long-time component in the decay curves which corresponds to $\tau = 14$ ns or 8 ns. The most important fact in the kinetic measurements seems to be that the additional long-wavelength band shows a substantially slower decay than the main band.

Finally, we include in this review the polarization-resolved measurements (Fig. 11). The main band shows a degree of polarization growing from 0.17 to 0.28 when the excitation wavelength increases from 360 to 530 nm. These data are in qualitative and quantitative agreement with the properties of AO monomers. A small degree of polarization for the additional band is consistent with the known tendency towards depolarization of fluorescence with aggregation. Moreover, a negative polarization of this band, pronounced for $\lambda_{ex} \geq 510$ nm, leads to the model of an AO dimer based on the simple exciton theory. This theory predicts a splitting of the S_1-state into two levels. A partly forbidden radiative transition from the lower excited dimer state has a lower probability, i.e. a longer decay time. In addition, its polarization is orthogonal to that of the transition to/from the highest excited state.

Thus all the peculiarities of the fluorescence revealed for AO in glass matrices can be understood if a part of molecules is assumed to be coupled in dimers. Our data on absorption saturation in the same structures [93] provide an estimate of the relative monomer/dimer concentrations (approximately 9:1).

Generally we may state that inorganic glasses heavily doped with organic dyes may be considered as exciting model objects that providing an opportunity to analyze the properties of aggregates under conditions in which other manifestations of concentrational effects such as absorption broadening and fluorescence quenching are substantially reduced due to the stable fixation of molecules in the solid matrix.

5.1.5 Fluorescence of Acridine Orange in Inorganic Glass Matrices

Inorganic glasses doped with organic dyes are considered as promising materials for quantum electronics, nonlinear optics and solar-energy devices [93]. Unlike liquid dye solutions, for practical applications it is extremely important that these structures can be impregnated by a dye to a very high concentration without a noticeable quenching of fluorescence. The concentration quenching in liquid solutions that originates from molecular collisions is strongly reduced when the molecules are imbedded in a rigid matrix. The fewer collisional effects make it possible to analyse other aspects of molecular interactions at high concentrations. By means of absorption spectroscopy Rodhamine 6G in porous silica gel was recently shown to exhibit a tendency to dimerization with growing concentration of the dye in the matrix. Similarly, the absorption spectrum of acridine orange (AO) in inorganic glasses at concentrations near 10^{-3} mol l^{-1} was found to be substantially broader than that of a liquid solution and to exhibit selective bleaching under excitation by tunable narrow-band laser radiation. Spectral peculiarities in absorption and bleaching were tentatively attributed to the presence of molecules in the form of dimers [93].

5.2 Cationic Fluorescein in Sol-Gel Glass

Fluorescein may exist as six different molecular structures, depending upon pH. In strongly acidic solutions the cationic form is predominant, absorbing at about 430 nm and emitting at about 460 nm. The neutral form occurs at about pH = 2, while mono- and di-anionic forms appear consecutively above pH = 2, with absorption maxima between 460 and 490 nm and emission maxima at about 510 nm.

We recently studied the nonlinear optical behavior of fluorescein and its derivative mercury-fluorescein in boric acid in heavy lead-tin-fluoride glass, in composite sol-gel glass and in pure sol-gel glass [84–86]. In these studies we found that the magnitude of the nonlinear optical effect of fluorescein depends strongly on the position of the absorption and emission peaks and on the lifetime of delayed luminescence. The strongest nonlinear effects were observed in boric acid, where the absorption peaks at about 430 nm and lifetime of delayed luminescence is about 3 s, and in heavy glass, where the absorption peaks at about 440 nm and lifetime of delayed luminescence is about 200 ms. In these media, fluorescein retains its cationic form. In a "normal" sol-gel glass (i.e. prepared by a weakly acidic catalysis), where the position of the absorption peak indicates a slightly basic environment the nonlinear effects are weak. It is therefore a matter of considerable importance to be able to manufacture acidic fluorescein in sol-gel glass in order to obtain high nonlinear effects in this material. We present below a method for preparing acidic fluorescein in glass and its spectroscopy.

The following materials were used for preparing the sol-gel films doped with cationic fluorescein: tetramethoxysilane (TEOS), Fluka, ethanol, hydrochloric acid, sulfuric acid, Frutarom, the surface active agent Triton X-100BDH and disodium fluorescein from Kodak.

For the cationic form, the starting solution contains TEOS, water and ethanol in the molar ration of 1:16:21 and variable amount sulfuric acid 0.06–0.4 mol l^{-1}. Five drops of Triton X-100 were added to 25 ml of this solution. Fluorescein was dissolved in ethanol and added to the precursor solution to yield a concentration of about 2.9×10^{-2} mol l^{-1}.

For the anionic form, the starting solution usually contains hydrochloric acid in molar ratio to TEOS of 1:0.05.

The films were prepared by the dip-coating technique. After the dipping the samples were immediately placed on hot plate at 100–120 °C.

The thickness of the films on the glass substrate varies between 0.1 and 0.3 mm, as determined from interference patterns [95]. The absorbance of the films was measured using a double-beam Perkin-Elmer Lambda 5 spectrophotometer and the fluorescence was measured using a Jasco FP-770 spectrofluorimeter.

5.2.1 Spectroscopy of Cationic Fluorescein in Glass Films

Fig. 12 presents absorption spectra of thin glass films doped with fluorescein prepared in the presence of sulfuric acid (dotted line) and prepared by catalysis under weakly acidic conditions (full line). The peak at 438 nm corresponds to the cationic form of fluorescein. This absorption peak also observed in the case of

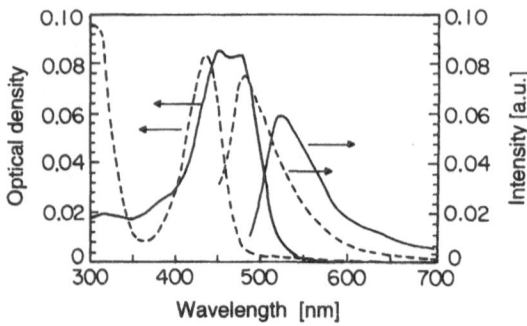

Fig. 12. Absorption and emission spectra of cationic (*broken curve*) and monoanionic (*full curve*) of fluorescein in thin glass films

cationic fluorescein in an alcoholic solution of fluorescein at pH = 2, which emits at about 480 nm. The absorption spectrum of the fluorescein under weakly acidic conditions is red-shifted in respect to this peak (solid line), and its emission centers at about 520 nm. As mentioned above, the nonlinear properties of the anionic form are much weaker than those of the cationic form.

It should be noted that fluorescein exists in stable cationic and mono-anionic forms at room temperature and that small changes of conditions can significantly shift the equilibrium, which is accompanied by a change in color of the film.

The following experiment demonstrates the changes due to ammonia treatment on a film containing the cationic form. The sample was placed at a distance of 10 cm from a drop of 25% ammonia in water in a closed container and its absorption and emission spectra were recorded as a function of time at one-minute intervals. Figure 13 demonstrates the decrease in intensity of the emission peak at 489 nm due to cationic form and the increase in intensity of the peak at 580 nm due the anionic form. Both emissions were excited at 452 nm. The situation is reversible when the ammonia treated sample is exposed to HCl vapor whereby the cationic form is changed to the anionic-form.

A similar situation is detected in the absorption spectrum presented in Fig. 4 for a bulk glass, where the cationic form changes to the anionic under ammonia treatment. Here we see a decrease of the absorption at 430 nm and an increase of the absorption at 480 nm.

5.3 Malachite Green in Thin Glass Films

The triphenylmethane colorant Malachite Green (MG) incorporated in sol-gel glasses or other polymers recently attracted attention because of its unusual optical properties, which are also manifested on very short time scale. Recently [96] a gel glass composed of SiO_2 and ZrO_2 containing MG^+ was shown to become reversibly transparent at incident energies in the order of microjoules and with a recovery time in the order of picoseconds. Therefore, incorporation

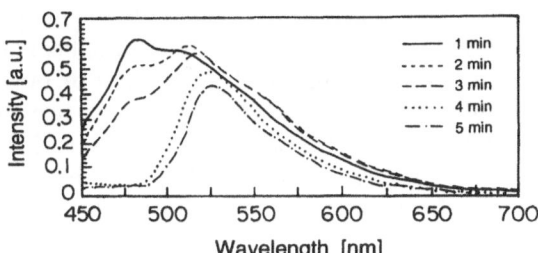

Fig. 13. The change of emission of cationic fluorescein in a glass film under the influence of ammonia vapour

of MG into glass allows new materials to be prepared for stable saturable absorbers, used for producing femtosecond pulses by Q-switching.

A systematic study of absorption and in particular the novel emission spectra showing high luminescence yields as a function of the preparative procedure [97, 98] have encouraged attempts to rationalize the unusual spectroscopy based on a general distinction between the coordination numbers $N = 3$ and $N = 4$ of carbon in aromatics. Of particular interest is the monocationic form of Malachite Green MG^+ and its protonated form in gel glasses [98], in which the luminescent ions are situated in a small isolated cavity. In Malachite Green the colourless form is attributed to the four-coordinated carbon atom in the carbinol and the coloured forms of the monocation and the dication to three-coordinated carbon atom.

The strongly coloured cations ($N = 3$) all have a central carbon coordinated to the three C_6 hexagons. However, at a given pH, slow equilibria have been established with almost all colourless carbinols (HO)MG and protonated $(H_2O)MG^+$ has been extensively studied [111–114], also for different forms of MG that are substituted at several positions on the remaining phenyl ring. A way to explain the above behaviour may involve Lewis acidity [99] where a definite (possibly proton-free) Lewis acid forms a strong adduct with a base (with a reactive pair of electrons in the Lewis paradigm [100]). Contrary to the proton affinity of a base, which can be described according to Brönsted by one pK-value in a given solvent (able itself to be protonated), the affinity between a Lewis acid and a given base, cannot generally be described by the sum of the logarithms of the acid and base strengths. A striking case is $B(OH)_3$, which shows strong affinity to some bases, allowing the coordination number to increase from 3 to 4. In water, $B(OH)_4^-$ is formed [101] with an effective pK of 9 for $H_2O + B(OH)_3$. The coordination to boron has been shown in fluorescein (here flu^0, existing as two tautomers in different solvents) which can be protonated to $H(flu)^+$ at low pH (pK is 2.2) having a quite characteristic absorption spectrum. Molten boric acid (dehydrated to approximately $B_{11}O_{16}OH$) dissolves flu^0, as a similar species, showing a highly unusual luminescence. At liquid air temperature, the first triplet state emits in the orange with a long lifetime. Warming up to room temperature, this emission is replaced by a greenish blue band, still with the observed lifetime of several (up to 20) seconds, at the same energy as the exceedingly rapid (10^{-8} s) emission. The phenomenon is called "delayed fluorescence", the triplet state being (very slowly) thermally excited to the first excited singlet of flu^0 bound to boric acid glass. In other situations, including sol-gel glass hosts flu^0 loses two protons (one derived from a carboxylic substituent) with pK = 4.4 and 6.7 in aqueous solution, providing broad-band green emission (between the first excited singlet and the ground state) with a quantum yield above 0.9.

5.3.1 Preparation of Samples Containing Malachite Green

The sol-gel glasses with Malachite Green were prepared from tetramethoxysilane (TMOS), $Si(OCH_3)_4$ containing varying concentrations of water and methanol, and using as catalysts either methacrylic, acetic, hydrochloric, phosphoric, hydrochloric, nitric or sulfuric acid, or dilute sodium hydroxide, and in one experiment without any catalyst. The detailed ratio between components and the sequence of preparation may be found in Ref. 97. The drying temperature varied between 50 and 100 °C. Glasses prepared using either methacrylic, acetic or hydrofluoric acids, or sodium hydroxyde, as catalysts showed the blue-green colour of the monocation MG$^+$ (predominant at pH 4 to 6 in aqueous solutions), with absorption peaks close to 615, 420 and 312 nm and emission typically at 650 nm (Fig. 15). It turns out that these transitions vary only 620.7 ± 4.4, 425.6 ± 2.2 and 315 ± 1.5 in ten different solvents and polymers.

When the sol-gel glass is prepared under very strongly acidic conditions, or is impregnated with a strong acid, only the yellow H(MG)$^{2+}$ is perceived. This has an absorption maximum at between 420 and 450 nm and it emits at about 570 nm. A striking effect of ammonia on MG$^+$ is also presented in Fig. 14 where it is shown that if a glass that has been prepared with a molar ratio of TMOS/water/methacrylic acid of 1/16/0.0001 and containing 10.6 mol l^{-1} of

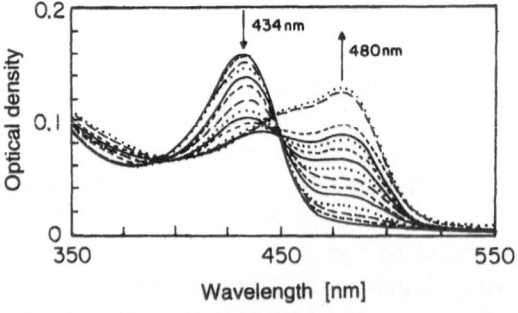

Fig. 14. The change of absorption of cationic fluorescein in fluorescein-doped sol-gel bulk as a function of exposure to ammonia vapour at 1-minute intervals. The 434-nm peak decreases, the 480-nm peak increases

MG^+ oxalate, is immersed in 0.5 M ammonia solution at pH 11.5, its MG^+ absorption decreases by a third, whereas the quantum efficiency of fluorescence at 670 nm increases from 0.003 to 0.2. A possible explanation for this unexpected behaviour may be that the ammonia in solution washes out those MG^+ species that were dissolved in the water within the gel cavities. The remaining MG^+ molecules are chemically bound to the silicate groups, forming a complex of the type $R_3C + OSi(O...)_3^-$, which possess low nonradiative relaxation rates and a high quantum yield of fluorescence.

When the sol-gel glass is impregnated at sufficiently low pH, only the yellow $H(MG)^{2+}$ is perceived, having a band close to 430 nm (an extreme case being 450 nm) and emitting around 570 nm.

5.3.2 Excited State Relaxation and Photochemical Bleaching

The fact that aqueous green MG^+ solution slowly fades at pH > 6 (as mentioned above) may be simply due to a reaction with water that produces HO(MG). Experimental results on the excited-state sol gel glasses prepared from $Si(OCH_3)_4$ and containing blue green MG^+ show three absorption bands, typically between 605 and 616 nm due to the first excited singlet S_1, a weaker maximum between 411 and 421 nm due to a second singlet S_2 (in aqueous solutions at 618 nm and 424 nm respectively) and a third band at 312 nm ascribed to S_3. Not only does the MG^+ cation emits a broad band at 650 nm (representing a Stokes shift of S_1) but light absorption in S_2 provides an additional emission band at 470 nm, and in S_3 even a third band at 360 nm. Hence, a cascade transition allows emission of Stokes-shifted S_2 and S_1 from S_2 excitation, and from all three S_n when illuminated at 300 to 320 nm, with the highest energy centered around 380 nm.

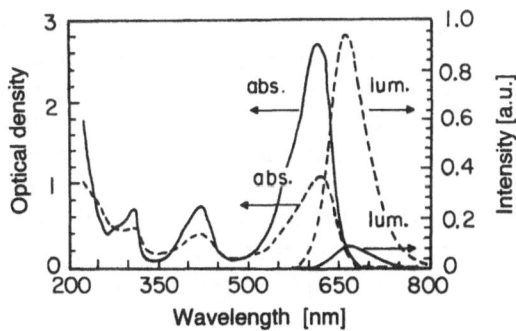

Fig. 15. Absorption and luminescence spectra of MG^+ in sol-gel bulk before (*full curve*) and after (*dashed curves*) immersion in 0.5 molar aqueous ammonia for 24 hours. The glass was prepared using a solution of $Si(OCH_3)_4$, water and methacrylic acid (ratio 1:16:0.001), and originally containing 7.4 micromolar MG^+ oxalate. Both optical intensity and luminescence intensity are plotted vertically in arbitrary units

Malachite Green exists in glasses, as well as in other viscous materials, as a strongly coloured monocation or as a weaker coloured dication; both cases the central carbon atom has three carbon neighbours. Upon illumination or when a base is added, the cations are changed reversibly to the colourless carbinol form in which the central atom acquires fourfold coordination. The dark green monocation in glasses exhibits fluorescence from S_3, S_2 and S_1 excited states. The quantum efficiency of the S_1 luminescence is about 20% at room temperature, indicating relatively low nonradiative losses in this medium.

Under spectral experimental conditions, mainly using strong acids as catalysts for the formation of gel glasses, a purple form (possibly a complexed monocation) is formed, which may also produce a shoulder close to 550 nm in the absorption spectrum of the green form (unless this is a vibronic excitation of S_1, or an excited singlet with a molar extinction coefficient, ε, of about 10^4, situated between S_1 and S_2). In the ethanolic solution it may be bound to oxalate anions [101].

6 Second-Order Nonlinear Materials

Most of this section deals with the sol-gel optics research devoted to NLO materials [102] was first related to third order processes which are compatible with the isotropy of amorphous sol-gel matrices. Semiconductor, metal particles and organic molecules inside amorphous sol-gel matrices are in general randomly oriented, thus ruling out the emission of second harmonic. The second order nonlinearities can be only achieved in a noncentrosymmetric environment' due to symmetry consideration, and organic chromophores can be oriented in hybrid sol-gel metrices by using electrical field induced second harmonic (EFISH) or corona electrical field poling techniques [102]. In the work performed by the groups of Livage [102], Sanchez and Zyss [103–105] organic molecules were chemically bonded to the oxide backbone of gels. The chemical bonding of the dye to the sol-gel matrix allowed to increase the dye concentration without any crystallization effects. SHG responses of 1.6 p,/V were obtained. Consequently a land of opportunity for the synthesis of optical sol-gel devices with efficient second harmonic properties was opened. Then, over the past two years, there has been increasing interest in second order NLO materials synthesized via sol-gel chemistry. Among the strategies used to, improve the NLO response of the hybrid coatings the control of the chromophore relaxation by increasing the matrix rigidity is without no doubts the most important point to be solved in order to be able to make efficient NLO devices.

6.1 Example of the Synthesis of Hybrid NLO Materials

TSDP (N- [3-Triethoxysilyl-propyl]-2,4 Dinitrophenyl amine) was used [102] as the siloxane network precursor carrying the NLO chromophore while tetramethoxylane (TMOS) was used as a crosslinking reagent to increase the network rigidity. Sol-gel coatings with different TSDP:TMOS molar ratios (1, 0.8, 0.6, 0.4, 0.2) were prepared as follows:

The precursors, TSDP and TMOS, were mixed in THF and co-hydrolyzed with acidic water (HCl; pH = 1). The H_2O:Si molar ratio was 2:1. The solution was then stirred for 30 minutes and the resulting sols were aged for several hours. From these sols, hydrophobic transparent films several micrometers thick were produced without crack and failure on ordinary soda-lime glass-sheets that had been previously cleaned and dried. The resulting hybrid xerogels were then characterized. In this paper, the samples will be labelled Tx/Qy, where T stands for TSDP, Q for the silicon added from TMSO, and x and y are the molar percentages (x + y = 100) of the two precursors. Details on the recording of the NLO and spectroscopic data are reported in Ref. 102.

Poled hybrid siloxane-oxide coatings synthesized through hydrolysis and condensation of $Si(OMe)_4$ and TDSP precursors leads to SHG values of 10 pm/V. The glass transition temperatures of the hybrid coatings increase with the TMOS content and the thermal curing time. These modifications are related to an increase in the degree of condensation and crosslinking of siloxane and silica species; they have a strong influence on the relaxation behavior of the NLO chromophores grated into these hybrid networks.

The use of organic and dopant molecules for crosslinking in the sol-gel process has also been described by Kador et al. [115].

7 Conclusions

7.1 Quantum Confinement

Nanocluster materials have already had a significant impact in the area of photonic-electronic devices. The miniaturization of electronic and optical devices has revolutionized response times, energy loss, and transport efficiency. In addition, the presence or absence of a few atoms and the geometrical disposition of each atom can significantly modify electronic and photonic properties as one approaches the nanosize regime. This control can be further supplemented by "packaging" assemblies of atoms or molecules into thin-film or nanocomposite bulk material so as to define surface states, cluster environment and geometry, intercluster interactions, and consequently a wide tunable range of optical and charge-carrier responses.

One specific example attracting commercial interest is optical data storage. In writing an optical disk, it is desirable to obtain the highest possible resolution and optical density. Laser beam collimation is one technique used to optimize the optical density. Nanosized semiconductors and metals display a peculiar nonlinear optical property: they will absorb light if it is not too intense, but will transmit light if the number of photons surpasses a certain threshold value. By using a thin-film mask containing the semiconductor cluster during writing, only the center portion of the beam, which has a photon flux above a certain intensity cutoff, is transmitted, resulting in a dramatic ($\sim 50\%$) improvement in both optical density and recording quality. The use of shorter wavelengths could further improve the resolution. Fortunately, the optical absorption and emission bands of a semiconductor are shifted to shorter wavelengths if the semiconductor atoms are in appropriately small clusters. These effects are consequences of the fact that the electrons in the semiconductor are quantum-confined. Obviously, the desired size of the cluster is the first property that needs to be considered when creating a nanocomposite material containing the semiconductor nanoclusters described above. In principal this can be arbitrarily adjusted.

7.2 Off-Resonance Nonlinear Optics

Nonlinear optics is concerned with the way in which the electromagnetic field of a light wave interacts with the electromagnetic field of matter and with other light waves. The interaction of light with a nonlinear material will cause the properties of the material to change, and the next photon that arrives will "see" a different material. As light travels through a material, its electromagnetic field interacts with other electromagnetic fields within the material. These internal fields are a function of the time-dependent electron density distribution in the material and the electric fields of the other light waves, if, for example, two or more light sources are used. In a nonlinear optical (NLO) material, strong interactions can exist among the various fields. These interactions may change the frequency, the phase, the polarization, or the path of the incident light. The chemist's aim is to develop material that will control this mediation so that photons can be modulated or combined (wave-mixing). In addition, it is necessary to fine-tune both the magnitude and the response time of the optical processes.

Porous glasses, organic media, and layered material play an important role in nanocomposite inclusion chemistry and some comments concerning their use are briefly presented.

Potential and feasible applications of quantum-confined materials have been demonstrated in the areas of photocatalysis [106], phase-conjugate optical systems, optical switching for parallel data processing, resonant tunneling and field effect transistors, low-gain lasers and frequency mixing [107, 108]. The primary limitation is the ability of the synthetic-materials chemist, to create and package nanophases in such a way as to obtain uniformly precise sizes, intercluster separations, optical densities, and suitable process abilities. Thus the sol-gel processing of nonlinear materials may play a key role in achieving the proper materials.

7.3 Glasses and Disordered Media

Porous glasses prepared by the sol-gel method are promising in that they offer the advantages of providing a large range of pore sizes, ease of optical characterization, and the potential for use as thin films and monoliths in optical devices. Photo- and thermal stability are crucial if the nanocomposites are to be used in laser device applications. In optical computer or optical switching applications, for example, the nanocomposite should be able to perform trillions of switching operations per second for years at a time. As inorganic hosts, porous glasses have been shown to greatly increase the lifetime of organic dye lasers which have been built into their pores [19]. This is an important property of the inorganic host–organic guest combination that will undoubtedly be pursued further.

Acknowledgements. The author is grateful to C.K. Jørgensen and C. Klingshirn for most helpful discussions and to M. Eyal, D. Brusilovsky, H. Minti and D. Shamrakov for providing many of the experimental results. The grant from the Swiss National Science Foundation was instrumental for the efficient collaboration between the groups in Geneva and in Jerusalem.

8 References

1. Shen YR (1984) The Principles of Nonlinear Optics. Wiley, New York
2. Prasad PN, Williams DJ (1984) Introduction to Nonlinear Optical Effects in Molecules and Polymers. Wiley, New York
3. Chemla DS, Zyss HJ (1987) Nonlinear optical properties of organic molecules and crystals. Academic Press, New York
4. Reisfeld R, Nanoparticles in Amorphous Solids and their Nonlinear Properties, Presented at the Eleventh Course of Nonlinear Spectroscopy of Solids; Advances and Applications (1994) Di Bartolo B, (Ed.), Plenum, NATO ASI Series B Physics, 399: 491
5. Klein L (1993) Sol Gel Optics:"Processing and Applications" (Kluwer Academic Publisher, Boston)
6. Reisfeld R (1995) "Nanoparticles in Amorphous Solids", in Advanced Materials in Optics, Electro-Optics and Communication Technologies, Eds. Vincenzini P, Righini RG. Proceedings of Topical Symposium VII on Advanced Materials in Optics, Electro-Optics and Communication Technologies, Techna Publisher, Faenza, pp 3–13
7. Reisfeld R, Jørgensen CK (1992) Optical properties of colorants or luminescent species in sol-gel glasses, Structure and Bonding, Springer-Verlag (1991), Reisfeld R, Jørgensen CK eds., 77: 207–256
8. Avnir D, Levy D, Reisfeld R (1984) J Phys Chem, 88: 5956
9. Avnir D, Kaufman V, Reisfeld R (1985) J Noncrystalline Solids, 74: 395
10. Schmittrink S, Miller DAB, Chemla DC (1987) "Theory of the linear and nonlinear optical-properties of semiconductor microcrystallites", Phys Rev B 35: 8113–8125; Schmittrink S, Chemla DS, Miller DAB (1989) "Linear and nonlinear optical-properties of semiconductor quantum wells". Review, Adv Phys 38: 89–188
11. Wang Y, Herron N, Mahler W (1989) "Linear-optical and nonlinear-optical properties of semiconducter clusters", Suna A, J Opt Soc Am B 6: 808–813
12. Stucky Galen D, Mac Dougall James E (1990) "Quantum confinement and host gest chemistry-probing and new dimension", Science, 247: 669–678
13. Anderson DZ, Bull MRS (1988) 13, 30, Glass AM, Ed., "Photonic Materials", ibid., no. 8

14. Brennan JG, Siguist T, Canoll PJ, Stuczyns SM, Brus LE, Steigerwald ML (1989) "The preparation of large semiconductor cluster via the pyrolysis of a molecular precursor. Note", J Am Chem Soc, 111: 4141–4143

15. Ekimov AI, Efros AL, Onuschchenko AA (1985) "Quantum size detection in semiconductor microcrystals", Solid State Commun, 56: 921–924

16. Brus L (1991) "Quantum Crystallites and nonlinear optics", Appl Phys A 53: 465–474

17. Mohan V, Anderson JB (1989) "Effect of crystallite shape of exciton energy. Quantum Monte-Carlo calculation", Chem Phys Lett 156: 520–524

18. Efros AL (1982) "Interband absorption of light in semiconductor sphere", Sov Phys Semicond 16: 772–775

19. Brus LE (1984) "Electron-electron and electron-hole ineractions in small semiconductors crystallites. The size dependence of the lowest excited electronic state", J Chem Phys 80: 4403–4409

20. Fojtik A, Weller H, Koch U, Henglein A (1984) "Photo-chemistry of colloidal metal sulfides. 8. Photo-physics of extremely small CdS and magic agglomeration numbers", Ber. Bunsenges, 88: 969–977

21. Steigerwald ML, Brus LE (1989) "Synthesis, stabilizations, and electronic- structure of quantum semiconducture nanoclusters. Review", Ann R Mater, 19: 471–495

22. Brus L (1987) "Size dependent development of band-structure in semiconductor crystallites", New J Chem (France), 11: 123–127

23. Reisfeld R (1992) SPIE 1758 Sol-Gel Optics II, p 546

24. Minti H, Eyal M, Reisfeld R, Berkovic G (1991) "Quantum dots of cadmium sulfide thin glass films prepared by sol-gel technique", Chem Phys Lett, 183: 277–282

25. Reisfeld R, Minti H, Eyal M (1991) "Active glasses prepared by the sol-gel method including islands of CdS or silver", SPIE Proc 1513: 360–67

26. Reisfeld R (1991) "Semiconductor quantum dots in amorphous materials", in "Optical Properties of Excited States in Solids". Di Bartolo B (Ed.), Plenum NATO, ASI Series B, 301: 601–621

27. Reisfeld R, Minti H, Eyal M, Chernyak V (1993) Nonlinear properties of semiconductor quantum dots and organic molecules in glasses, prepared by the sol gel method. J Nonlinear Optics, 5: 339–360

28. Mackenzie JD, Ulrich DR (eds) (1988) Proceedings of the Third International Conference on Ultrastructure Processing. Wiley, New York

29. Brinker CJ, Clark DE, Ulrich DR (eds) (1986) Better ceramics through chemistry II [MRS Symposium Vol 73] and (1988) Better ceramics through chemistry III [MRS Symposium Vol 121] Materials Research Soc, Pittsburg

30. Dislich H (1988) Thin films from the sol-gel process, in: Klein LC (ed), Sol-gel technology for thin films, fibers, preforms, electronics and speciality shapes, Noyes Publishers, Park Ridge NJ, 50–79

31. Sanchez C, Livage J (1990) "Sol-gel chemistry from metal alkoxide precursor", New J Chem 14: 513–21

32. Sol-Gel Science and Technology, Edited by Aegerter MA, Iafelicci M, Souza DF, Zonotto ED (1989) World Scientific, Singapore

33. Reisfeld R (1989) in: Sol-Gel Science and Technology, Edited by Aegerter MA, Jafelicci M, Souza DF, Zanotto ED, World Scientific, Singapore, p 323

34. Reisfeld R (1990) J Non-Cryst Solids, 121: 254

35. Reisfeld R (1990) Theory and applications of spectroscopically active glasses prepared by the sol-gel method, Sol-Gel Optics, SPIE Proc. 1328: 29

36. Reisfeld R, Zusman R, Eyal M (1988) Chem Phys Lett 147: 142

37. Fricke J, Reichenauer G (1987) "Structure investigation of SiO_2 aerogels", J Non-Cryst Solids 95: 1135–1142

38. Sorek Y, Reisfeld R, Tenne R (1994) The Microstructure of Titanium-Modified Silica Glass Waveguides Prepared by the Sol-Gel Method, Chem Phys Letters 227: 235–242

39. Sorek Y, Reisfeld R, Finkelstein I, Ruschin R (1995) Light Amplification in a Dye-Doped Glass Planar Waveguide, Appl Phys Lett 66: 1169–1171

40. Sorek Y, Reisfeld R, Weiss AM (1995) Effect of composition and morphology on the spectral properties and stability of dyes doped in a sol-gel glass waveguides, Chem Phys Lett 244: 371–378

41. Takada T, Yano T, Yasumori A, Yamane M, Mackenzie JD (1992) "Preparation of quantum-size CdS-doped Na_2O-B_2O_3-SiO_2 glasses with high nonlinearity", J Non-Cryst Solids 147: 631–635

42. Reisfeld R, Jørgensen ChK (1992) in: Structure and Bonding 77: 207, Springer-Verlag Berlin
43. Schmidt H (1989) Sol-Gel Science and Technology, Edited by Aegerter MA, Iafelicci M, Souza DF, Zonotto ED, World Scientific, Singapore, p 432
44. Reisfeld R, ibid p 323
45. Reisfeld R (1990) J Non-Cryst Solids, 121: 254
46. Reisfeld R, Brusilovsky D, Eyal M, Miron E, Burstein Z, Ivri J (1989) Chem Phys Lett 160: 4344
47. Reisfeld R, Chernyak V, Eyal M, Weitz A (1988) Laser and spectroscopic characterization of thin films. Proc Int'l Conf on Optical Science and Engineering, Optical Materials Technology for Energy Efficiency and Solar Energy Conversion VII, Solar Collecting Devices, SPIE Proceedings 1016: 240–246
48. Shamrakov D, Reisfeld R (1993) Superradiant Laser Operation of Red Perylimide Dye Doped Silica-Polymethylmethacrylate Composite, Chem Phys Letters 213: 47–54
49. Schmidt HM, Weller H (1986) "Photochemistry of colloidal semiconductors. 15. Quantum size effects in semiconducture crystallites. Calculations of the energy spectrum for the confined exciton", Chem Phys Lett 129, 615–618
50. Nogami M (1994) Semiconductor-Doped Sol-Gel Optics, Sol Gel Optics, Ed. Lisa Klein Kluwer, pp 329–344
51. Li C-Y, Kao YH, Hayashi K, Takada T, Mackenzie JD, Kang K-I, Lee S-G, Peyghambarian N, Yamane M, Zhang G, Najafi SI, "Improving CdS dot materials by the sol-gel method, SPIE 2288 Sol-Gel Optics III, pp 151–162
52. Guglielmi M, Martucci A, Righini G-C, Pelli S (1994) CdS- and PbS-doped silica-titania optical waveguides, SPIE 2288 Sol-Gel Optics III, pp 174–182
53. Reisfeld R, Minti H (1994) Nonlinear Properties of Semiconductor Quantum Dots in Glasses Prepared by the Sol-Gel Method, J Sol-Gel Science and Technology 2: 641–645
54. Masumoto Y, Wamura T, Iwaki A (1989) "Homogeneous width of exciton absorption-spectra in CuCl microcrystals", Appl Phys Lett 55: 2535–2537
55. Nogami M, Zhu Y-Q, Tohyama Y, Nagasaka K (1991) J Am Ceram Soc, 74: 238
56. Itoh T, Furumiya M (1991) "Size-dependent homogeneous broadening of confined excitons in CuCl microcrystals", J of Luminiscence 48&49: 704–708
57. Lifshitz E, Yassen M, Bykov L, Dag I, Chaim R (1994) "Nanometer Sized Particles of PbI_2 Embedded in SiO_2 Films", J Phys Chem 98: 1459–1463
58. Lifshitz E, Yassen M, Bykov L, Dag I, Chaim R (1995) Photodecomposition and Regeneration of PbI_2 Nanometer Sized Particles, embedded in Porous Silica Films", J Phys Chem 99: 1245–1250
59. Jiang Zhonghong, Ye Hui (1993) World Optical Conference, Shanghai PRC 133
60. Tseng JY, Li CY, Takada T, Lechner C, Mackenzie JD (1992) SPIE 1758, Sol-Gel Optics II, p 612.
61. Mie G (1908) Ann Phys (Leipzig) 25: 377
62. Maxwell-Garnett JC (1904) Philos Trans R Soc London 203: 385; (1906) 205: 237
63. Doremus RH (1964) J Chem Phys 40: 2389; (1965) 42: 414
64. Doyle WJ (1958) Phys Rev 111: 1067
65. Kawabata A, Kubo R (1966) J Phys Soc Jpn 21: 1765
66. Genzel L, Martin TP, Kreibig U (1975) Z Phys B 21: 339
67. Johnson PB, Christy RW (1972) Phys Rev B 6: 4370
68. Agrawal GP, Cojan C, Flytzanis C (1977) Phys Rev Lett 38: 711
69. Hache F, Richard D, Flytzanis C (1986) J Opt Soc Am B 3: 1647–1655
70. Agrawal GP, Cojan C, Flytzanis C (1977) Phys Rev Lett 38: 711
71. Johnson PB, Christy RW (1972) Phys Rev B 6: 4370
72. Yoshiyuki Asahara. Materials Research Laboratory, Hoya Corporation 3-3-1 Musashino, Akishima-shi Tokyo 196. Nonlinear glass materials, lecture presented at Eighth Cimtec World Ceramic Congress, Florence, June 28–July 4, 1994.
72a. Sakka S, Kozuka H, Zhao G (1994) "Sol-gel preparation of metal particle/oxide nanocomposites", SPIE 2288 Sol-Gel Optics III, pp 108–119
73. Menning M, Schmitt M, Becker U, Jung G, Schmidt H (1994) "Gold colloids in sol-gel derived SiO_2 coating in glass and their linear and nonlinear optical properties", SPIE 2288 Sol-Gel Optics III, pp 130–139
74. Horan P, Blau W, Byrne H, Berglund P (1990) "Simple setup for rapid testing of third-order nonlinear optical materials" Appl Opt 29: 31–36
75. Weyl WA (1951) Coloured Glasses, The Society of Glass Technology; Sheffield, (new print)

76. Brusilovsky D, Eyal M, Reisfeld R (1988) Chem Phys Lett 153: 203
77. Reisfeld R, Eyal M, Brusilovsky D (1988) Chem Phys Lett 153: 210
78. Reisfeld R, Minti H, Eyal M (1991) In glasses for Optoelectronics II, Proc SPIE, 1513, ed. by Righini G, SPIE, Bellingham, p 360
79. Menning M, Spanhel J, Schmidt H, Betzholz S (1992) "Photoinduced formation of silver colloides in a ferrosilicate sol-gel system", J Non-Cryst Solids, 146 & 148: 326–330
80. Hinsch A, Zastrow A (1992) "The production of small colloidal silver particles in a thin SiO_2 sol-gel glass layers", J Non-Crystal Solids, 147: 579–581
81. Innocenzi P (1994) "Methyltriethoxysilane derviced coatings for optical applications", SPIE, 2288 Sol-Gel Optics III, pp 87–95
82. Tompkin WR, Boyd RW, Hall DW, Tick PA (1987) "Nonlinear-optical propertie of lead tin fluorophosphate glass containing acridine-dyes", J Opt Soc Am B4: 1030–1034
83. Kramer MA, Tompkin WR, Boyd RW (1986) Phys Rev A34: 2026
84. Reisfeld R, Eyal M, Gvishi R, Jørgensen CK (1987) Photochemical behavior of luminescent dyes in sol-gel and boric acid glasses. Proc 7th Int'l Symp on the Photochemistry and Photophysics of Coordination Compounds, Elmau, Germany, March 29–April 2, 1987. Springer-Verlag, Heidelberg-New York, pp 313–316
85. Reisfeld R, Eyal M, Gvishi R (1987) Spectroscopic behavior of fluorescein and its di(mercury acetate) adduct in glasses. Chem Phys Lett 138: 377–383
86. Eyal M, Reisfeld R (1988) High yield singlet-triplet transfer for efficient saturable absorbers. Proc Int'l Conf on Luminescence (ICL'87), Beijing, China J Luminescence, 40–41: 539–540
87. Graham S, Renner R, Klingshirn C, Schrepp W, Reisfeld R, Brusilovsky D, Eyal M (1989) Pump- and probe beam measurements in organic materials. Paper presented at Int Conf Materials for Non-linear and Electro-optics, Cambridge, 1989. Inst Phys Conf Ser (Bristol, New York) No 103; Section 2.2, 157–162
88. Graham S, Eyal M, Thoma M, Brusilovsky D, Reisfeld R, Klingshirn C (1991) Nonlinear absorption and laser induced gratings in glasses doped with acridine orange and methyl orange. J Luminescence 48–49: 325–328
89. Graham S, Thoma M, Klingshirn C, Eyal M, Brusilovsky D, Reisfeld R, Gaponenko SV, Yu Lebed V, Zimin LG (1991) "Laser Induced Gratings and Nonlinear Optics in Organic Materials", in Organic Materials for Nonlinear Optics II, Edited by Hann RA and Bloor, Royal Society of Chemistry, Thomas Graham House, Cambridge 142–148
90. Gaponenko SV, Germanenko IN, Stupak AP, Eyal N, Brusilovsky D, Reisfeld R, Graham S, Klingshirn C (1994) Fluorescence of acridine orange in inorganic glass matrices, App Phys B 58: 283–288
90a. Graham S, Klingshirn C, Gaponenko S, Germanenko I, Gribkovskii V, Zimin L, Malinovskii I, Stupak A, Eyal M, Brusilovsky D, Reisfeld R (1991) "Nonlinear Optical and Luminescent Properties of Acridine Orange in porous glasses", published in the NATO ASI Series, "Proc Symp Physics and Chemistry of Finite Systems: From Clusters to Crystals", Richmond, USA, October: 8–12
91. Yariv E, Reisfeld R, Aryeh Weiss M (1992) Optical nonlinearities of methyl red in various solid matrices, SPIE Vol. 1972 on the 8th meeting on Optical Engineering in Israel, pp 14–16, 46–54
92. Todorov T, Nikolova L, Tomova N, Dragostinova V (1986) IEEE J Quant Electron QE-22: 1262
93. Graham S, Thoma M, Klingshirn C, Eyal M, Brusilovsky D, Reisfeld R, Gaponenko SV, Yu Lebed V, Zimin LG (1991) "Laser Induced Gratings and Nonlinear Optics in Organic Materials", in Organic Materials for Nonlinear Optics II, Edited by Hann RA, Bloor D, Royal Society of Chemistry, Thomas Graham House, Cambridge, pp 142–148
94. Speiser S, Chisena FL (1988) "Optical Bistability in Fluorescein Dyes", Appl Phys, B45: 137–144
95. Reisfeld R, Chernyak V, Eyal M, Weitz A (1988) Laser and spectroscopic characterization of thin films. Proc Int'l Conf on Optical Science and Engineering, Optical Materials Technology for Energy Efficiency and Solar Energy Conversion VII, Solar Collecting Devices, SPIE Proceedings 1016: 240–246
96. Canva M, LeSaux G, Georges P, Brun A, Chaput F, Boilot JP (1991) "Time-resolved saturated absorption recovery in malachite green-doped xerogel", Chem Phys Lett, 176: 495–498
97. Chernyak V, Reisfeld R (1991) Chem Phys Lett 181: 39

98. Reisfeld R, Chernyak V, Jørgensen CK (1992) "Photophysical behaviour of Malachite Green in solid and liquid media", Chimia, 46: 148–151

99. Lewis GN, Lipkin D, Magel TT (1941) J Am Chem Soc 63: 3005

100a. Jørgensen CK (1984) "The problems for two-electron bond in inorganic compounds; analysis of the coordination number N", Top Curr Chem 124: 1–32

100b. Jørgensen CK (1989) "Are atoms significantly modified by chemical bonding?", Top Curr Chem 150: 1–46

101. Ingri N, Lagestrom G, Frydman M, Sillen LG (1957) Acta Chem Scand 11: 1034

102. Sanchez C, Lebeau B, Bruno Viana (1994) "NLO and Luminescent Properties of Hybrid Siloxane-Oxide Coating", SPIE 2288 Sol-Gel Optics III, pp 227–238

103. Toussaere E, Zyss J, Griesmar P, Sanchez C (1991) "Second harmonic generation from poled organic molecules incorporated into sol-gel matrices", Non Linear Optics I, p 349

104. Sanchez C, Griesmar P, Toussaere E, Puccetti G, Ledoux I, Zyss J (1992) "Nonlinear optical properties of organically doped metal-oxide based gels", Non Linear Optics, 4, p 245

105. Griesmar P, Sanchez C, Pucetti G, Ledoux I, Zyss I "Second harmonic generation from organic molecules incorportated in sol-gel matrices", Molecular Engineering, 1(3), p 205, 1991

106. Fox MA, Pettit TC (1989) "Photoactivity of zeolite-supported Cadmium-sulfide. Hydrogen evolution in the presence of sacrificial donors", Langmuir, 5: 1056–1061

107. Marder SR, Sohn JE, Stucky GD (Eds.) (1991) Materials for Nonlinear Optics: Chemical Perspectives , ACS Symposium Series, Vol 455

108. Kubena RL, Joyce RJ, Ward JW, Garvin HL, Stratton FP and Brault RG (1987) "Dot lithography for zero-dimensional quantum wells using focused ion-beams", App Phys Lett 50: 1589–1591

109. Kozuka H, Zhao G, Sakka S (1994) J of Sol Gel Science and Technology 2: 741

110. Flytzanis C (1991) "Impact of quantum confinement on optical nonlinearities of metal and semiconductor crystallites in glasses" in Materials for Photonic Devices, ed. A. D'Andrea, A. Lopicciriello, G. Marletta, S. Viticoli, World Scientific Publishing Co., Singapore, p 359.

111. Cigen R, Eckstrom CG, Acta Chem Scand 17 (1963) 2083; ibid 18 (1964) 157

112. Eckstrom CG, Acta Chem Scand 19 (1965) 1381; ibid 20 (1966) 444

113. Bengtsson G, Acta Chem Scand 21 (1967) 1138 and 2544

114. Ritchie CD, Sager WF, Lewis ES (1962) J Am Chem Soc 84: 2349

115. Kador L, Fischer R, Haarer D, Kasemann R, Brück S, Schmidt H and Dür H (1993) "Optical Second-Harmonic Effects of Sol-Gel Inorganic–Organic Nanocomposites", Adv Mater 5: 370–373

Sol-Gel Chromogenic Materials and Devices

Michel A. Aegerter*

Instituto de Física de São Carlos, Universidade de São Paulo, Caixa Postal 369, 13560-970, São Carlos, SP, Brasil
*Present address: Institut für Nene Materialien – INM, Im Stadtwald, Gebäude 43, D-66123 Saarbrücken – Germany

In the last few years the sol-gel process has turned into an interesting and promising method of synthesizing materials for obtaining thin or thick films with definite functions. The techniques of film preparation such as dip and spin coating are simple and allow us to prepare coatings with smooth optical surfaces with controlled stoichiometry, structure and texture. In this paper we give an up to date overview of what has been achieved in the field of chromogenic materials such as anodic or cathodic electrochromic coatings, counter or ion storage electrodes, transparent electron conductors and ionic conductors to be used in electrochromic (EC) devices. We also review the sol-gel research in the related areas of photochromic, thermochromic and electrooptic sol-gel materials whose properties are essentially used to modulate the luminous or solar energy as well as sol-gel materials prepared in the form of nanoparticles proposed recently for the development of a new type of solar cell. Finally we stress the future developments in these fast growing fields.

1 Objective

The objective of this paper is to give essentially an up to date review of what has been achieved in the field of *Chromogenic Materials and Devices* prepared by the *Sol-Gel Process*. All the references obtained through a STN International File Search (22.3.1994) with keywords sol, gel and electrochromism or photochromism or thermochromism as well as more recent ones are cited so that the reader should have a complete overview of this fast growing field which has gained importance in science as well as in industry during the last few years.

2 Introduction

International interest in solar energy research has strongly fluctuated during the last two decades. It was high during the latter half of the 1970s as a consequence of the oil crisis but faded somewhat in the 1980s. Today it is recognized worldwide that the technologies involved in energy efficiency and solar energy could alleviate the environmental crisis manifested by global heating through the greenhouse effect. The hazardous increases of UV radiation through the holes in the ozone layer also contribute to the pressure to limit use of fossil and nuclear fuels because of their environmental impacts. However this revival is also due to the increasing interest of other specialized markets such as automotive, aerospace, defense, toys, etc. which look very promising at short term, with sales impact exceeding several billion dollars.

A wide class of materials, called *chromogenic* materials, are known to change in a persistent but reversible manner their optical properties such as optical transmission, absorption, reflectance and/or emitance in response to changes in ambient conditions [1].

For instance Cd or Ag doped glasses and organic dye doped polymers or organic dyes incorporated in inorganic porous matrices change their color when exposed to UV or visible light. This *photochromic* property is widely used today in commercial products such as lenses. A few materials, such as VO_2, are known to exhibit an analogous effect when heated at a defined temperature and are known as *thermochromic* materials. Others, called *barochromic* materials, change their color when exposed to a change in ambient pressure; this is exhibited for instance by samarium sulphide. Other devices used *liquid crystals or polymer dispersed* materials incorporated in a liquid; the application of an electric field changes the orientation of these molecules altering the optical absorption or scattering of the layer.

Technologically, however, the most promising chromogenic effect is called *electrochromism* and is defined as the persistent but reversible optical change produced electrochemically. A wide range of materials are known today to exhibit a reversible coloration by the application of an electric field or more precisely by the passage of an electric current and ions through them. These materials are required to exhibit a mixed electronic and ionic conductivity and consequently are often amorphous and porous to provide an open network for rapid ionic diffusion. A common feature of such materials, unlike the liquid crystals used in displays, is that once the material is colored, the applied voltage can be switched off and the color retained, making the electrochromic (EC) devices more energy efficient.

Commercially, hearing aid battery powdered electrochromic sunglasses and other systems are already on the market. However, owing to the more than 30 million new cars and trucks produced worldwide each year and to the recent advance of technology, in the near future the greatest opportunities for EC technology is without any doubt in the automotive industry [2,3]. Small to medium area transmissive or reflective EC devices have already been success-fully commercialized for automobile rear and side view mirrors (more than one million sold in 1993) and automotive sunroofs.

This success represents the first stepping stone towards large area devices for application in architectural glazing that will allow active optical response to changing environmental conditions. This technology will save on cooling and lighting energy costs and, at the same times, provide glare control and improved thermal comfort as shown in Fig. 1 [4, 5].

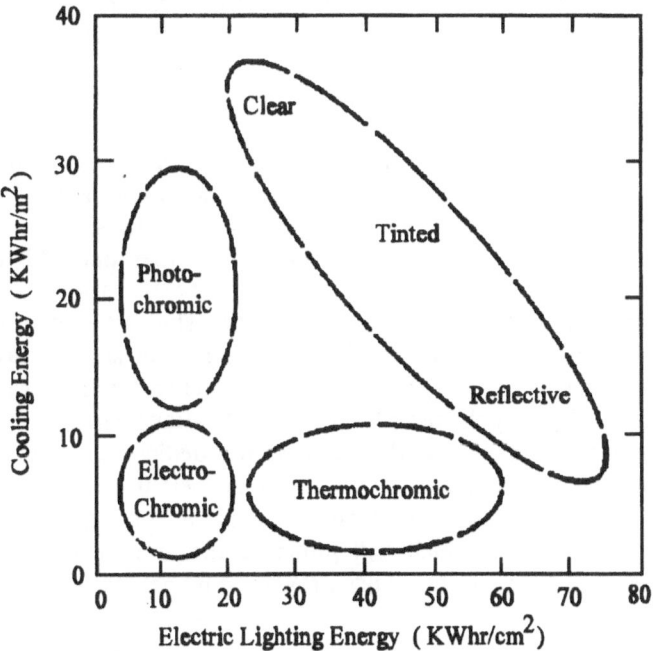

Fig. 1. Schematic diagram of projected cooling and electric lighting energy for a building using different window panes [4]

3 Electrochromic Devices

Electrochromic devices have been classified into several categories depending on their application, their spectral transmission variation or their configuration. Reilly et al. [6] classified *ideal* electrochromic materials into three types according to their transmittance, which could vary typically between 0.1 and 0.8, and switches over different spectral ranges such as the entire solar spectrum, the visible subspectrum only or the solar infrared subspectrum only (Fig. 2). As each of these spectral types could change transmittance by a corresponding change in either absorption or reflection, six basic electrochromic materials can be considered. On the other hand, Cronin and Agrawal [3], in a more pragmatic way, classified the devices according to their configuration. As electrochromism has never been observed in an isolated compound and requires a simultaneous double injection of electrons and ions, the device configuration involves the development of an electrochemical cell consisting of multiple layer stacks that are sequentially deposited or laminated. Figure 3 shows a few typical examples; the simplest device uses a liquid solution between two electronically conductive electrodes (Fig. 3a) and the optical modulation results in a redox reaction at the liquid/electrode interface when an appropriate potential is applied to the conductive electrodes. For transmissive devices both conductive electrodes must be transparent. For large area devices, to improve the safety of the device and to avoid the effect of hydrostatic pressure and the necessity of a seal, the liquid solution can be replaced by a solid-state electrochromic film such as a polymer. In this case the electrochemically active chromophores are incorporated into the matrix by chemical bonding.

More sophisticated configurations are also shown in the figure. These EC systems can be assembled on a glass substrate by sequential deposition where at least one of the coatings is an electrochromic layer (Fig. 3b) or, alternatively (Fig. 3c, d) an electrochromic film can be deposited on a conductive substrate (left part) and then assembled to the other part. All these devices may use a sealed liquid electrolyte as shown in both examples or, as preferred for large area applications, a solid electrolyte. The last system (Fig. 3d) makes use of a counter electrode which serves as an ion storage layer which can also be electrochromic in order to achieve a higher dynamic transmission range. These are only typical examples and modifications can be made in order to improve the performance of the systems for dedicated applications or to take into account the particular conditions of their use. The reader should consult the specialized literature for more insight on these questions [3–5, 7]. The performance factors which will influence the market acceptance of electrochromic windows have recently been reviewed by Selkowitz et al. [5] and Sullivan et al. [8].

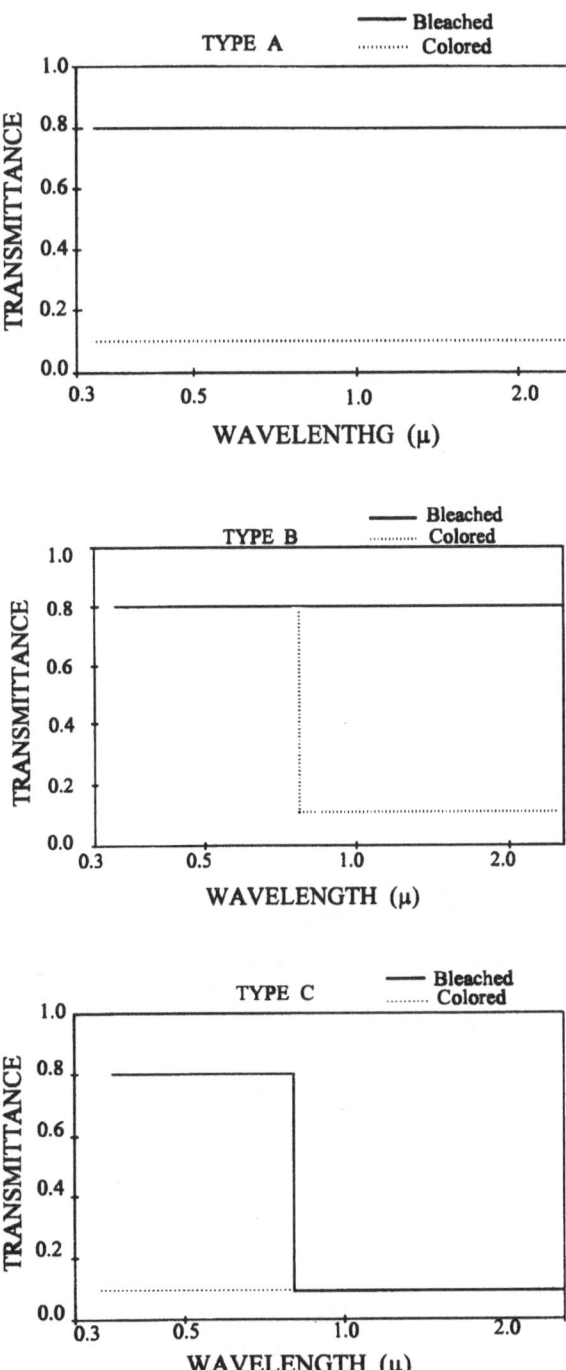

Fig. 2a–c. Ideal electrochromic spectra: **a** uniform switching over the entire solar spectrum; **b** switching over the infrared spectrum only; **c** switching over the visible spectrum only [5]

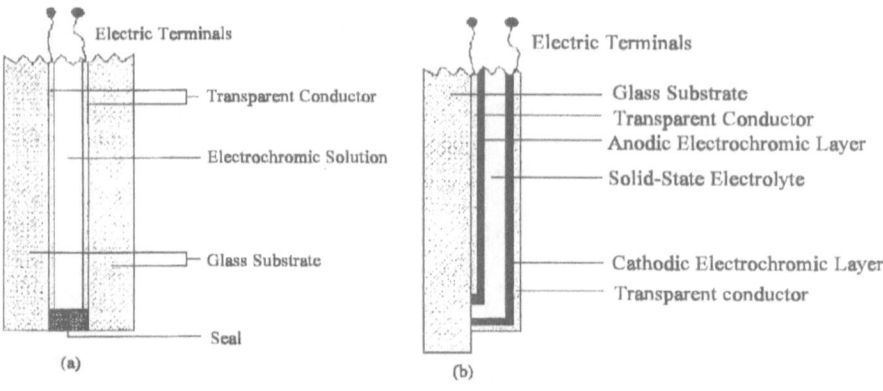

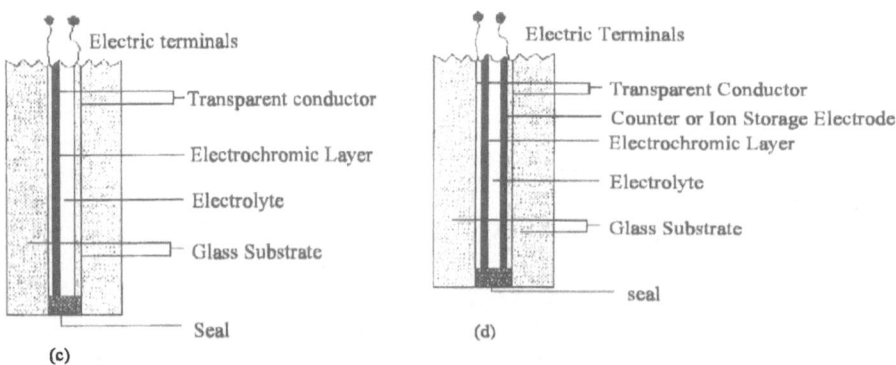

Fig. 3a–d. Typical designs of electrochromic devices [3]: **a** EC device with a liquid solution between two electronically conducting electrodes; **b** EC device which uses an EC thin film and a solid state electrolyte and which can be assembled on one glass substrate by sequential deposition; **c** EC device which uses a liquid or solid electrolyte (without the seal) and an EC coating; **d** EC device similar to **c** but which also uses a counter electrode (ion storage) which could also be electrochromic

4 Materials for Electrochromic Devices

4.1 Overview

Electrochromism has been known since 1953 when Kraus [9] discovered that a vapor-deposited WO_3 layer on a semitransparent metal layer (Cr, Ag) was intensely blue in color when cathodically polarized in 0.1 N H_2SO_4. However world wide research on this topic started only after the fundamental work of Deb [10, 11] on the same material two decades later. Today several other materials are known to exhibit such a property. Most of them are inorganic oxides of transition metals, to which we shall restrict this review, but several organic materials, mostly doped polymers, have been also discovered [12, 13]. However, despite hundreds of scientific and technical papers, the fundamentals of the phenomenon are still not well understood. Usually the coloration and bleaching of these materials is described schematically by

$$MeO_n + xI^+ + xe^- \leftrightarrow I_xMeO_n \tag{1}$$

where Me is a metal atom, I^+ is a singly charged ion such as H^+, Li^+, Na^+, K^+, Ag^+, e^- is an electron and n depends on the particular type of oxide. Some materials color when they are cathodically polarized, others when anodically polarized and a few of them color in both states. According to Granqvist [14], the electrochromic oxides can be divided into three main groups with regards to their crystalline structure: perovskite-like, rutile-like and layer and block structure, and the general occurrence of MeO_6 building blocks in these materials is of fundamental importance.

A partial list of the inorganic electrochromic materials used to formulate EC devices is given in Table 1 and, for sake of completeness, Tables 2 and 3 list the most important materials which can be used as transparent electronic conductor and electrolyte (H^+ and Li^+ conductors) respectively. In order to be useful for application, these materials should have not only outstanding physical and chemical properties but should also meet severe criteria regarding temperature range, cyclability, corrosion resistance, etc. For a definite application most of them may not yet meet these criteria or meet them only partially. As an example, Table 4 shows the desirable features of an electrochromic automobile sunroof [3]. For other applications some of the requirements are even more stringent [2, 4]. A complete description of device evaluation and test methods can be found in Czanderna and Lampert [15].

4.2 Sol-Gel Materials

Practically all the materials listed in Tables 1–3 were originally prepared by non sol-gel methods such evaporation, DC diode, magnetic or reactive sputtering, electrochemical deposition, etc. and no detailed explanations will be given here. The reader should consult the specialized literature and the references given in the tables should be helpful to start.

The sol-gel process will not be reviewed in this paper as several reference books are available today on this subject [103, 127, 128]. The method appears now to be one of the most promising technologies for preparing coatings with tailored properties [129]. In the field of chromogenic materials, it has until recently been used mainly to develop the chemistry of the precursor sols, showing that the method was adequate to obtain coatings with chromogenic properties. More recently, however, because of its intrinsic advantages over other techniques, which include the use of inexpensive coating equipment for

Table 1. List of inorganic electrochromic materials prepared from sol-gel processes and non-sol processes (only partial list)

Polarization	Material	State	Color	Reference (non sol-gel)	Reference (sol-gel)
Cathodic	WO_3	a, c	blue	[11, 16]	[3, 17–33]
Cathodic	K_xWO_3	c	blue	[34]	
Cathodic	WO_3–TiO_2	a	blue	[35]	[35]
Cathodic	WO_3–MoO_3				[36]
Cathodic	MoO_3	a, c		[37]	[38, 39]
Cathodic	CeO_2		UV		[35, 40, 41]
Cathodic	CeO_2–SnO_2				[41]
Cathodic	CeO_2–TiO_2	c, *	UV		[40, 42–50]
Cathodic	TiO_2		grey	[51, 52]	[53–58, 77]
Cathodic	TiO_2–Al_2O_3		blue		[53]
Cathodic	TiO_2–Cr_2O_3		blue		[53]
Cathodic	TiO_2–WO_3				[57, 59, 60]
Cathodic	TiO_2-viologen				[58]
Cathodic	Nb_2O_5	a, c	a-brown c-blue	[61–65]	[42, 55, 56, 66–74]
Cathodic	Fe_2O_3				[91–93]
Cathodic	Fe_2O_3–TiO_2				[92, 93]
Cathodic	Fe_2O_3–SiO_2				[92, 93]
Cathodic	SnO_2	*			[94]
Both	V_2O_5	c	green, yellow, red	[75, 76]	[39, 54, 75, 77–80]
Both	V_2O_5–Na_2O				[81, 82]
Both	V_2O_5–Ta_2O_5	powder	grey		[83]
Both	V_2O_5–Nb_2O_5	powder	grey		[83]
	V_2O_5–TiO_2	a	blue, green, yellow, reddish-brown		[84]
Anodic	IrO_2			[85]	
Anodic	NiO			[86, 87]	
Anodic	$Ni(OH)_2$			[88]	
Anodic	$Fe_4(Fe(CN)_6)_3$			[89]	
Both	Rh_2O_3			[90]	

a Amorphous
c Crystalline
* More useful for counter electrode

Table 2. Partial list of inorganic transparent conductors used for the fabrication of EC devices (reference non sol-gel [95–97])

	Materials Reference (sol-gel)
In_2O_3:Sn (ITO)	[98–100]
SnO_2	[101–110]
SnO_2:Zr	[107]
SnO_2:Ti	[107]
SnO_2:Sb	[111]
SnO_2:F	
V_2O_5	[112]
ZnO:Al	
Au (thin film)	

Table 3. Partial list of inorganic and organic electrolytes

H$^+$	Li$^+$	Reference (non sol-gel)	Reference (sol-gel)
Ta$_2$O$_5$		[113–115]	[116, 117]
ZrO$_2$		[114]	
HU$_2$PO$_4$4H$_2$O		[118, 119]	
SiO$_2$		[52]	
Aminosil (SiO$_2$)			[120]
ORMOCER-TiO$_2$	ORMOCER-TiO$_2$		[121]
Sol-Gel Polymer PBSS			[122]
PWA-TiO$_2$			[59]
HPA-SiO$_2$			[59]
	Li$_3$N	[123]	
	LiAlF$_4$	[124, 125]	
	LiNbO$_3$	[124]	
	LiTiO$_3$	[124]	
	LiAlSiO$_4$	[126]	
	Nasicon		[145]
	Li$_2$O–SiO$_2$–P$_2$O$_5$		[148, 154]
	(LiCl)$_2$–R$_2$O$_3$–SiO$_2$		[151, 152, 155–157]
	V$_2$O$_5$		[159]
	ORMOLYTE		[160]

Table 4. Desirable features for an electrochromic automobile sunroof [3]

Luminous Transmission Range	< 10% in colored range
	> 50% in bleached state
Solar Transmission	< 10% in colored state
Response time	< 15 min. to achieve 90% of coloration or bleach
Temperature Range	− 40 to 100 °C
Cyclic Durability	> 20 000 cycles
Lifetime	> 5 years
Others	Solid state, low power consumption, low cost

coating small and large areas, and especially its applicability to design chemically the molecular precursors to better control the texture and structure of the films [129, 130], this method was used to develop better oxide coatings and even new coatings with properties which have never been obtained previously. EC devices built partly or entirely with sol-gel materials have been extensively tested since 1991.

In the next sections each group of sol-gel materials is reviewed separately.

4.3 Sol-Gel Electrochromic Coatings

The oxide coatings which have been prepared by the sol-gel process and whose electrochromic characteristics have been reported and discussed involve practically eight groups: WO$_3$, MoO$_3$, TiO$_2$, Nb$_2$O$_5$, V$_2$O$_5$, CeO$_2$, Fe$_2$O$_3$ and SnO$_2$ and mixed compounds of these materials.

An extensive review on the same subject has been produced by Agrawal et al. [131]. Limited descriptions of some topics have also been given [29, 78, 132–134].

4.3.1 WO_3 and WO_3 Doped with TiO_2 or MoO_3

WO_3. WO_3 is the most studied electrochromic material and is considered today the best for EC applications. It is a cathodic EC material and it changes its color from transparent or yellow to deep blue with a large optical modulation when it is reduced by H^+ or Li^+:

$$WO_3 + xM^+ + xe^- \rightleftharpoons M_xWO_3 \tag{2}$$

where $M^+ = H^+$, Li^+, etc.

The bronze can also be formed when sodium, potassium or silver are used. However as the size of the ions increases the rate of diffusion decreases, diminishing the rate of optical modulation [17, 18].

At least four basic sol-gel routes have been developed for the preparation of the WO_3 sols:

1) acidification of sodium tungstate [19–21];
2) use of peroxypolytungstic acid [22–25];
3) hydrolysis of alkoxides [22, 26–28];
4) reaction of tungsten chloride and oxychloride with alcohols [30, 32, 39, 135].

The first method is one of the earliest efforts to produce sol-gel WO_3 films and its major advantage is the formation of WO_3 at room temperature with no formation of decomposition products. It is therefore possible to produce thick and unstressed films. However the stability of the solutions (which is an important parameter for industrial production) and the adhesion of the coatings to the substrate were found not to be adequate, although the sol stability could be slightly improved by complexing it with appropriate additives such as dimethysulfoxide [20].

In the second method, tungsten and tungsten carbide powder are digested by an aqueous solution of hydrogen peroxide. The product, a peroxytungstic acid, is isolated and then dissolved in a polar solvent such as alcohol or water. The acid is decomposed into tungsten oxide during the heat treatment of the film at low temperature (100–200 °C).

The tungsten alkoxide (third method) is the classical sol-gel route for any kind of oxide but is expensive and consequently not useful for industrial application. Crack-free films could only be prepared with small thickness (< 50 nm). It is therefore necessary to repeat the deposition process to obtain thicker films to get a sufficient optical contrast.

According to Livage [29] the reaction of tungsten oxy-chloride ($WOCl_4$) with isopropanol is the best method of preparation as it is a cheap technique and leads to a sol that is stable for several months because of the formation of molecular oligomeric species $(WOCl_{4-x}(O^iPr)_x)_n$ [30]. In this method, tungsten chloride is reacted with an anhydrous alcohol and the resulting material is diluted with more alcohol to obtain the precursor sol. The films have a high uniformity, better than those prepared via a colloidal route. The same precursor has also been used to deposit WO_3 by spray pyrolysis [31]. Micrometer-thick hydrated films of composition $WO_3 \cdot nH_2O$ are easily obtained, with the control of the amount of water and adequate heat treatments controlling the film morphology, essential for these applications. The electrochromic properties were found to be dependent upon the values of n. It was also found that the switching time and the long term memory decrease when the amount of water increases due to faster ion diffusion through the gel network [32]. When Li^+ ions are used the best electrochemical stability was encountered for n = 0.5 as higher amount of water leads to a gradual decrease of the current, an indication of the occurence of an irreversible process [21].

Finally two recent interesting results concerned with the improvement of WO_3 have appeared. Cronin et al. [3, 18] have patented the preparation of a WO_3 electrochromic coating solution prepared by reacting metallic W with a mixture of a peroxy acid such as hydrogen peroxide and an organic acid such as acetic or propionic acid at between -10 and $+12\,°C$ for 16–26 h. The resulting product, a W-peroxy acid, was then esterified by reacting with low boiling one to three carbon alcohol to produce a peroxyester-W derivative (PTE). The sol was found to be stable if stored below $10\,°C$. The deposited coating was converted to an electrochromically active layer by removing the volatile organics by firing at a temperature as low as $100\,°C$. It is claimed that the optical transmission of the coatings prepared in accordance with the disclosures decreases from 85% to less than 15% in a matter of seconds. Higher firing temperatures will only increase the film density, making the film tougher and more resistant to scratching. Such a method can be extended to other transition metals such as Mo, Mn, Cr, Rh, Ir, Ni, etc. Due to the low temperature of the film heat treatment, it is without doubt an important technique for obtaining films for industrial applications.

Denesuk et al. [33] have presented a novel synthesis of a sol-gel WO_3 precursor which produces films with different microstructures. The precursor sols were prepared using the above method in anhydrous ethanol in which was added 32 mol% oxalic acid dihydrate (referenced to the W-metal content in PTE). The coatings, prepared by the dip-coating method, were heat treated at $250\,°C$ in ambient atmosphere. The films prepared without the addition of oxalic acid were homogeneous and amorphous. Those prepared with the oxalic acid had an inhomogeneous amorphous/crystalline hybrid structure (with no remaining trace of the additive) and contained small (~ 5 nm) regions of increased electron density. The optoelectrochemical data indicate that the intercalation capacity and the dynamic optical efficiency of these hybrid films are much larger and essentially independent of cycling (Fig. 4).

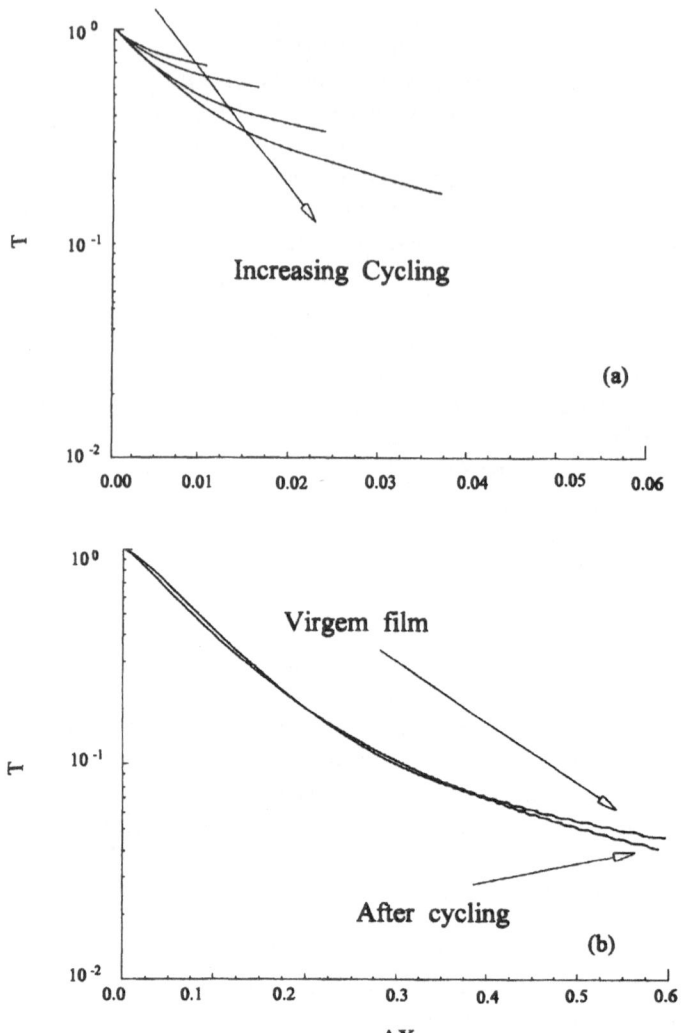

Fig. 4a, b. Log-linear plot of the transmission T vs the change in intercalation $\Delta \times$ for a WO_3 film: **a** prepared without oxalic acid dihydrate; **b** prepared with 32% of the same additive. In **a** one observes a strong cycling effect while in **b** the effect does not exist. This latter film possesses a larger average optical efficiency than the former [33]

These two last results are especially promising for the preparation of commercial EC products and show that the sol-gel method has now reached a state of the art such that it may supplant conventional technologies.

WO_3–TiO_2. Using a sol prepared with a mixture of tungsten chloride (WCl_6) tetraisobutylorthotitanate ($Ti(OBu)_4$) and ethanol, stable for weeks, Göttsche et al. [35] have claimed a better electrochemical stability under Li insertion for films dipped four times (thickness ~ 150–190 nm) and then tempered in air at 200 °C for 1 h (the same property has also been found in sputtered deposited films). However, the addition of TiO_2 (up to 50 mol%) was found to reduce the number of tungsten active sites and to lead to a decrease in the reversible optical transmittance variation with a corresponding reduction of the coloration efficiency from about 70 cm^2/C for an undoped film to 35 cm^2/C for a 33% Ti doped one. On the other hand, the incorporation of Ti was found to prevent the crystallization of tungsten oxide so that the films can be solidified at higher temperatures. The addition of 10–15 mol% of TiO_2 resulted in coatings with maximum cyclic lifetimes.

WO_3–MoO_3. WO_3–MoO_3 films have been deposited from colloidal solutions prepared by mixing H_2WO_4 and H_2MoO_4. After filtration through a cation-exchanger column, the sol was kept at a constant temperature to promote the polymerization [36]. A blue shift of the absorption band compared to pure WO_3 was observed which was maximum for a molar ratio W: Mo of 7.3.

4.3.2 MoO_3

This compound also forms a number of bronzes with alkali metals. It colors when reduced by ion insertion but its ability to color is lower than that of WO_3. Sol-gel MoO_3 has been prepared by Moser and Lynam [39] using halogenated compounds as precursors. Wang et al. [38] have used molybdenum chloro-ethoxide as precursor hydrolyzed in water with a mole ratio H_2O/Mo of two to four. The films were found to be amorphous below 350 °C. At higher temperatures the films crystallized in air into MoO_3. At 600 °C Mo_9O_{26} was also found and other oxides were found if the heat treatment were made under Ar atmosphere or vacuum.

4.3.3 TiO_2 and TiO_2 Doped with Al_2O, Cr_2O_3, WO_3 or Coated with Viologen

TiO_2. Such coatings have been prepared from the classical alkoxy route by Doeuff and Sanchez [53], Nabavi et al. [54], and Ozer et al. [55, 56]. The last authors have prepared amorphous gel coatings by a spin coating technique by using a drop of Ti $(OBu)_4$–BuOH solution. When amorphous (T < 400 °C), the insertion of Li^+ ions was found to occur at random and no peak was observed in voltammetry but instead a continuous reduction and oxidation curve was obtained. After heating to 400 °C the films were found to be crystalline (anatase) and the voltammograms for Li insertion (20 mC/cm^2) showed two cathodic and anodic waves corresponding to defined sites in the TiO_2 structure. In both cases the color-bleaching cycles were reversible. In pure compound the color was grey but was blue for doped materials (Al, Cr). Ozer et al. [55], using sols prepared with Ti (isoPrO)$_4$, Ti (isoBuO)$_4$ and acetic acid as catalyst, have found that the coatings were electrochromic under certain preparation conditions.

Bell et al. [57] have also reported electrochromism in a film of pure TiO$_2$ prepared from a sol made with Ti isopropoxide and ethanol or isopropanol, showing a neutral greyish color under Li insertion.

TiO$_2$ gels have also been synthesized from modified alkoxide precursors using acetic acid which is known to slow down the hydrolysis-condensation process of Ti alkoxides and thereby preventing precipitation [77]. The electrochemical and electrochromic properties of 0.5-μm thick film prepared in this way are quite similar to those obtained without acetic acid, but their color was found to be blue instead of grey. This was attributed to the presence of some acetate groups which remained bounded to the metal. The ligand field around the titanium is therefore different and changes the color. The presence of impurities or organic groups may therefore change the cosmetic aspect of the film after ion insertion and may be a tool to adjust the color of the coating in the colored state.

A new approach has recently been proposed by Hagfeld et al. [58]. Thick (3.5–4 μm) coatings of *nanocrystalline* TiO$_2$ particles (which can easily be prepared by the sol-gel process [103]) have been made by spreading a paste of 15-nm size colloidal TiO$_2$ on SnO$_2$:F conducting glass. After autoclaving at 200 °C and firing at 450 °C in air for 30 min the films were found to be crystalline with the anatase crystal structure. These films are highly porous from the outer layer to the conducting back contact with a roughness factor of 540. Cyclic voltammetry performed in LiClO$_4$/acetonitrile electrolyte shows a cathodic and anodic peak and the charge stored during the intercalation process was very high, 0.11 C/cm^2 for both the forward and reverse scan showing that the

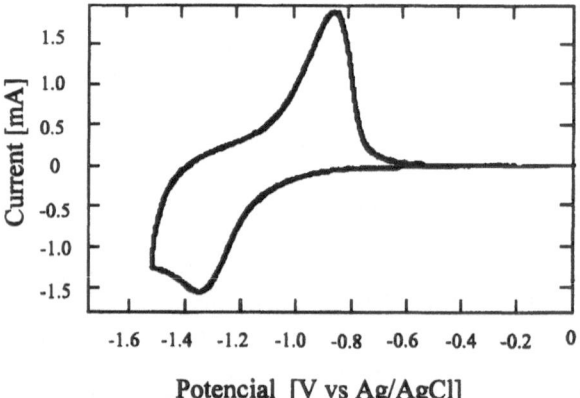

Fig. 5. Cyclic voltammetry of a transparent nanocrystalline TiO$_2$ electrode. Charge passed is $Q_{cathodic} = 0.11$ C/cm^2 and $Q_{anodic} = 0.11$ C/cm^2. The film thickness is 3.5 μm and electrode area 1 cm^2. Ar purged 1 mol/l LiClO$_4$ in acetonitrile is used as electrolyte. The reference electrode is Ag/AgCl (sat. KCl in water) and the counter electrode is glassy carbon. Voltage is scanned from 0 V to − 1.50 V to 0 V at a scan rate of 5 mV/s [58]

intercalation is strictly reversible (Fig. 5). The Li^+ insertion is accompanied by an intense color change from transparent to dark blue as shown in Fig. 6, which develops in about 25 s after the application of a step potential. On switching back the potential to the initial potential value, the coloration disappears on the same time scale. Stability tests performed with a sample of 0.44 cm^2 geometric area show a decline of the cathodic wave of $\sim$ 25% during the first three cycles. No other change had been observed in up to 110 cycles.

This behavior contrasts with that of anatase films prepared by conventional methods which are unable to intercalate Li ions to any significant extent. Apparently the nanoporous morphology of these new devices greatly facilitates the reversible Li intercalation.

This recent investigation establishes the advantage of using nanocrystalline coatings for Li^+ intercalation due to the unique morphology and surface structure of the coating.

$TiO_2:Al_2O_3$. Doeuff and Sanchez [53] have also reported some properties of TiO_2 doped with aluminum. This dopant was introduced in the colloidal titanium sol in the form of aluminum-*sec*-tributylate $(Al(OBu^s)_3)$. The films, heat treated at 400 °C for 2 h, were slightly green-brown but transparent. Under Li^+ insertion they turned blue due to a large unstructured optical absorption band centered at 750 nm probably caused by intervalence transfer between Ti(III) and Ti(IV). This is in agreement with ESR measurements which confirm the presence of Ti^{3+} in distorted octahedral symmetry in the reduced compound. The change in color, grey for pure TiO_2 to blue for doped TiO_2, is attributed to the modification of the localization of Ti^{3+} due to the coulombic field created by the impurity.

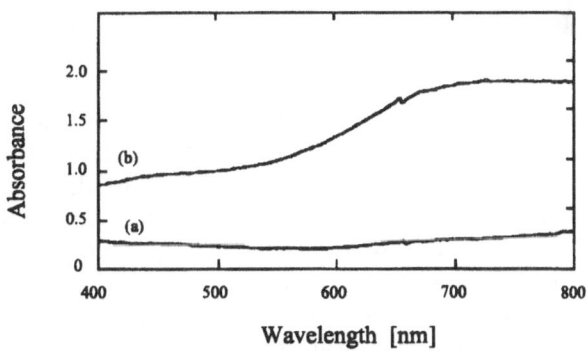

Fig. 6. UV-VIS observed during electrochromic switching of nanocrystalline TiO_2. *Line a* – bleached state ($-$ 0.64 V vs Ag/AgCl); *line b* – colored state after polarizing to $-$ 1.64 V vs Ag/AgCl for 10 s. 1 mol/l LiClO$_4$ in acetonitrile, film thickness 3.5 μm. The observed background absorption in the bleached state is due to light scattering. Then, 10 s after applying the potential step, the absorbance increases to a value above 1 in the whole visible region with a broad maximum of absorbance close to 2 in a wavelength interval of 660–880 nm. Charge passed in 0.10 C/cm^2 [58]

$TiO_2:Cr_2O_3$. The same authors [53] also reported similar conclusions with Cr^{3+} doping.

TiO_2-WO_3. Dip coated films containing 0–100% WO_3 have recently been prepared by Bell et al. [57] from precursors containing Ti isopropoxide or Ti propoxide diluted in ethanol and a tungsten alkoxide precursor. The films have been fired in air at 300 °C. The results are in general agreement with those obtained by Göttsche et al. [136] with respect to the coloration performance of the mixed $WO_3:TiO_2$ films. They exhibit coloration by absorption. The authors found that the absorption decreases with the TiO_2 content but that the bleaching was improved at the same time. A wide range of composition (at least 67–100 mol% W) gives a good coloration efficiency. The color of these films is somewhat different from that found in pure WO_3 due to the presence of an additional absorption band in the 2.2–3.2 eV range. They have a more greyish appearance in the colored state.

Recently Stangar et al. [59] have incorporated phosphotungstic acid $H_3PW_{12}O_{40}.xH_2O(PWA)$ to TiO_2 xerogel with a ratio PWA/Ti = 7:100. This compound is known as an electrochromic and ion conductor material and belongs to a large family of heteropoly compounds of high molecular weight with Keggin structure of the anion. The optical density change was only about 40% in the visible due to a large structureless band having its maximum in the UV region and decreasing slowly at longer wavelength. The maximum amount of injected proton was 25 mC/cm^2 and the kinetics was slow. However, this compound has interesting properties. Since protons are already present in the PWA structure there is no need for an additional ion conductive material to be in close contact with the electrochromic material to produce its coloration (Fig. 3b). Therefore the layer can be reduced at the transparent conductor (SnO_2:F) interface and oxidised at the PWA/counter electrode interface. However the main drawback is its strong acidity and solubility in water and some organic solvents, which should limit its use for commercial products.

TiO_2-Viologen. An interesting approach has recently been proposed by Hagfeld et al. [58]. Instead of using a doped material, the authors have coated TiO_2 nanoparticles with viologen attached to them by carboxylic groups. These organic compounds have a low redox potential, show significant reversibility, exhibit in the visible a large change of the extinction coefficient following reduction, and possess a first reduction potential that is essentially pH independent. Figure 7 shows the UV-Visible spectral change observed for such a system having a high coloration efficiency of 85 cm^2/C. Stability tests indicate a gradual decrease of the charge exchanged which levels after about 100 cycles to about 75% of the initial value. This new approach looks promising as it combines the effect of a large active area obtained by the use of nanosize particles with the electrochromic properties of organic molecules.

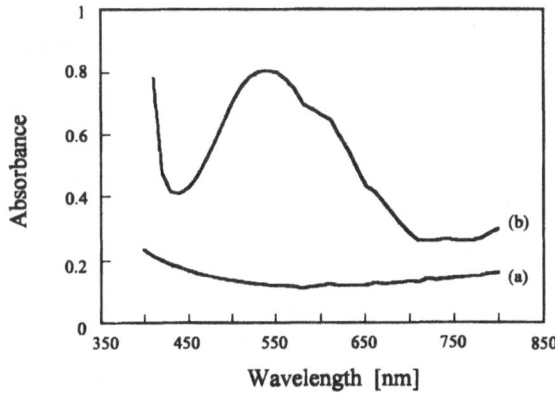

Fig. 7. UV-VIS spectral changes observed during electrochromic switching of a viologen coated nanocrystalline TiO_2 electrode. *Line a* – bleached state under open circuit condition; *line b* – colored state after polarizing to -0.86 V. 0.1 mol/l $TPAClO_4$ in propylene carbonate is used as electrolyte and the electrode area is 1.0 cm^2. The observed background absorption in the bleached state is due to light scattering. Charge passed during coloration is 8.0 mC/cm^2 [58]

4.3.4 V_2O_5 and V_2O_5 Doped with Na_2O, Nb_2O_5, Ta_2O_5 or TiO_2

V_2O_5. The vanadium oxides exist in different stoichiometry, but V_2O_5 is the most used for EC devices. Since the change in the optical spectrum is not as large as in WO_3, it has been used mainly as ion storage electrodes in conjunction with WO_3 as the electrochromic layer, or as electrodes in electrochemical energy storage systems [80]. The reversibility for Li^+ ions is good and is better when V_2O_5 is in the crystalline state [75]. These coatings are yellow in the transparent (bleached) state and therefore slightly reduce the maximum light transmission.

V_2O_5 electrochromic gels have mainly been synthesized by Livage's group using the alkoxy route [54, 77, 78] and by Yoshino et al. [79] using, for instance, alkoxy vanadate VO$(OAm^i)_3$ hydrolyzed with an excess of water. The chemical structure of the precursors (VO$(OAm^i)_3$ and VO$(OPr^i)_3$) have been studied by ESR by Nabavi and Sanchez [137]. The nature of the final material was found to depend on the hydrolysis ratio H_2O/V. For a ratio smaller than three the alkoxy ligands are not all hydrolyzed, leading to the formation of oxy polymers in which organic groups remain bonded to vanadium. These gels exhibit an orange color and give rise to transparent films when dried at room temperature [77]. With higher ratio (H_2O/vanadium > 100) the layer (0.5 µm thick) is mainly a hydrous oxide gel (V_2O_5 1.8 H_2O) with a pale yellow color. The gels have mixed electronic and protonic conductivity and Li ion insertion was found to be reversible.

The study of the structure of these gels shows that the material is made from V_2O_5 ribbons separated by solvent molecules where the vanadium ions are in a square pyramidal environment; the structure inside the ribbons is close to that observed in orthorhombic V_2O_5. When the colloidal solution is deposited on a substrate, the layers are found to be anisotropic due to the turbostratic stacking of the ribbons perpendicular to the substrate. The electrochemical insertion of Li^+ is reversible, showing two peaks for reduction and oxidation, a behavior which indicates that Li^+ insertion occurs at rather well defined sites and not at random as in an amorphous oxide. After insertion the layers are green due to a broad absorption band whose maximum is at about 1450 nm, typical of intervalence transfers between V^{4+} and V^{5+}, and the kinetics of the coloring and bleaching is fast (a few seconds). Using thicker films (about 10 µm) such devices show multiple colors, turning red to yellow to green.

Alkoxyhalides of vanadium have also been used [39]. In this method vanadium trifluoride oxide, vanadium oxytrichloride or vanadium tribromide have been reacted with an anhydrous alcohol. The fluorinated precursor gave the largest optical modulation.

V_2O_5–Na_2O. $Na_{0.32}V_2O_5$ bronzes have been reported by Pereira-Ramos et al. [82] and Bach et al. [81]. A colloidal solution of V_2O_5 was first prepared using an alkoxy route. Then the solution was passed through an ion exchange column to incorporate sodium. Unlike pure V_2O_5 these materials were found to have a 3-D tunnel structure enabling them to insert Li^+ and Na^+ ions by diffusion without structural change.

V_2O_5–Nb_2O_5. $NbVO_5$ bronzes have been prepared by Amarilla et al. [83] from solutions of vanadyl tri-*tert*-butoxide and $NbCl_5$ in isopropyl alcohol. Li^+ extraction and insertion was verified chemically on powders. The bronzes change their color from yellow to dark grey on reduction.

V_2O_5–Ta_2O_5. The same route was used to obtain $TaVO_5$ bronze [83]

V_2O_5–TiO_2. Mixed V_2O_5–TiO_2 thin coating have been prepared by Nagase et al. [84] from sols prepared from a mixture of $VO(O-iso-C_3H_7)_2$ and $Ti(O-iso-C_3H_7)_4$ alkoxides diluted in isopropanol. The sol was stabilized by the addition of acetylacetone and excess of acetic acid. Dip coated films present an electrochromism which is strongly dependent on the atomic ratio $x = Ti/(V + Ti)$ and the calcination temperature. After heat treatment at 400 °C for instance, a two step coloration blue ↔ green ↔ yellow for $0 < x < 0.17$ was observed, while Ti rich coatings $(0.60 < x < 0.67)$ exhibited a reddish brown color at potential -0.8 V vs SCE attributed to the presence of V^{3+} ions. The coatings were essentially amorphous with small crystallites of TiO_2 (anatase). At 500 °C the coatings were found to have properties similar to pure V_2O_5.

4.3.5 Nb_2O_5

Sol-gel Nb_2O_5 films are new, very promising candidates for electrochromic coatings. Very few studies have been reported on the electrochromic properties of Nb_2O_5. Reichman and Bard [61] showed the occurrence of such effects with a 15-μm thick coating produced on the surface of a niobium metallic disk by heating at ~500 °C for about 10 min. A coloring effect, chemically stable and with fast kinetics (1–2 s), was seen in reflexion under either H^+ or Li^+ insertion. Gomes et al. [63, 64] have studied in detail the protonic electrochromic properties of 20-μm thick opaque coating prepared in the same way and Alves [65] has confirmed the possibility of inserting Li ions in a 1-mm thick Nb_2O_5 ceramic prepared from commercial CBMM powder sintered at 800 °C.

The first attempt to fabricate sol-gel Nb_2O_5 for electrochemical purpose has been reported by Lee and Crayston [66] who have spin coated an ITO coated glass electrode with a mixture of $NbCl_5$ dissolved in EtOH. Hydrolysis and gelation were completed in 1 mol/dm^3 H_2SO_4 solution. After drying at room temperature the result was a 5–10 μm thick film with substantial cracking (10-μm islands) and peeling due to important shrinkage. Cyclic voltammograms in $LiClO_4$–MeCN electrolyte showed a blue coloration with a fast coloration (~6 s) and bleaching (~3 s) kinetics and a 6 cm^2/C coloring efficiency. However the durability of the electrochromic response was only a few cycles. The quality of the film has been slightly improved by adding a trialkoxysilane (Glymo) to the precursor sol in order to obtain a Nb–Si Ormocer. Recently Faria and Bulhões [67] have prepared niobium pentoxide films 2.8-μm thick dip-coated on ITO glass from a sol made by dissolving 40 wt% citric acid in 60 wt% ethylene glycol at 60 °C to which was added $NH_4H_2\{NbO(C_2O_4)_3\}3H_2O$. Under Li insertion the films showed a blue color with a transmission change of 82 to 30% at 631.8 nm (5 mC/cm^2).

Colloidal Nb_2O_5 sols have been prepared in Aegerter's laboratory [68, 71–74] using an alkoxide route. Pentabutoxide of niobium ($Nb(OBu^n)_5$) was first synthesized following the Na process described by Mehrotra [138] for other metals. This precursor was then mixed with glacial acetic acid (CH_3COOH) with molar ratio 1/2 resulting in a sol stable at room temperature for several months. The size distribution of the sol particles measured in a light scattering experiment has a z average mean of 16.7 nm. The films had an excellent microstructure with no cracks and defects even at the microscopic scale. They were found to be amorphous up to ~520 °C and after Li^+ insertion their color was brown. At 560 °C the material is crystalline and its structure was identified as the TT phase corresponding to $Nb_{16}O_{38}X_4$ or $Nb_2(OX)_{5+n}$ in which some oxygen atoms have been replaced by other monovalent species such as OH^-, Cl^-, vacancies, etc. The electrochemical properties, tested with 250-nm thick films up to 2000 voltammetry cycles indicate that the cycles are reversible with a maximum charge inserted of 22 mC/cm^2 corresponding to a change in transmission of 80 to 20% in the optical range 400–1200 nm. The color of the coating is deep blue, similar to WO_3 (Fig. 8). No variation of the charge has been noted. The kinetics is rapid, of the order of a few seconds for bleaching and coloring. Protons were also inserted but the lifetime of the coatings did not exceed a few hundred cycles due probably to corrosion problem at the interface EC layer-electrolyte. These new sol-gel coatings of excellent optical and electro-

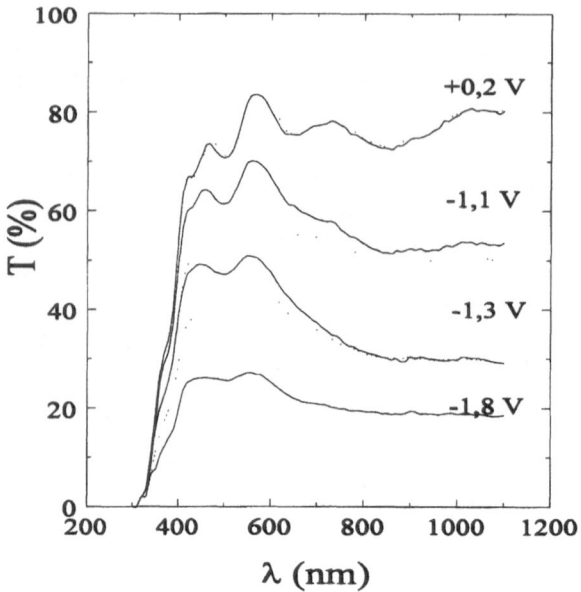

Fig. 8. Optical transmission spectrum of Nb_2O_5 thin film heat treated at 560 °C for 2 h in O_2 atmosphere and measured after different step potentials during a complete voltammetry cycle [68, 71– 74]

chemical quality and stability appear quite adequate as substitute WO_3 coatings.

Ozer et al. [69] have also prepared Nb_2O_5 coatings by hydrolysis and polymerization reactions of polymeric niobium ethoxide solution in ethanol with addition of acetic acid. The films were prepared by a spin coating technique and then heat treated at 150 °C only, so that their structure was amorphous. The cathodic and anodic charge obtained at a sweep rate of 20 mV/s were approximately 20 mC/cm² for Li insertion and the films were found quite stable over 1000 cycles without significant change or degradation. As the films were amorphous their color was brown and the change in transmittance was only of the order of 30% in the visible range, with rather slow kinetics for coloring (30 s) but a fast rise time for bleaching (3 s). Such films are proposed as counter electrodes in anodically coloring EC nickel oxide devices.

4.3.6 CeO_2 and CeO_2 Doped with TiO_2 or SnO_2

CeO_2. Pure and doped CeO_2 coatings are promising materials to be used as optically passive counter electrodes (ion storage). This compound has two stable valencies available (III and IV) and is transparent in the spectral range 0.35–30 µm.

Only three works have been reported on their preparation. Atkinson and Guppy [139] prepared pure CeO_2 gel coatings from cerium hydroxide which were peptized with HNO_3 with a nitrate to ceria mole ratio of 27:100. However the authors only reported on the mechanical stability of the films in relation to crack formation. Recently Stangar et al. [40] have prepared pure CeO_2 sols by thermal decomposition of $Ce(NH_4)_2(NO_3)_6$. Peroxycomplexes were made by adding H_2O_2 to the yellow solution of the starting compound. A brown Ce(IV) complex decomposed into a yellow gelatinous precipitate which was used after peptization with HNO_3 for making very stable sols and homogeneous, crack free coatings with a particulate texture. The films heat treated at 520 °C are crystalline but exhibit a more amorphous structure when heat treated at 300 °C. The electrochemical properties show a high reversibility for Li^+ insertion (better then CeO_2–TiO_2, see below), stable after five cycles.

The total charge exchanged during cycling was found to be dependent on the thickness of the coatings and varied from 1 mC/cm^2 for a 25-nm thick film to 9 mC/cm^2 for a 250-nm thick coating (measured after 70 s). The interesting thing about these films is that they show only a coloration in the UV region typically below 370 nm (Fig. 9). It was claimed that their storage capacity was superior to CeO_2–TiO_2 coatings but their time response was inferior (see below).

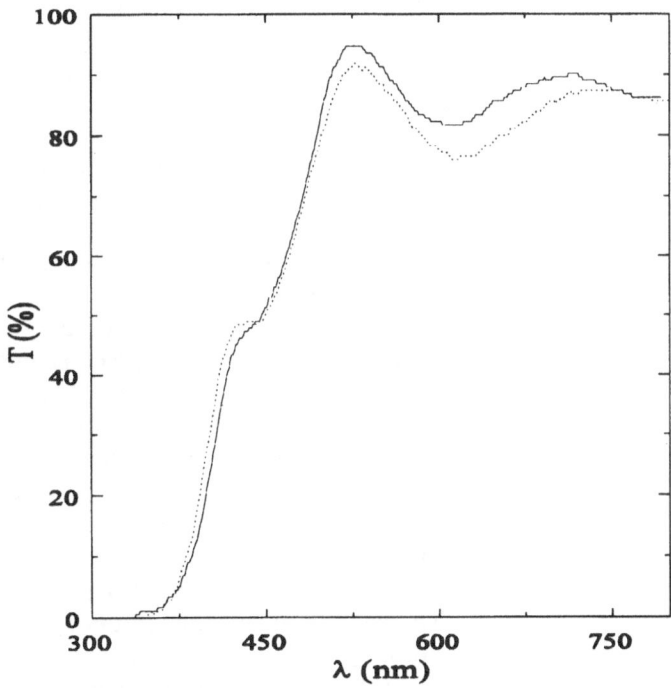

Fig. 9. Optical transmission of the system glass/ITO/CeO_2–TiO_2(sg) (——) before Li^+ insertion, (– – –) after Li^+ insertion [44, 46, 47]

Much better results have been reported by Orel and Orel [41] using sols made with the same precursor $Ce(NH_4)_2(NO_3)_6$. In this method precipates have been obtained by addition of NH_4OH at pH = 9. After washing with bidistilled water in order to remove NH_4^+, Cl^-, and NO_3^-, peptization was performed by adding an equimolar quantity of HNO_3 to obtain a colloidal sol. Coatings have been obtained by a dip coating technique followed by heat treatment at 300–500 °C. The authors report that the amount of charge exchanged increases with the CeO_2 oxide concentration in the sol and the thickness of the films (up to 8 dips corresponding to 280 nm). The highest value was about 20 mC/cm^2.

CeO_2–TiO_2. These compounds have similar (but better) electrochemical properties than pure CeO_2. Their preparation was first reported by Baudry et al. [43] and a series of other papers [44–48]. The sols have been prepared by dissolving cerium amonium nitrate, $Ce(NH_4)_2(NO_3)_6$, and titanium isopropoxide in ethanol, isopropanol or 2 methoxy-ethanol. The sol had to be aged for about seven days. In order to get the best electrochemical activity the dip coated film had to be fired at 450 °C. At this stage the structure of the films was essentially amorphous with the presence of small CeO_2 crystallites. Similar films obtained from sols prepared by mixing two alkoxides, $Ce(OBu^s)_4$ and $Ti(OBu^n)_4$, have later been reported by Kéomany et al. [49]. In this case the coating were found to be amorphous for concentrations of CeO_2 below 50% and the size of the CeO_2 nanocrystallites included in the TiO_2 amorphous matrix increases from ~ 1 nm for 5% CeO_2 to 5 nm for pure CeO_2 [49]. Contrary to the work of Stangar et al. [40], the best electrochemical Li^+ insertion has been found for an equimolar mixed oxide [43, 49] with a total charge exchanged at 50 mV/min as high as 20 mC/cm^2 [49], almost a factor of two higher than that claimed by Macedo [50]. This is probably due to the fact that Kéomani et al. have prepared films with extremely small particles (or sites or hanging bond or defects structure which can receive the Li^+ ion), and the amorphous porous TiO_2 matrix probably appears as a more continuous structure for the Li^+ ions to reach the surface of each crystallite or site.

CeO_2–TiO_2 coatings have also been prepared by Stangar et al. [40] and Orel and Orel [41] using the route proposed by Makishima et al. [140] in which $CeCl_3$ was used in combination with titanium isopropoxide $Ti(OiPr)_4$ with a Ce/Ti mole ratio of one to one. Their overall properties were similar to those found by the Brazilian and French groups but their electrochemical properties were inferior. No comments about the CeO_2 nanoparticles have been reported.

CeO_2–SnO_2. Along the same direction Orel and Orel [41] have reported the preparation of SnO_2-doped CeO_2 coatings using sols made with $Ce(NH_4)_2$ $(NO_3)_6$ and $SnCl_4$. The method of sol preparation is similar to that described above by the same authors for pure CeO_2. As before, those films are not colored in the visible region and significant improvement of charge transfer was obtained with films prepared with 17 mol% SnO_2, varying from 10 mC/cm^2 for a 60-nm thick film to 22 mC/cm^2 for a 280-nm thick film (eight coatings) to be compared to 4 mC/cm² and 16 mC/cm² for pure CeO_2 coatings. However the charge capacity was found to decrease slightly with the number of cycles (up to 400). The total charge exchanged was also found to depend on the CeO_2 concentration in the sol. With a coating of eight films it typically varied from 13 mC/cm² for a concentration of 4.8×10^{-3} mol/20 ml to 20 mC/cm² for a concentration of 9.6×10^{-3} mol/20 ml.

All these sol-gel coatings prepared with doped CeO_2 now look very promising for use as a counter electrode for any EC all solid state devices. The possibility of controlling the various parameters at our disposition during the preparation of the sols or films has practically triplicated the amount of change since the original work of Baudry et al. [43].

4.3.7 Fe_2O_3 and Fe_2O_3 doped with TiO_2 or SiO_2

FeO_3. Only three recent works have been reported on the preparation and the characterization of iron oxide electrochromic coatings. Moser and Lynam [91] have used ferric nitrate-ethyl acetate solution and recently Orel et al. [92, 93] have precipitated $FeCl_3 \cdot 6H_2O$ with $(NH_3)_{aq}$ followed by peptization of the precipitate with glacial acetic acid. Amorphous γ-Fe_2O_3 films obtained at 300 °C show electrochromic properties under Li^+ ions insertion (anodic coloration, $\Delta T = 60\%$ at 300 nm) but their transformation into α-Fe_2O_3 at 500 °C turns them inactive. The presence of OH^- in the coatings and small grain size particles were found of fundamental importance to develop electrochromism. The adherence of the films to the electronic ITO coating was poor and the mechanical stability deteriorated with cycling.

$Fe_2O_3 : TiO_2$ and $Fe_2O_3 : SiO_2$. The electrochemical stability of the iron oxide was modified by admixing (nonabsorbing Ti and Si oxides prepared with the $FeCl_3 \cdot 6H_2O$ precursor in combination with 3-aminopropylmethoxysilane (3-APMS) or ferric (III) nitrate alcohol solution in combination with $Ti(^iPr)_y$, respectively by Orel et al. [92, 93]. Mixed Fe/Ti and Fe/Si oxide films color and bleach in a narrower spectral range (300–400 nm) than pure Fe_2O_3 and the optical transmission variation drops to 15–30%. However the mixed oxides exhibit extremely large charge intercalation (up to 60 mC/cm²), much higher than most of the known electrochromic coatings. This property makes these coatings interesting for use as counter electrodes (ion reservoir).

4.3.8 SnO_2

Olivi et al. [94] developed a new method for the synthesis of SnO_2 by using 50 wt% citric acid in ethylene glycol at 60 °C to which was added tin citrate in a 3/1 molar ratio (acid/tin) and a few drops of concentrated nitric acid in order to solubilize tin citrate and catalyze the esterification reaction performed at

110 °C. The films, calcined at 500 °C, are highly transparent in the 350–800 nm range, homogeneous and crystalline. No optical changes have been observed either in the oxidized or reduced forms using Li electrolyte. The charge associated with the process at a sweep rate of 10 mV/s was however very low, 4 μC/cm^2, but such a system also appears as an attractive material for transparent counter electrodes in transmissive EC devices using non-aqueous electrolytes.

4.4 Sol-Gel Conductive Coatings

Little work has been done on using the sol-gel process to obtain electronic and ionic conductive coatings with explicit functions for EC devices. In these systems the electronic conductors have the function of making electric contact between the external source and the device and should be able to insert and extract electrons from the EC coatings. However these coatings should also present a combination of optical, mechanical, chemical and aesthetic properties which have to be fulfilled at the same time. For instance, for EC devices used as energy saving windows, it is mandatory to have high IR reflection (9.5 μm), high transmission for solar and visible light energy and long term stability.

The electrolyte which is in direct contact with the EC layer (Fig. 3c,d) should exchange ions such H$^+$, Li$^+$, Na$^+$, etc. with it. Depending on the configuration of the device it may also serve as the ion storage (Fig. 3b, c) or unite the EC layer to the counter electrode (ion storage coating) and allow only the ions to migrate through it (Fig. 3d). To fabricate EC devices with good response times, the ionic conductivity should be higher than 10^{-7} S cm^{-1} and at least 10^{-4} S cm^{-1} for display devices.

4.4.1 Electronic Conductive Coatings

ITO. Several years ago Arfsten et al. [98, 99] developed sol-gel indium tin oxide (ITO) coatings prepared by dip coating with heat treatment between 400 and 500 °C in reduced atmosphere, to be used either alone as heat mirror coatings or in EC devices. The authors did not however disclose their preparation method but only their physical properties which are given in Table 5.

More recently Takahashi et al. [100] got similar results using a mixture of tin isopropoxide and indium acetate precursors stabilized with diethanolamine (HOCH$_2$CH$_2$)$_2$NH, or a mixture of indium alkoxide dissolved in isopropanol stabilized by triethanolamine to obtain In$_2$O$_3$ coatings and ITO coatings when adding Sn isopropoxide. This stabilizer was found to suppress the alkoxide hydrolysis leading to stable sols without the occurrence of gel formation and precipitation and to enhance the solubilities of alkoxides and even metallic acetates into alcoholic solvents.

Table 5. Optimized properties of ITO sol-gel for application in displays coatings with different thicknesses (a), (b), and (c) for windows

Properties	(a)	(b)	(c)
Film thickness (nm)	20–30	80–100	270
Density of carriers (10^{20} cm^{-3})	4	5–6	5–6
Mobility (cm^{2}V^{-1}s^{-1})	15–20	60–70	35–40
Conductivity (S cm^{-1})	800–1200	5000–6000	3000–3500
Sheet resistance (Ω cm^{-2})	500	~25	10

SnO$_2$. Sn(IV) oxide is a wide gap semiconductor (Eg = 3.97 eV) with transmittance cut-off at 335 nm. When doped with F, Sb or Mo it becomes electrically conductive while its optical solar transmission is invariant (85–95%). Due to the presence of free charge carriers introduced by doping, the films start to reflect thermal infrared radiation for $v > 2000$ cm^{-1}. Because of these properties SnO$_2$ has a widespread use in various fields.

Different organic and inorganic precursors have been used for making SnO$_2$ powders and coatings via the sol-gel route. Up to now more studies have been devoted to the alkoxide route using Sn and Sb alkoxides [101–105, 107, 141]. Yet the drawback of the method is that alkoxides are expensive, extremely sensitive to heat, moisture and light and their preparation is time consuming.

The inorganic route relies on the mixing of soluble salts of Sn, such as SnCl$_4$ and SnCl$_5$, in combination with Sb salts such as SbCl$_3$ and SbCl$_5$. Gels of SnO$_2$ prepared entirely via the inorganic sol-gel route were reported by Goodman and Gregg [108]. Undoped SnO$_2$ gels were also made by Giesekke et al. [109] and recently by Hiratsuka et al. [106] who succeeded in decreasing the time for gelling the sols by peptizing the precipitates with NH$_4^+$ and OH$^-$ ions. They also found that an excess of Cl$^-$ ions retarded the gelling time. SnO$_2$–Sb$_2$O$_5$ semiconducting glaze [111] was also made from SnCl$_4$ and SbCl$_5$ precursors, using H$_2$O$_2$ in combination with an alcohol as solvent for the preparation of Sn-based gels.

Maddalena et al. [107] reported electrically conductive Zr- and Ti-doped SnO$_2$ films prepared from the SnCl$_2$ precursor in combination with Zr(OC$_3$H$_7$)$_4$ and Ti(OC$_4$H$_9$)$_4$. The alcoholic solution of the precursor underwent gelation when left in an open vessel for 5 days at 30 °C.

Orel et al. [110] have prepared an undoped SnO$_2$ thin solid coating by dip-coating techniques using an aqueous SnCl$_4$. 5H$_2$O precursor peptized with (NH$_3$)$_{aq}$. Sn-doped coatings have been achieved by adding SbCl$_3$ to the sol in concentrations of 1–10 mol%. Analysing the plasma frequency determined by FTIR reflection spectroscopy, they reported that the concentrations of free carriers in their sample were smaller compared to those obtained with SnO$_2$ prepared by spray pyrolysis or RF sputtering, despite the large concentration (10%) of the dopant species introduced. The films have a visible near infrared transmittance of about 75% and exhibit, when measured without post heat treatments, (i.e. as deposited), a conductivity of about 83 S cm^{-1}. However it decreases with the film thickness because the crystallization is favored at greater coating thicknesses. The dip coating process could be repeated up to 20 times.

V_2O_5. V_2O_5 films have also been proposed as conductive coatings. Pozarnsky et al. [112] have obtained thin films from sols prepared by ion-exchange of sodium metavanadate solution. They found that their thickness and surface morphology were directly related to the age of the sol. The last property was found to change from that of a featureless surface after one day to a continual coverage of micron-sized fibers as the sols aged. The conductivity of the coating, which is of the order of 2 to $3 \times 10^{-4}\,S\,cm^{-1}$, was unaffected by the aging process.

4.4.2 Ionic Conductive Coatings

An ionic conductor is an electrolyte which is able to conduct ions such H^+, Li^+, Na^+, K^+, etc. In EC devices H^+ and Li^+ are most used as they have a higher mobility and Li^+ ionic conductors are preferred because of their low corrosive effect at the adjacent coatings. For most EC devices the electrolyte must be transparent and preferentially of solid state to avoid any leakage or pressure gradient observed with liquid ones. Although the literature on ion conductors is vast (several reviews on inorganic ion conductors suitable for EC devices and other applications can be found in [7] and [86]), only a few works have been produced on sol-gel process and the description of their properties in EC devices is rare. In fact, to date, all commercial devices use liquid or polymer type electrolytes. This is essentially due to their position in the EC devices (Fig. 3) which makes fabrication of all solid state devices rather difficult, as the chemistry and heat treatments required to place one more solid state layer and maintain the interface and the structure of the underlying layers (EC electronic conductor/glass and storage coating/electronic conductor/glass) become rather restrictive. In addition one would have generally to charge one of the electrodes with the ions before completing the stack and ensure that the nature of the bonds between these ions and the matrix is not altered or the ions do not migrate during further processing. A major limitation with the sol-gel process, which has been observed and yet not overcome, is the inability to produce a material that can solidify without releasing volatiles or that does not change either through aging or during the cycling process over long periods of time. Usually reactions, such hydrolysis and condensation, continue after device assembly and separate the interfaces with the formation of bubbles.

According to Grandqvist [142] the requirements for such coatings should be:

(i) high ionic conductivity, between 10^{-4}–$10^{-7}\,S/cm$ depending of the application;

(ii) electronic conductivity smaller than 10^{-12} S/cm;

(iii) long cycling durability at operation temperature;

(iv) good adhesion with the adjacent layers and optical transparency in the spectral range of interest.

Most of the research in this field has been realized in order to obtain sol-gel conductive coatings for H^+ and Li^+ to which we shall restrict our discussion. In both cases there are basically two classes of materials which may serve for this purpose. The first one is based on the fabrication of inorganic oxide material and the other, more promising from our point of view, on the preparation of organic-inorganic hybrids which combine the better conductive properties of polymer type material with the better mechanical strength of inorganic backbone. Also, from the point of view of the fabrication of complete devices, the Ormocer type coating will be easier to use. However to our knowledge none of these coatings are presently employed in commercial products which are all built with liquid electrolytes.

H^+ *Conductor.* The only inorganic H^+ conductor coating which has been described for such applications is Ta_2O_5. Ling et al. [116] have reported the preparation of spin-coated Ta_2O_5 thin films, pure or doped with Al, using the alkoxide route (Ta ethoxide). The films have been studied with heat treatment up to 1200 °C. As their goal was to improve the storage of dielectric capacitor, they did not report any conductivity data.

Recently Ozer et al. [117] have prepared Ta_2O_5 from the same route using a mixture of tantalum ethoxide $(Ta(OC_2H_5)_5)$, acetic acid and water. Spin coated films at 2500 rpm on ITO glass or ITO coated with WO_3 on glass have then been heat treated at 150 °C for 1 h. The preparation of thick films required 10–15 repetitions of the above procedures, and the films were found to be hard, durable and stable. From a.c. impedance and electrochemical measurements, the authors conclude that the sol-gel films have higher ionic conductivity and better properties than Ta_2O_5 films obtained by other methods. Depending on the deposition conditions, ionic conductivities as high as 5×10^{-6} S cm^{-1} have been obtained, i.e. almost a factor two higher than RF sputtered films. This increase in conductivity is thought to come from the fact that the sol-gel films are more porous (packing density $\rho = 3.2$ g/cm^3) than the films deposited by other techniques, thus providing a larger surface area for absorbing the protons. Tests performed on half an EC window glass/ITO/WO_3/Ta_2O_5 (190-nm thick) with a H_2SO_4pH = 2 solution as electrolyte resulted in strong transmission variation between the colored and bleached states. The photopic weighted change and the solar weighted transmittance change are $T_p = 85.1\%$–21.9% and $T_s = 75.6\%$–14.2% respectively. The voltammograms do not present degradation up to 150 cycles.

Charbouillot et al. [120] were the first to propose a promising new family of advanced materials which they called AMINOSILS. These compounds were prepared by hydrolysis and condensation of organo-tri-alkoxide-silane

R'Si(OR)$_3$ (R' = amino group) in presence of a strong mineral acid – HClO$_4$, HCF$_3$SO$_3$, HCH$_3$SO$_3$, HCl, HNO$_3$, HCH$_3$COO or H$_3$PO$_4$, and water. The silica backbone provides mechanical strength while the organo amino groups offer solvation properties with respect to guest ionic species. Non-porous films about 10 μm thick were found to be stable in atmospheric atmosphere and up to 180 °C. At room temperature their conductivity is about 10^{-5} S cm^{-1}.

At the same time Judeinstein et al. [121] have proposed an organic modified TiO$_2$ gel made from a sol prepared from a mixture of acetic acid and pure Ti(OBu^n)$_4$ to which was added glycerol (CH$_2$OHCH–OHCH$_2$OH) in a molar ratio glycerol/Ti ranging between 12 and 30 to 1. During the reaction it is believed that the metal alkoxide reacted with the glycerol forming organic bridges between Ti atoms. The hydrolysis of the alkoxy group was performed in excess water. The resulting viscous gel was stable up to 80 °C. The conductivity was deduced from the intercept of the arc of a circle with the real axis of the a.c. electrical impedance plotted in the complex plane and was found to follow an Arrhenius behavior $\sigma T = \sigma_0 \exp(-Eg/kT)$ in the temperature range -40 to $+60$ °C. Room temperature conductivity varied between 3×10^{-6} and 10^{-5} Ω^{-1} cm^{-1}. It increases with the amount of acetic acid while the activation energy increases with the amount of glycerol. This coating was tested with satisfactory results in a complete sol-gel EC window (see Sect. 4.5).

A new proton-conducting polymer electrolyte, poly (benzylsulfonic acid) siloxane (PBSS) was developed by a sol-gel process by Sanchez et al. [122]. The hydrolysis and condensation reactions of triethoxybenzylsilane were studied under different catalytic conditions. PBSS was found to be thermally stable up to 300 °C in air and high conductivities ranging from 2×10^{-3} to 10^{-2} S cm^{-1} at room temperature were measured by a.c. impedance spectroscopy. The same laboratory [143] had also previously developed a proton-conducting polymer based on the copolymerization via the sol-gel process of sulfonamide-containing groups partially deprotonated and an internal plasticizer (polyethylene oxide (I) segments). All the organic groups are attached to trialkoxysilanes which act as the silica-based backbone. The films were found to be flexible, homogeneous, transparent, thermally stable up to 220 °C and electrochemically stable up to 2 V. The highest conductivities, 2×10^{-7} S cm^{-1} at 30 °C and 10^{-5} S cm^{-1} at ~ 84 °C, have been obtained with $\sim 15\%$ sulfonamide deprotonated groups and $\sim 10\%$ of (I) plasticizer (weight average molecular weight 2000).

Another interesting approach, discussed earlier (Sect. 4.3.3), has been reported by Stangar et al. [59] who developed a combined electrochromic ion conductor based on phosphotungstic acid (PWA) doped TiO$_2$ gels, which already contained protons. Using a.c. impedance spectroscopy the authors found that, at room temperature, PWA/Ti gels have a higher proton conductivity ($\sigma = 1.7 \times 10^{-4}$ S cm^{-1}) than pure TiO$_2$ gels ($\sigma = 4.6 \times 10^{-6}$ S cm^{-1}) or even aminosils [120]. Higher values ($\sigma = 10^{-2}$ S cm^{-1}) were found with PWA/Si gels by Tatsumisago et al. [144] and for water soluble PWA doped polyvinyl alcohol (PA) gels (σ up to 9.8×10^{-2} S cm^{-1}).

Stangar et al. [59] have also prepared silica gel from 3-aminopropyl-triethoxysilane with $R = HPA/Si = 0.1$. Their conductivity was about 2.2×10^{-5} S cm^{-1}, higher than that of aminosils doped with acid having small ions. The authors claimed that the small polarizability of the large Keggin's ion decreases the influence of ion-ion interactions leading to an increase in proton conductivities.

Li$^+$ conductor. Lithium ion conducting amorphous solid state inorganic materials such as glasses and glass ceramics are materials to be considered for solid state batteries and related EC devices. They exhibit isotropic ionic conductivity but their preparation by conventional melting techniques requires high temperatures which limits the alkali concentration. Therefore the preparation of fast ion Li or Na conductors by the sol-gel process is certainly an asset. This field has recently been reviewed by Boilot and Colomban [145], Klein et al. [146] and Klein [147].

Earlier works on these materials essentially concentrated on the preparation process itself. It was found that the lithium gels were more stable than their sodium and potassium counter-parts and that the glasses and glass-ceramics derived from lithium silicate gels are more stable and more durable in corrosion tests. Among the different compositions studied, the family of sodium zirconium-silico-phosphate gels known as Nasicon (Na Super Ionic Conductor) has been prepared from a mixture of alkoxides [145], and room temperature conductivities as high as 10^{-6} S cm^{-1} for gels and up to about 10^{-3} S cm^{-1} for glasses have been achieved. In the last case the conductivity was found to be similar to those of equivalent polycrystalline ceramics with, however, a higher activation energy.

More promising results seem to have been obtained in systems containing other network formers such as P_2O_5 [148–150], Al_2O_3 [151], Ga_2O_3 [152] and V_2O_5 [153].

Hayri and Greenblatt [148, 154] have studied the fabrication of Li_2O (Na_2O)-SiO_2-P_2O_5 xerogels and found that the conductivity increases with the concentrations of both lithium and phosphorus, obtaining relatively high values up to 2×10^{-3} S cm^{-1} at 300 K. However the conductivity was found to be associated with protons which always exist in different forms in xerogels (surface hydroxyls, hydrogen bonded water, etc.) as the process is carried out in aqueous media. Nevertheless several studies have been performed on these systems in order to understand their chemistry, molecular structure and microstructure (see [147] for references).

Wang et al. [151, 152, 155–157] have studied the ionic conductivities of ($LiCl$)$_2$-R_2O_3-SiO_2 with R = B, Al, Ga xerogels. The materials have been prepared by mixing a solution of TEOS dissolved in methanol with a solution of aluminium nitrate dissolved in methanol to which was added acid and water and then LiCl. Dip coated films ~ 0.8 μm thick have been dried and then heat treated at 300 °C. XPS measurements indicate the presence of residual organics with possible atmospheric carbon contamination and show that the oxygen has three forms: bridging (BO), non-bridging (NBO) and O = C–O–. The highest lithium conductivity (measured by a.c. impedance) was found in the lithium alumino silicate xerogels and was attributed to the relatively large concentration of mobile lithium ions associated with tetrahedral AlO_4 units and/or with non-bridging oxygens in the silica network [151]. Linear behavior was observed in the temperature range from 100 to around $\sim 300\,°C$. The presence of organics or carbon residues is claimed to have an insignificant effect on the ionic conductivity. At temperatures below $100\,°C$, σ is lower than $10^{-7}\,S\,cm^{-1}$, with an activation energy of $0.75\,eV$, a value which is much lower than that of polymers and which is not acceptable for the fabrication of EC devices. However these films have the advantages of a better stability at higher temperature and good mechanical strength.

Solid state batteries of the type Ni/oxide cathode (Li_yMnO_2)/lithium aluminosilicate gel/Li/Ni have recently been realized [158], having initial open circuit voltages up to $3.4\,V$.

In lithium silicate gels, a.c. conductivity is determined by the pore size and density, the water content and the precursor salt in terms of the anion. At room temperature its value is low, of the order of $10^{-7}\,S\,cm^{-1}$ but increases with temperature according to an Arrhenius law (as the process is thermally activated).

V_2O_5 has also been tested by Trifonova [159] to be used as an electrode material for solid state rechargeable lithium batteries. The sol was made using a decavanic acid solution prepared by passing a $0.25\,mol/l$ sodium vanadate solution through a cation (H^+) exchanger in order to eliminate completely the Na ions. The sol polymerized spontaneously on standing. The gel was spread manually on Ni foil and dried. Composite electrodes of $V_2O_5 \cdot nH_2O$ and PEO were also prepared to improve the mechanical properties of the films. The electrochemical characterization was performed in an all solid state cell with a polymer as electrolyte $Li/PEO–LiCF_3SO_3/Li_xV_2O_5 \cdot nH_2O$. The maximum energy density was $460\,Wh/Kg$ obtained for $Li_{1.35}V_2O_5 \cdot 0.5nH_2O$, a modest value, and the cycling was found to be fully reversible only if all the loosely bound water is eliminated by heat treatment (dry gels). Up to now none of the inorganic systems have been used for EC devices.

The most promising low temperature Li^+ ion conductors materials seem to be lithium doped ORMOCER. In a recent study ORMOLYTES (ORganically MOdified electroLYTES) have been prepared and characterized by diverse techniques by Judeinstein et al. [160] from mixtures of tetraethoxysilane, tetraethylene glycol (PEG_n) and lithium salt $(LIClO_4)$. Transparent monoliths and films have been obtained. Their structure was found to be diphasic with silica clusters (observed by SAXS) providing the mechanical properties and with an organic phase allowing the dissolution of large quantities of salt. Their thermal stability was checked between -25 and $+95\,°C$. The values of the ionic conductivity measured at $300\,K$ in anhydrous materials treated in vacuum and after heat treatment increased drastically with the amount of lithium salt and varied between 10^{-8} and $6 \times 10^{-5}\,S\,cm^{-1}$, much larger than that of the pure matrix ($< 10^{-10}\,S\,cm^{-1}$), proving that the conductivity came from the dissolved salt. As a function of temperature, deviation from an Arrhenius law has been observed revealing two different slopes and the activation energies were found to be between 0.5 and 0.9 eV.

In earlier work, Judeinstein et al. [121] incorporated $LiClO_4$ in the modified hybrid TiO_2 gel. This material, which is already a good proton conductor, exhibits even higher conductivity with lithium ions, up to $4 \times 10^{-4}\,S\,cm^{-1}$ at room temperature. The process is also thermally activated (E = 0.45 eV).

4.5 Sol-Gel Electrochemical Devices

A few groups have developed EC devices in which either some or all active layers have been prepared from gels. The first windows were reported by Judeinstein et al. [121] for which both the EC layer and the electrolyte were made from gels. They had the following configuration: non-symmetric $SnO_2/WO_3(sg)/TiO_2$–$LiClO_4(sg)/SnO_2$ or symmetric always colored $SnO_2/WO_3(sg)/TiO_2$–$LiClO_4(sg)/WO_3(sg)/SnO_2$. The non-symmetric window was initially transparent. The main characteristics are given in Table 6.

Various small size windows have been mounted in Aegerter's Advanced Materials Laboratory to test some of the sol-gel coatings made there. The CeO_2–TiO_2 storage layer was tested in a cell ITO/Li_xWO_3 (evaporation)/POE–$LiN(CF_3SO_2)_2/CeO_2$–$TiO_2(sg)/ITO$ showing a transmission from 60 to about 20% (at 750 nm) and reasonable kinetics, limited by the insertion process of Li^+ into the CeO_2–TiO_2 layer [45, 48, 50] (Fig. 10). Similar results have been obtained substituting the polymer electrolyte by a cellulose polyacetate polymer, a proton conductor. Its kinetics were faster but the window lifetime was shorter due to corrosion problems at the WO_3 coating by the acid (protonic) conductor [46, 50, 161]. Sol-gel TiO_2 protonic conductors similar to that proposed by Judeinstein et al. [121] have been tested in symmetric and non-symmetric windows of configuration [44, 47, 50]:

1) ITO/CeO_2–$TiO_2(sg)/TiO_2(sg)/CeO_2$–$TiO_2(sg)/ITO$;

2) $ITO/H_xWO_3(sg)/TiO_2(sg)/CeO_2TiO_2(sg)/ITO$;

3) $ITO/H_xWO_3(evap.)/TiO_2(sg)/CeO_2$–$TiO_2(sg)/ITO$.

The first window does not present any coloration (see Sect. 4.3.6) and no degradation has been observed up to 30×10^3 voltammetry cycles. The optical transmission of the second one can be seen in the first and 360th cycles (Fig. 11) where the effect of the corrosion at the $WO_3(sg)/TiO_2(sg)$ interface is clearly observed. During the first cycles the performance of the windows was found to be similar to those made by conventional processes. Better results should however be obtained by using Ti butoxide instead of Ti isopropoxide for the preparation of the electrolyte with rigid control of the amount of water in the preparation of the WO_3 (sg) layer.

Table 6. Characteristic of all-gel electrochromic cells (6 cm²) corresponding to a change in optical density of 0.8 [121]

	Non-symmetric	Symmetric
Coloration voltage	− 3.5 V	− 2.5 V
Response time	50 s	50 s
Bleaching voltage	+ 2.0 V	+ 2.5 V
Response time	50 s	50 s
Lifetime (cycles)	$> 3 \times 10^3$	$< 4 \times 10^4$
Open circuit	4 months	> 6 months
Memory		many months

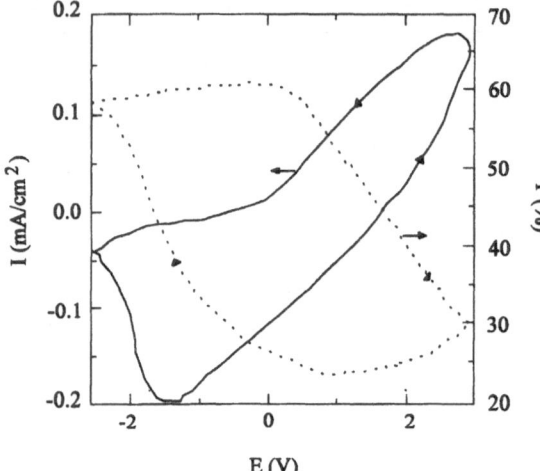

Fig. 10. Opto-electronic cyclic voltammetry of an EC window $ITO/Li_xWO_3(evap)/POE\text{-}LiN(CF_3SO_2)_2/CeO_2\text{-}TiO_2(sg)/ITO$ measured at $\lambda = 750$ nm [45, 48, 50]

Ozer et al. [56] have built an "all-gel" device with the configuration $ITO(sg)/TiO_2(sg)/electrolyte(gel)/ITO(sg)$ in which the electrolyte was a gel prepared from poly (vinylbutyral), glycol ether and LiCl. The response time was ∼ 50 s under applied voltage of ± 2.8 V with open circuit memory as long as 5 months.

The performance of a transmissive EC device with the configuration $ITO/WO_3/H_3PO_4$-polyvinyl alcohol (PVA)/SnO_2:Sb(sg), where the last layer acts as an ion storage and conductive electrode, has been briefly reported by Orel and Orel [41], who reported large optical transmission variations mainly in the near-infrared region. The coloration kinetics were adequate but the bleaching kinetics lasted from 1 to 3 min.

Stangar et al. [59] have also tested a thin PWA/Ti xerogel film in a semi-liquid electrochromic EC cell with reflectance modulation. The gel was found to change color irreversibly from transparent yellow to blue. Reversibility was achieved by putting a thin Ag film acting as a counter electrode in contact with

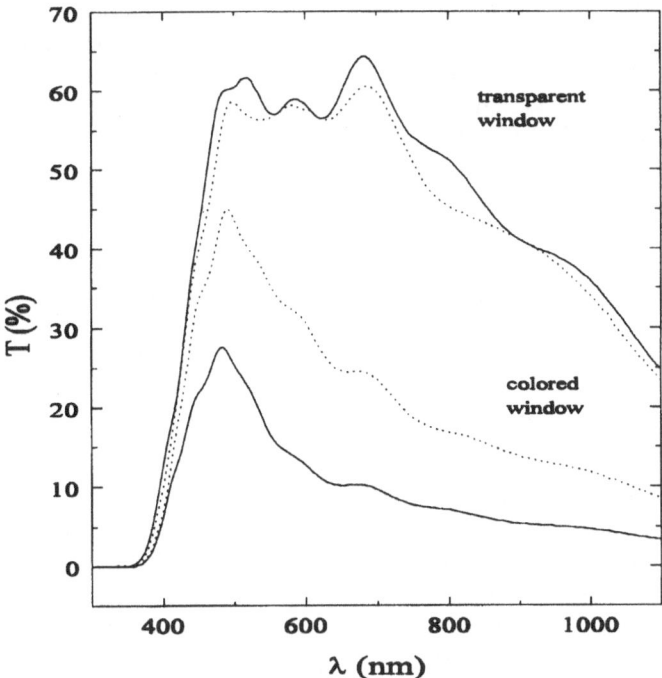

Fig. 11. Optical transmission of an EC window ITO/H$_x$WO$_3$(sg)/TiO$_2$(sg)/CeO$_2$–TiO$_2$(sg)/ITO measured during the first cycle (——) and the 360th cycle (– – – –) [44, 47, 50]

one of the ITO electrodes. The electrochromism behavior of PWA/Ti solid films was verified in an all solid state device ITO/PWA:Ti(sg)/H$_3$PO$_4$–PVA/ SnO$_2$:Sb(sg). Reversible response was reported with well defined anodic and cathodic peaks. However the transmittance changes were smaller than 15% due to the small thickness (200 nm) of the PWA/Ti EC film into which only 4 mC/cm^2 could be inserted. The reversibility and cyclability were claimed to be satisfactory.

Other systems involving at least one sol-gel layer can be found in [162, 163]. Developments are also being made in different industrial laboratories but no public disclosures are yet available.

The fabrication of all solid state EC devices is quite a difficult task. Nevertheless several sol-gel films have been shown to have adequate properties to be used for such a purpose. CeO$_2$–TiO$_2$ layer (which was also prepared recently using conventional techniques) looks promising as a storage coating but its charge capacity should be slightly improved. Nb$_2$O$_5$ is a promising EC layer but WO$_3$ coating is today the best candidate from the scientific and technological points of view and recent results indicate that it will be used soon for EC rear-view mirrors in automobiles. The main challenge remains the development of an adequate Li$^+$ solid state electrolyte. With our present knowledge inorganic glasses or glass-ceramics still have too low ionic conductivity and cannot be used in EC devices which have to tolerate temperatures as low as $\sim -20\,°C$. They can be useful for other applications such as high temperature batteries. The best systems to be contemplated are hybrid organic-inorganic materials (ORMOCER, ORMOSIL or ORMOLYTES). The results obtained up to now are promising but need to be improved. Another important problem which is rarely mentioned is the physical and chemical compatibility of these different materials at their interface and research is needed in this direction, bearing in mind parameters such as long term corrosion and cyclic durability and the large temperature variation to which EC devices will be submitted (-20 to $100\,°C$).

5 Materials for other Chromogenic Devices

As mentioned in the introduction, besides electrochromic materials, other materials present chromogenic properties which can be used to develop devices for optical transmission modulation with regards to luminous and solar radiation. The sol-gel researches in the field of *photochromism, thermochromism and electrooptics* are relatively recent and the sol-gel method looks very promising and particularly well adapted to incorporate molecules or nanoscale crystallites in an inorganic or mixed inorganic-organic matrix which can change their optical properties by light irradiation, temperature variation or application of an electric field respectively.

5.1 Sol-Gel Photochromic Materials

The feasibility of incorporating a photochromic dye (Aberchrome 670) into SiO_2 gels was first reported by Kaufman et al. in 1986 [164]. Later Levy et al. [165, 166] introduced spiropyrane molecules in silica matrices and studied their photochromic behavior during the transition sol → xerogel, showing the importance of the local environment near the molecules on such properties and the effect of the precursor materials used for the preparation of the sols. Analogous results have been obtained by Matsui et al. [167]. Photochromism of spiropyranes has also been investigated in aluminosilicate gels derived from diisobutoxy-aluminoxy-triethoxysilane by Preston et al. [168]. No reverse photochromism was observed and both the photochromic response and the color change rate decreased during the drying process, ceasing after a weight loss of around 75%.

Nogami and Sugiura [169] have studied the effect of the Al_2O_3 content of spiropyrane molecules in Al_2O_3–SiO_2 prepared from alkoxide precursor and found that the photochromism response decreased with increasing Al_2O_3 content. No more response was observed for 33% Al_2O_3.

More recently Yamamaka et al. [170] incorporated 2,3-diphenylindenone oxide (DPIO) into aluminosilicate, silica (TMOS) and ormosil sols prior to gelation. After gelation these materials are colorless. Upon exposure to ultraviolet light the organic molecule isomerizes to form 1,3-diphenyl-2-benzopyrilium-4-oxide (DPBO) and the materials turn red. The interconversion back to the colorless state was facilitated by exposing the molecule to visible light of wavelength greater than 435 nm. The fading of the absorption after blocking the UV irradiation is characterized by long lifetime components (~ 20 s) in all three types of materials but in addition, a short one (~ 5 µs) was found in aluminosilicate samples. All matrices showed a gradual decrease in their lifetime with time (measured up to 120 days after the gel preparation) which, according to the authors, may result from the increase of the polarity in the gel environment during aging.

Photo isomerization of 4-methoxy-4'-(2-hydroyethoxy) azobenzene (MHAB) in SiO_2 gel films was investigated by Ueda et al. [171] and their results have been compared with those obtained using the same dye incorpored in poly(methylmetacrylate) (PMMA) films. The cisfraction in the photostationary state in the sol-gel films was smaller than in PPMA films due to the higher rigidity of the sol-gel matrix. Later Ueda [172] reported on the effects of the introduction of one or two trimethoxysilyl groups (TES) into the AB molecules on the photochromism. The introduction of one TES group has only a slight influence on the photoisomerization and thermal reversion, but the incorporation of two TES groups depressed these parameters in the sol-gel films.

Spirooxazine molecules (SO) were also incorporated in ORMOCER by Hou et al. [173], prepared from organically modified silicon alkoxides $R'Si(OR)_3$ such as methyltrimethoxysilane (MTMS), and 3-glycidyloxypropyl-trimethoxysilane (GPTMS) which have longer GP chains to create more flexible matrix for the dye and give additional polymerization between the epoxy groups. However the gelling time of these compounds is too long (> 1 month). When mixed together these precursors produced an interesting matrix which possesses thermal and photochemical stabilities comparable to SO-doped PMMA material with, however, much better photochromic response and faster color-change rate. However the fast decrease of the light induced absorption with temperature (up to 35 °C) still remains a problem.

Hou et al. [174] have also introduced photochromic dyes in aluminosilicate gels and monitored their photochromic properties during the sol-wetgel-xerogel transformation. The photochromic activity of the aluminosilicate gels decreases rapidly and even vanishes in the wetgel-xerogel stage while that of ORMOCER was found to level off at a reasonably high photochromic intensity which remains almost unchanged for over 4 months. The color-fading speed was, however, found to be similar to that in ethanol but the photostability was considerably improved.

Along the same direction, Judeinstein [175] has initiated research using polyoxymetallates (POM) as precursors prepared by reacting trichloro (or trialkoxy) silane with lacunar $K_4SiW_{11}O_{39}$. The POM are small compact oxide networks about 1 nm size based on the sharing of MO_6 structural unit (M = W, Mo, V...) with a very high electronic density and interesting redox properties. The final material has been modified by adding radical such as vinyl, allyl, styryl, methacryl which have then been linked together via radical polymerization or by hydrosilation reactions between silane and vinyl derivatives or direct reaction of POM with bis (trichlorosilyl) derivatives. Transparent films of 0.3–1 µm thickness have been deposited. They turn blue upon UV irradiation or electrochemical reduction due to the occurrence of a wide absorption band in the visible-IR range characteristic of an intervalence transition W^{5+}–W^{6+} of the polymetallate.

It appears therefore that the synthesis of mixed organic-inorganic materials with a mixing of the components at the molecular scale leads to a great flexibility to reach materials with improved specific properties for all kinds of chromogenic applications and also for catalysis. These advanced compounds, nanocomposites or dye incorporated ORMOCER, open a brand new field with promising prospects.

Finally, to our knowledge, only one piece of work has been successfully carried out with inorganic photochromic materials. Photochromic glass coatings 1.5 µm thick were synthesized by Mennig et al. [176] by infiltration of Ag^+ into a predried Na–Al–B–Si gel layer. The formation of small Ag colloids was initiated by a soft heat treatment and then converted to AgCl crystallites of about 40 nm diameter by a HCl vapor treatment. Such coatings turn brown-violet under Hg-Xe light or Ar laser irradiation due to the formation of Ag crystallites of about 5 nm. The reverse process was realized by heating at 400 °C. It is claimed that no decay was observed after numerous cycles. Holograms have been obtained.

5.2 Sol-Gel Thermochromic Materials

Although several transition metal oxides or sulfides are known to undergo a sudden metal-to-nonmetal transition over a discrete temperature range [177] with pronounced and reversible changes in the optical (thermochromism), electrical conductivity and magnetic properties, very few works have been reported on the sol-gel preparation of such materials and the characterization of such properties.

Vanadium dioxide, VO_2, is known to undergo such a thermally induced semiconductor-to-metal phase transition at 68 °C [177]. Early works of Geffcken and Berger [178] and Schroeder [179, 180] are now being revived as today the state-of-the-art of the sol-gel process offers distinct advantages over the vacuum deposition of thin metal oxide films. Potember and Speck [181] have synthesized VO_2 film from a sol prepared from $V(Ot-Bu)_4$ dissolved in isopropanol to which was added W and Mo impurities in the form of oxytetraethoxide in order to modify the transition temperature. After hydrolysis and condensation under N_2 atmosphere, glass slides were dip coated and heat treated at 600 °C under N_2 to prevent the oxidation of the vanadium compound. The product formed is $V_{1-x}M_xO_2$ and the reduction of VO_2 was performed in 1 mTorr vacuum at 600 °C (or in nitrogen atmosphere). Figure 12 shows the spectral transmission of a 100 nm thick VO_2 film above and below the transition temperature and Fig. 13 shows the temperature dependence of the optical

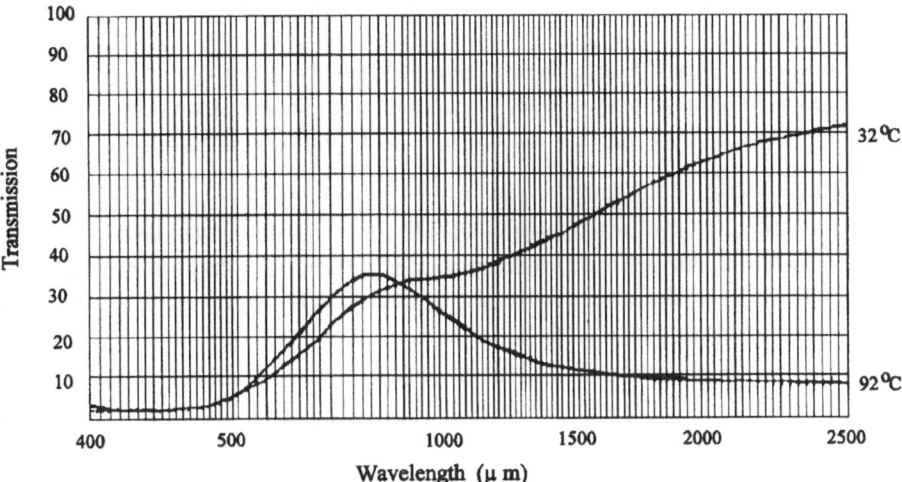

Fig. 12. Spectral transmission of a 100 nm thick sol-gel VO_2 film above and below the transition temperature [181]

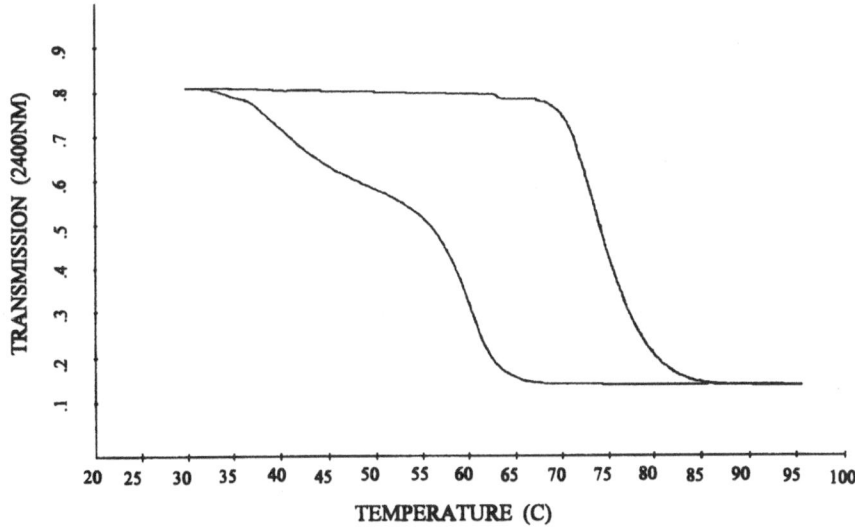

Fig. 13. Temperature dependence of the optical transmission of the same film at 2.4 µm, showing hysteresis effect (the transmission T increases at a temperature 10 °C lower on cooling) [181]

transmission at 2.4 µm. Stoichiometry, film thickness, type of substrate and crystalline orientation were found to affect the thermochromic properties.

5.3 Sol-Gel Electrooptical Materials

Electrooptical effects have been demonstrated by Levy's group [182 –186] by trapping dispersed microdroplets of liquid crystals (LC) in silica gel-glass (GDLC).

A mixture of methyltriacethoxysilane, K15 (4′-pentyl-4-biphenylcarbonitrite) and bidistilled water was polymerized at room temperature. The addition of Ti isopropoxide has also been used as dopant to increase the gel-glass refractive index and to reduce the refractive index differences between LC and matrix. This new class of doped sol-gel materials was sandwiched between two glass plates coated with transparent conductive electrodes. The device shows a reversible switching from opaque to clear transparent state when a voltage is applied to the conductive electrodes and acts as an optical shutter which, unlike other LC

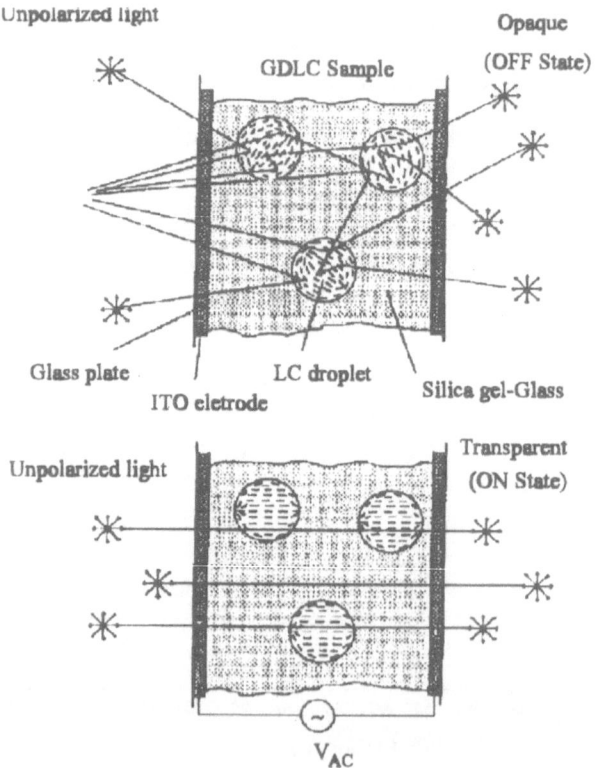

Fig. 14. Illustration of the switching principle of a GDLC device [184]

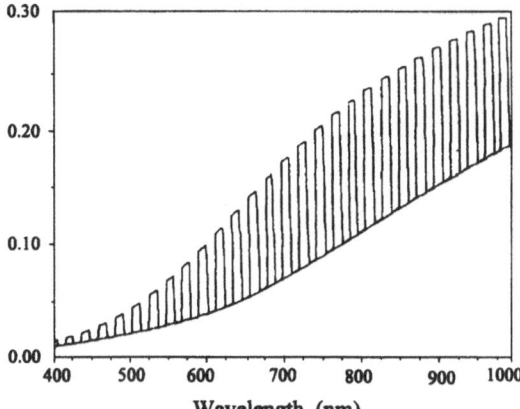

Fig. 15. Transmission spectra of a doped 30 μm thick GDLC (1:57 TIPO, 3.4:1 ethanol molar ratios relative to silane), switching peak voltage 150 V (saturation) [184]

devices, does not require polarizers. The optical switching is based on the dielectric anisotropy and birefringence of LCs where the applied electric field produces a torque that reorients the LC molecules, modifying in turn the refractive index encountered by the incoming light. Figure 14 illustrates the switching effect of such a device and Fig. 15 shows a typical spectral change of optical transmission. The dynamic behavior of such a device (typically in the range of 1 to 10 ms) was found to depend on the size of the LC droplets.

6 Sol-Gel Materials for Solar Cells

The sol-gel process can produce nanometer sized particles of several oxides [103] which can be deposited on a conducting glass support leading to transparent nanocrystalline films. When the materials have semiconducting properties (TiO_2, Fe_2O_3, Nb_2O_5, etc.) they should find a variety of interesting applications based on the fact that very large internal surface areas, with roughness factors of the order of 1000, are easily obtained. If the particles are sintered together to allow electronic contact, they can allow for electronic charge carrier conduction. A first practical embodiment uses such films for conversion of light to electricity (solar cell) which, in contrast to a conventional photovoltaic cell, separates the function of light absorption and carrier transport. The light harvesting is carried out by a sensitizer coated on the surface of the particles which initiates electron transfer events in the semiconducting particles leading to charge separation. The principle of the device, originally developed by Graetzel several years ago (a review can be found in [187]) is shown in Fig. 16.

When derivatized with suitable chromophore such as ruthenium complexes where one of the ligands is 4,4'-dicarboxy-2,2'-bipyridyl, Graetzel [188] has

principle of the nanocrystalline solar cell

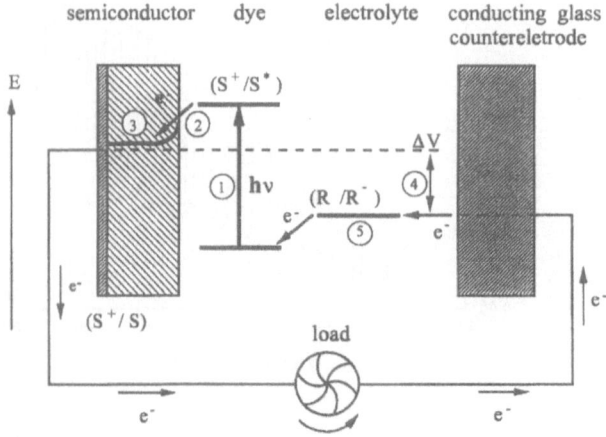

Fig. 16. Schematic representation of the principle of new photovoltaic cell indicating the electron energy level in the different phases. The cell voltage observed under illumination corresponds to the difference in the quasi-Fermi level of TiO_2 under illumination and the electrochemical potential of the electrolyte. The latter is equal to the Nernst potential of the redox couple (R/R^-) used to mediate charge transfer between the electrodes. The *layer 3*, 10 μm thick, is made of nanoscale TiO_2 particles [188]

shown that TiO_2 films prepared with 10–20 nm size particles give extraordinary efficiencies for the conversion of incident photons into electric current, in some cases exceeding 90%. Photovoltaic devices with 18.2 mA/cm^2 photocurrent and overall light to electric energy conversion yield up to 10% under simulated AM 1.5 solar radiation (96.4 mW/cm^2) have been obtained.

Other sol-gel semiconducting materials may also be used. In a prelimanary work, Barros Filho et al. [189] have studied the UV action spectrum of 20–30 nm colloidal Nb_2O_5 particles and obtained promising results.

These new developments, which are somewhat outside the scope of this paper but for which the structure of the devices is quite similar to those of EC devices, appear to be extremely promising and the sol-gel process, once again, is one of the principal methods to be used to obtain outstanding coatings with specific functions.

7 Future Developments and Conclusion

Sol-gel processing of coatings for EC devices and other related systems like solar cells has been reviewed including report literature up to mid-1994. Up to few years ago most of the research has been concentrated on developing the chemical processing of the sols in order to produce optically acceptable thin films and to show that these layers had the specific properties required for EC devices. Long term physical and electrochemical behavior, fundamental properties for technical applications, as well as microscopic studies of the phenomena have been rarely addressed. Nevertheless the good results obtained with the sol-gel processing has prompted more rigorous studies trying to fill these gaps and even complete devices have been tested in the last four years. Small size mirrors (up to 100 cm^2) are likely to be commercialized soon using at least one EC sol-gel layer. Devices with more than one sol-gel layer can be expected in the near future but more research and tests are necessary. Large scale devices such as smart windows require still more research in order to overcome the many technical difficulties encountered during the scale-up development.

From the physics and chemistry points of view there is a definite trend, observed in the last few years research, that the coatings which intercalate ions (H^+, Li^+, etc.) such as the EC and ion storage electrodes, or the coatings whose properties rely essentially on the attachment of specific molecules which modify their physical properties (mainly optical) under ambient stimuli (light, electric field, etc.), have to be made with colloidal nanoparticles linked together by some adequate thermal treatment. These coatings have quite high surface areas (up to several hundreds of square meters) which allows linking or binding of high amounts of ions or molecules to the active sites and a relatively high porosity allowing for diffusion of a large amount of ions (H^+, Li^+, etc.) On the other hand, due to their small thickness ($\leq \mu m$) and nanoscale size of their constituents, these coatings retain adequate optical appearance. In this respect the sol-gel process is also definitively very promising as it allows preparation of the layers in a relatively simple and cheap way.

Another development which is also quite peculiar to the sol-gel process is the possibility of preparing hybrid organic-inorganic materials. This field is relatively new and has not yet been fully exploited for the development of such devices. The principal application should be in the realization of solid but still viscous electrolytes in order to substitute the polymeric or liquid systems in the 20–100 °C range.

We firmly believe that the future of the sol-gel process for the preparation of thin or thick layers for EC or related devices (such as solar cells) is bright with high probability of seeing, in the short to medium term, technical applications of high interest in these fields.

8 References

1. Lampert CM, Grandqvist CG (1990). In: Large-area chromogenics: materials and devices for transmittance control, SPIE IS4, 2, SPIE, Bellingham, Washington, USA
2. Lynam NR, Agrawal A (1990). In: Large-area chromogenics: materials and devices for transmittance control, SPIE IS4, 46, SPIE, Bellingham, Washington, USA

3. Cronin JP, Agrawal A (1994). In: Perspectives on glass science and technology. Symposium in honor of the 90th birthday of Prof. N. Kreidl, Triesenberg, Liechtenstein, to be published

4. Selkowitz SE, Lampert CM (1990). In: Large-area chromogenics: materials and devices for transmittance control, SPIE IS4, Optical Engineering Press, Bellingham, Washington, USA, 22

5. Selkowitz SC, Rubin M, Lee ES, Sullivan R (1994). In: Optical materials technology for energy efficiency and solar energy conversion XIII, SPIE 2255, 226, SPIE, Bellingham, Washington, USA

6. Reilly S, Arasteh D, Selkowitz SE (1991) Solar energy mater 1: 22

7. Agrawal M, Cronin JP, Zhang R (1992). In: Sol-gel optics II, SPIE, 1758, 300, SPIE, Bellingham, Washington, USA

8. Sullivan R, Lee ES, Papamichael K, Rubin M, Selkowitz S (1994). In: Optical materials technology for energy efficiency and solar energy conversion XIII, SPIE 2255, 443, SPIE, Bellingham, Washington, USA

9. Kraus T (1953) unpublished report (Balzers, Liechtenstein)

10. Deb SK (1969) Appl Opt Suppl 3: 192

11. Deb SK (1973) Phil Mag 27: 801

12. Inganäs O (1990). In: Large-area chromogenics: materials and devices for transmittance control, SPIE IS4, 328, SPIE, Bellingham, Washington, USA

13. Yang SC (1990). In: Large-area chromogenics: materials and devices for transmittance control SPIE IS4, 335, SPIE, Bellingham, Washington, USA

14. Grandqvist CG (1994) Solar energy materials and solar cells 32: 369

15. Czanderna AW, Lampert CA (1990) Solar Energy Research Institute SERT Golden, Col, USA TP 255-3637 U Category 316 DE 90000 334

16. Haas TE, Goldner RB (1990). In: Large-area chromogenics: materials and devices for transmittance control, SPIE IS 4, 170, SPIE, Bellingham, Washington, USA

17. Cronin JP, Tarico DJ, Tonazzi JCC, Agrawal A, Kennedy SR (1992). In: Sol-Gel Optics II, SPIE 1758, 343, SPIE, Bellingham, Washington, USA

18. Cronin JP, Tarico DJ, Tonazzi JCC, Agrawal A, Kennedy SR (1993) Solar energy materials and solar cells 29: 371

19. Chemseddine A, Morineau R, Livage J (1983) Solid State Ionics 9–10: 357

20. Xu G, Chen L (1988) Solid State Ionics 28–30: 1726

21. Judeinstein P, Livage J (1989) Materials Science and Engineering 133: 129

22. Yamamaka K (1981) Jpn J Applied Physics 20: 1307

23. Oi J, Kishimoto A, Kudo T (1992) J Solid State Chemistry 96: 13

24. Yamanaka K, Ohkawoto H, Kidon H, Kudo T (1986) Jpn J Applied Physics 25: 1420

25. Itoh K, Okamoto T, Wakita S, Niikura H, Murabayashi M (1991) Appl Organomet Chem 5: 295

26. Unuma H, Tonooka K, Suzuki Y, Furusaki T, Kodaira K, Matsushita T (1986) J Mat Lett 5: 1248

27. Takase A, Miyakawa K (1991) Jpn J Appl Phys Part 2 30: L1508

28. Bell JM, Green DC, Patterson A, Smith GB, MacDonald KA, Lee K, Kirkup LD, Cullen JD, West BO Apoccia L, Kenny MJ, Wilunski LS (1991) in: Opt. Mater. Technol. Energy Effic. Energy Convers. SPIE 1536, 29, SPIE, Bellingham, Washington, USA

29. Livage J (1992) Solid State Ionics 50: 307

30. Judeinstein P, Livage J (1991) J Mater Chem 1: 621

31. Craigen D, Mackintosh A, Hickman J, Colbow K (1986) J Electrochem Soc 133: 1529

32. Judeinstein P, Livage J (1990). In: Sol-Gel Optics – SPIE, Bellingham, Washington, USA, 1328, 344

33. Denesuk M, Cronin JP, Kennedy SR, Law KJ, Nielson GF, Uhlmann DR (1994). In: International Symposium on Optical materials technology for energy efficiency and solar energy conversion XIII, SPIE 2255, 52, SPIE Bellingham, Washington, USA

34. Joo SK, Raistrick JD, Huggins RA (1985) Solid State Ionics 17: 313

35. Göttsche J, Hinsch A, Wittwer P (1993) Solar Energy Materials and Solar Cells 31: 415

36. Yoshino T, Baba N, Yasuda K (1988) Nippon Kagaku Kaishi 9: 1525

37. Donnadieu A (1990) In: Large-Area Chromogenics: Materials and Devices for Transmittance Control, SPIE IS 4, 191, SPIE, Bellingham, Washington, USA

38. Wang B, Cheng J, Zhon W (1992) Huadong Huagong Xueynan Xuebao 18: 48

39. Moser FH, Lynam NR, US Patent (1989) 4, 855, 161

40. Stangar UL, Orel B, Grabec I, Ogoreve B, Kalcher K (1993) Solar Energy Materials and Solar Cells 31: 173

41. Orel ZC, Orel B (1994). In: Optical Materials Technology for Energy Efficiency and Solar Energy Conversion XIII, SPIE 2255, 285, SPIE, Bellingham, Washington, USA
42. Aegerter MA, LaSerra ER, Martins Rodrigues AC, Kordas G, Moore G, (1990). In: Sol-Gel Optics, SPIE 1328, 391SPIE, Bellingham, Washington, USA
43. Baudry P, Rodriguez ACM, Aegerter MA, Bulhões LOS (1990) J Non Crystal Solids 121: 319
44. Macedo MA, Dall'Antonia LH, Aegerter MA (1992). In: Sol-Gel Optics II – SPIE 1758, 320, SPIE, Bellingham, Washington, USA
45. Tonazzi JCL, Valla B, Macedo MA, Baudry P, Aegerter MA (1990). In: Sol Gel Optics, SPIE 1328, 375, SPIE, Bellingham, Washington, USA
46. Macedo MA, Dall'Antonia LH, Valla B, Aegerter MA (1992) J Non-Cryst Solids 147/148: 792
47. Macedo MA, Aegerter MA (1994) J Sol-Gel Science and Technology 2: 667
48. Valla B, Tonazzi JCL, Macedo MA, Dall'Antonia LH, Aegerter MA, Leones MAB, Bulhões LOS (1991) in: Optical Materials Technology for Energy Efficiency and Solar Energy Conversion X, SPIE 1536, SPIE, Bellingham, Washington, USA
49. Kéomany D, Poinsignon C, Deroo D (1995) Solar Energy Material and Solar Cells, 36: 397
50. Macedo MA (1994) PhD Thesis University of São Paulo
51. Ottaviani M, Panero S, Morzilli S, Scrosati B, Lazzari M (1986) Solid State Ionics 20: 197
52. Sata Y, Fujiwara R, Shimizu I, Inoue E (1982) Jpn J Appl Phys 21: 1642
53. Doeuff S, Sanchez C (1989) C R Acad Sci Ser 2 309: 351
54. Nabavi M, Doeuff S, Sanchez C, Livage J (1989) Mater Sci Eng B3: 203
55. Ozer N, Chen DG, Simmons JH (1991) Ceram Trans Glasses Electron Appl 20: 253
56. Ozer N, Tepehan F, Bozkurt N (1992) Thin Solid Films 219: 193
57. Bell JM, Barczynska J, Evans LA, MacDonald KA, Wang J, Green DC, Smith GB (1994). In: Optical materials technology for energy efficiency and solar energy conversion XIII, SPIE 2255, 324, SPIE, Bellingham, Washington, USA
58. Hagfeld A, Vlachopoulos N, Gilbert S, Grätzel M (1994). In: Optical materials technology for energy efficiency and solar energy conversion XIII, SPIE 2255, 297, SPIE, Bellingham, Washington, USA
59. Stangar UL, Orel B, Hutchins MG (1994). In: Optical materials technology for energy efficiency and solar energy conversion XIII, SPIE 2255, SPIE, Bellingham, Washington, USA, 261
60. Orel B, Stangar UL, Hutchins MG, Kalcher K (1994) J Non Cryst Solids 175: 251
61. Reichman B, Bard AJ (1980) J Electrochem Soc 127: 241
62. Yu PC (1991) Thesis, Tufts University, Dept. of Chemistry
63. Gomes MAB, Bulhões LOS, Castro SC, Damião AJ (1990) J Electrochem Soc 137(10): 3067
64. Gomes MAB, Bulhões LOS (1990) Electrochim. Acta 35(4): 765
65. Alves MC (1989) MSc Thesis, Federal University of São Carlos (Brazil)
66. Lee RG, Crayston JA (1991) J Mater Chem 1: 381
67. Faria RC, Bulhões LOS (1994) J Electrochem Soc 141: L29
68. Avellaneda CO, Macedo MA, Aegerter MA (1994). In: Proc. 38o. Congresso Brasileiro de Cerâmica, 109 Blumenau, SC
69. Ozer N, Barreto R, Büyüklinanl T, Lampert C (to be published) Solar Energy Materials and Solar Cells
70. Aegerter MA (1991) Patent pending No WO 91/02282 (PCT/BR90/00006)
71. Avellaneda CO, Macedo MA, Florentino AO, Barros Filho DA, Rabelo AA, Aegerter MA (1994). In: Proc 2nd conference "Sociedade brasileira de pesquisadores nikkeis", São Paulo, 17
72. Aegerter MA, Avellaneda CO (to be published). In: International Symposium on Sol-Gel Science and Technology, ACERS Pacific Coast Meeting, Los Angeles, USA
73. Avellaneda CO, Macedo MA, Florentino AO, Barros Filho DA, Aegerter MA (1994). In: Sol Gel Optics III, SPIE 2288, 422, Bellingham, Washington, USA
74. Avellaneda CO, Macedo MA, Florentino AO, Aegerter MA. In: Optical materials technology for energy efficiency and solar energy conversion XIII, SPIE 2255, 38, SPIE, Bellingham, Washington, USA
75. Cogan SF, Rauh RD, Plante TD, Nguyen NM, Westwood JD (1980). In: Physical electrochemistry division, The electrochemical society electrochromic materials, Pennington, New Jersey, 99
76. Talledo A, Andersson AM, Granqvist CG (1990) J Mater Res 5: 1253
77. Nabavi M, Sanchez C, Livage J (1991) Eur J Solid State Inorg Chem 28: 1173
78. Sanchez C (1992) Bol Soc Esp Ceram Vidrio 31: 191
79. Yoshino T, Baba N, Kouda Y (1987) Jpn J Appl Phys 26: 782

80. Desilvestro J, Haas O (1990) J Electrochem Soc 137: 50
81. Bach S, Pereira-Ramos JP, Baffier N, Messina R (1990) J Electrochem Soc 137: 1042
82. Pereira-Ramos JP, Messina R, Bach S, Baffier N (1990) Solid State Ionics 40–41: 970
83. Amarilla J-M, Casal B, Galvan J-C, Ruiz-Hitzky E (1992) Chem Mater 4: 62
84. Nagase K, Shimizu Y, Miura N, Yamazoe N (1993) J Ceram Soc Jpn 101: 1032
85. Hackwood S, Dayem AH, Beni A (1982) Phys Rev B 26: 471
86. Agrawal A, Habib HR, Agrawal RK, Cronin JP, Roberts DM, Popwich R-C, Lampert CM
    (1992) Thin Solid Films 221: 239
87. Svensson JSEM, Granqvist CG (1980) App Phys Lett 49: 1568
88. Lampert CM, Omstead TR, Yu PC (1985). In: SPIE, 562, 15, SPIE, Bellingham, Washington,
    USA
89. Miles MH, Stilwell DE, Hollins RA, Henry RA (1980). In: Physical electrochemistry division,
    The electrochemical society electrochromic materials, Pennington, New Jersey, 137
90. Gottesfeld S (1980) J Electrochem Soc 127: 272
91. Moser FH, Lynam NR (1990) US Patent, 4, 959, 247
92. Orel B, Macek M, Svege F, Kalcher K (1994) Thin Solid Films (in press)
93. Orel B, Macek M, Surca A (1994). In: Optical materials technology for energy efficiency and
    solar energy conversion XIII, SPIE 2255, 273, SPIE, Bellingham, Washington, USA
94. Olivi P, Pereira EC, Longo E, Varella JA, Bulhões LO (1993) J Electrochem Soc 140: L81
95. Chopra KL, Major S, Pandya DK (1983) Thin Solid Films 102: 1
96. Haacke G (1977) Ann Rev Mater Sci 7: 73
97. Lynam NR (1980). In: Physical electrochemistry division, the electrochemical society electroch-
    romic materials, Pennington, New Jersey, 201
98. Arfsten NJ (1984) J Non-Cryst Solids 63: 243
99. Arfsten NJ, Kaufman R, Dislich H (1984). In: Ultrastructural processing of ceramics, glasses
    and composites, 189, Wiley Interscience Publishers, New York
100. Takahashi Y, Hayashi H, Dhya Y (1992). In: Better Ceramics Through Chemistry V, MRS 271, 401
101. Takahashi Y, Wada Y (1990) J Electrochem Soc 137: 267
102. Gonzalez-Oliver GJR, Kato I (1986) J Non-Cryst Solids 82: 400
103. Brinker CJ, Scherer GW (1990). In: Sol-gel science: the physics and chemistry of sol-gel
     processing, Academic Press (ed) San Diego, 787p
104. Brinker CJ, Hurd AJ, Frye GC, Ward KJ, Ashley CS (1990) J Non-Cryst Solids 121: 294
105. Tsunashima A, Yoshimizu H, Kodaira K, Shimada S, Matsushito F (1986) J Mater Sci 21: 2731
106. Hiratsuka RS, Pulcinelli SH, Santilli CV (1990) J Non-Cryst Solids 121: 76
107. Maddalena A, DalMaschio R, Diré S, Raccanelli A (1990) J Non-Cryst Solids 121: 365
108. Goodman FJ, Gregg SJ (1960) J Chem Soc 237: 1162
109. Giesekke EW, Gutowsky HS, Kirkov P, Laitineu HA (1967) Inorg Chem 6: 1269
110. Orel B, Stanger UL, Crnjak-Orel Z, Bakovec P, Kosec M (1994) J Non-Cryst Solids 167: 272
111. Cocco G, Enzo S (1987) Mater Chem Phys 17: 541
112. Pozarnsky GA, Wright L, McCormick AV (1994) J Sol-Gel Science and Technology 3: 57
113. Lynam NR, Moser FH, Hichwa BP (1987). In: SPIE, 130
114. Hajimoto Y, Matsushima M, Ogura S (1979) J Electron Mater 8: 301
115. Uchikawa K, Niwa T (1987) US Patent 4, 652, 090
116. Ling HC, Yang MF, Rhodes WW (1986). In: Science of ceramic chemical processing, John
     Wiley and Sons, 285
117. Ozer N, He Y, Lampert CM (1994). In: Optical materials technology for energy efficiency and
     solar energy conversion XIII, SPIE 2255, 456, SPIE, Bellingham, Washington, USA
118. Howe AT, Sheffield SH, Childs PE, Shilton MG (1980) Thin Solid Films 67: 415
119. Takahashi T, Tanase S, Yamamoto O (1980) J Appl Electrochem 10: 415
120. Charbouillot Y, Ravaine D, Armand M, Poinsignon C (1988) J Non-Crystal Solids 103: 325
121. Judeinstein P, Livage J, Zarndiansky A, Rose R (1988) Solid State Ionics 28–30: 1722
122. Sanchez JY, Denoyelle A, Poinsignon C (1993) Polym Adv Technol 4: 89
123. Huggins RA (1977) Electrochim Acta 22: 773
124. Goldner RB, Haas TE, Seward G, Wong KK, Norton P, Foley G, Berera G, Wei G, Schulz S,
     Chapman R (1988) Solid State Ionics 28–30: 1715
125. Oi T, Miyauchi K (1981) Mater Res Bull 16: 1281
126. Raistrick ID, Ho C, Huggins RA (1976) Mater Res Bull 11: 953
127. Klein LC (1988) Sol-Gel Technology for Thin Films, Fibers, Preforms, Electronics and
     Specialty Forms, Noyes Publications (ed) New Jersey, USA, 407p

128. Aegerter MA, Jafelicci Jr. M, Souza DF, Zanotto ED (1989) Sol-Gel Science and Technology, World Scientific (ed) Singapore, 505p
129. Sakka S, Yoko T (1992). In: Chemistry, spectroscopy and applications of sol-gel glasses, Springer-Verlag Berlin, 89
130. Schmidt H (1992). In: Chemistry, spectroscopy and applications of sol-gel glasses, Springer-Verlag Berlin, 119
131. Agrawal A, Cronin JP, Zhang R (1993) Solar Energy Materials and Solar Cells 31: 9
132. Morineau R (1985) Vide, Couches Minces 40: 281
133. Uhlmann DR, Boulton JM, Teowee G, Weisenbach L, Zelinski BJ (1990). In: Sol-gel, SPIE 1328, 270, SPIE, Bellingham, Washington, USA
134. Livage J (1988) Chem Sci 28: 9
135. Habib MA, Glueck D (1989) Solar Energy Materials 18: 127
136. Göttsche J, Hinsch A, Wittwer V (1993). In: Large-area chromogenics: Materials and devices for transmittance control, SPIE IS4, 13, SPIE, Bellingham, Washington, USA
137. Nabavi M, Sanchez C (1990) C R Acad Sci Paris 310 – série II: 117
138. Mehrotra RC (1988) J Non-Cryst Solids 100: 1
139. Atkinson A, Guppy RM (1991) J Mater Sci 26: 3869
140. Makishima A, Kubo H, Wada K, Kitami Y, Shimohira T (1986) J Am Ceram Soc 69: C127
141. Giuntini JC, Granier W, Zanchetta TV, Taha A (1990) J Mater Sci Lett 9: 1383
142. Grandqvist CG (1993) Sol State Ionics 60: 213
143. Zea Bermudes VD, Baril D, Sanchez JY, Armand M, Poinsignon C (1992). In: Optical materials technology for energy efficiency and solar energy conversion XI: Chromogenics for Smart Windows, SPIE 1728, 180, SPIE, Bellingham, Washington, USA
144. Tatsumisago M, Kishida K, Minami T (1993) Solid State Ionics 59: 171
145. Boilot JP, Colomban P (1988). In: Sol-gel technology for thin films, fibers, preforms, electronics and specialty shapes, Noyes Publications, 303
146. Klein LC, Ho SF, Szu SP, Greenblatt M (1991). In: Applications of Analytical Techniques to the Characterization of Materials, Plenum Press, New York, 101
147. Klein LC (1993). In: CNRS European sol-gel summer school, Château de Bierville, France, 147
148. Hayri EA (1989) J Non-Cryst Solids 94: 167
149. Smaihi M, Petit D, Gourbillean F, Chaput F, Boilot JP (1991) Solid State Ionics 48: 213
150. Bozano D, Aegerter MA (1994) Private Communication
151. Wang B, Szu S, Greenblatt M, Klein CC (1992) Chem Mater 4: 191
152. Wang B, Szu S, Greenblatt M, Klein LC (1992) Solid State Ionics 53/56: 1214
153. Kuwano J, Naito Y, Kato M (1987) Yogyo Kyo Kaishi 95: 176
154. Hayri EA, Greenblatt M (1987) J Non-Cryst Solids 94: 387
155. Wang B, Szu S, Greenblatt M, Klein CC (1991) Solid State Ionics 47: 297
156. Wang B, Szu S, Greenblatt M, Klein LC (1992). In: The physics of non-crystalline solids, 203, Taylor & Francis London
157. Wang B, Greenblatt M, Yan J, Wu Y (1994) J Sol-Gel Science and Technology 2: 323
158. Ogasawara T, Klein LC (1994) J Sol-Gel Science and Technology 2: 611
159. Trifonova V (1994) J Sol-Gel Science and Technology 2: 447
160. Judeinstein P, Titman J, Stamm M, Schmidt H (1994) Chem Mater 6: 127
161. Macedo MA, Dall'Antonia LH, Aegerter MA (1992). In: Smart Materials Fabrication and Materials for Micro-Electro-Mechanical Systems MRS, 276, 125
162. Oihshi T, Maekawa S, Kato A (1992) Jpn Kokai Tokkyo Koho JP 04 242226 A2 920828 Heisi
163. Takahashi T, Nomura S (1993) Jpn Kokai Tokkyo Koho JP 05 177757 A2 930720 Heisei
164. Kaufman VR, Levy D, Avnir D (1986) J Non-Cryst Solids 82: 103
165. Levy D, Avnir D (1988) J Phys Chem 92: 4734
166. Levy D, Sinhorn S, Avnir D (1989) J Non-Cryst Solids 113: 137
167. Matsui K, Moroboshi T, Yoshida S (1989). In: MRS Int Meet Adv Mat, 12, 203
168. Preston D, Pouxviel JC, Novison T, Kaska W, Dunn B, Zink JI (1990) J Phys Chem 94: 4167
169. Nogami M, Sugiura T (1993) Mat Sci Lett 12: 1544
170. Yamamaka SA, Zink JI, Dunn B (1992). In: Sol-Gel Optics II, SPIE, 372
171. Ueda M, Kim HB, Ikeda T, Ichimara K (1992) Chem Mater 4: 1229
172. Ueda M (1993) J Non-Cryst Solids 163: 125
173. Hou L, Mennig M, Schmidt H (1994). In: Optical materials technology for energy efficiency and solar energy conversion XIII, SPIE 2255, 26, SPIE, Bellingham, Washington, USA
174. Hou L, Hoffmann B, Mennig M, Schmidt H (1994) J Sol-Gel Science and Technology 2: 635

175. Judeinstein P (1994) J Sol-Gel Science and Technology 2: 147
176. Mennig M, Krug H, Fink-Straube C, Oliveira PW, Schmidt H (1992). In: Sol-Gel Optics II, SPIE 1758, 387, SPIE, Bellingham, Washington, USA
177. Adler M (1968) Rev Mod Phys 40: 714
178. Geffcken W (1939). In: Jenaer Glaswerk Schott and Gen., Jena (ed) 411 (GDR Patent, 736)
179. Schroeder H (1962) Opt Acta 9: 249
180. Schroeder H (1969) Phys Thin Films 5: 87
181. Potember RS, Speck KR (1990). In: Sol-Gel Optics, SPIE 1328, 364 SPIE, Bellingham, Washington, USA
182. Levy D, Serna CJ, Oton JM (1991) Mat Letters 10: 470
183. Oton JM, A. Serrano A, Serna CJ, Levy D (1991) Liq Cryst 10: 733
184. Levy D, Serna CJ, Serrano A, Vidal J, Oton JM (1992). In: Sol-Gel Optics II, SPIE 1758, 476 SPIE, Bellingham, Washington, USA
185. Levy D, Quintana X, Covadonga R, Otón JM (1985). In: Sol-Gel Optics III, SPIE 2288, 529, Bellingham, Washington, USA
186. Levy D, Serrano A, Otón JM (1994) J Sol-Gel Science and Technology 2: 803
187. Graetzel M (1993) MRS Bulletin XVII 10: 61
188. Graetzel M (1994) J Sol-Gel Science and Technology 2: 673
189. Barros Filho DA, Maĉedo MA, Florentino A, Aegerter MA (1994). In: Proc. Congresso Brasileiro de Cerâmica, Blumenau, SC, Brasil, p 80

Luminescence of Cerium(III) Inter-Shell Transitions and Scintillator Action

Christian K. Jørgensen

Section de Chimie, Université de Genève, Sciences II, CH 1211 Geneva 4

Cerium(III) energy levels and photophysics represent an intermediate status between d-group behavior and the other lanthanides showing narrow absorption bands and luminescence bands reminiscent of $4f^q$ configuration atomic energy levels. Once the energetic separations of 0.5 to 2 eV of $5d$-like orbitals are recognized to provide huge Stokes shifts and rather wide emission bands, it is also obvious that Ce(III) does not have a standard symmetry of ligating atoms (e.g. coordination number $N = 9$ and D_{3h} known from aqua ions and anhydrous $LnCl_3$ and $LnBr_3$) but N is often 12, 11, 8, 7 or 6 (and rarely octahedral, as frequent in the d groups), 8 in irregular symmetries perhaps most typical. Scintillators for detecting *very* high kinetic energies (approaching the relativistic regime of photons and particles with very low rest-mass) involve closed shells, such as $4f$ and $5d$ in lutetium(III); barium $5p$, $5d$ and $4f$ in highly electrovalent barium(II) in transparent crystals; and perhaps $4f$ and $6s$ in hafnium(IV) and tantalum(V) fluorides.

1 Relations Between five *J*-levels below 11 eV of Gaseous Ce^{+3} and the Seven Kramers Doublets of ^{2}F (14 States), the 5d Shell, and the 6s Orbital

Although Cerium is the first lanthanide, its spectra, especially of compounds in transparent solvents, in crystals and in vitreous solids, are much simpler than the lanthanides from praseodymium(III) to thulium(III) containing from two to twelve 4*f* electrons in the lower set of excited levels; to a good approximation, these energy levels in condensed matter are bunched tightly in manifolds containing $(2J + 1)$ states [micro-states; mutually orthogonal many-electron wave-functions] corresponding for a given number q of 4*f* electrons, as well all the *J*-levels of $(14 - q)$ 4*f* electrons in agreement with Wolfgang Pauli's hole-equivalence:

$$q = (1 \text{ or } 13): 14; (2 \text{ or } 12): 91; (3 \text{ or } 11): 364;$$

$$(4 \text{ or } 10): 1001; (5 \text{ or } 9): 2002; (6 \text{ or } 8): 3003; (7): 3432 \qquad (1)$$

These numbers of states are the result of combinatoric distribution of q electrons on 14 available sites (the corresponding number of sites are $(2l + 1)$ 2 for other *l* values in the Periodic Table [0(*s*)], [1(*p*)], [2(*d*)], [3(*f*)], ...) [1–6]. It should be added that the individual sub-levels may contain $1, 2, 3, 4, ...$ states when q is *even* [nothing to do with the parity, decided by the total sum of *l* values in the electron configuration considered; the total parity is *odd*, if an odd number of electrons have *odd* *l* values (i.e. $1, 3, 5, 7, ...$)]. The general theorem of an odd-valued number of electrons (carrying one *l*) producing eigen-states with an even number of states, was shown by the Dutch-Danish physicist Kramers, and the resulting degeneracy of states [which can be resolved by "perturbations changing sign by a reversal of the arrow of time", i.e. in real life, magnetic fields] is called "Kramers degeneracy".

Chemical bonding effects can separate all the "sub-levels" discussed above, but not Kramers degeneracy without strong internal magnetic effects in the system. Chemical effects are here the concept of moderately weak separation of such energy that would have remained coinciding without the decreased symmetry induced by the neighbor atoms [a classical case is the broadening (and even splitting) of spectral lines of alkaline-metal or coinage metal *atoms* [7] or, gallium atoms in transparent films ("cool matrices") of solid argon, krypton and/or xenon [8], cooled by adjacent liquid helium].

Based on a model [9] elaborated by Hans Bethe in 1929, and refined for 4*f*^{q} systems by Hellwege [10] in 1948, the moderate splitting between 10 and 500 cm^{-1} in translucid lanthanide compounds [11, 12] the electrostatic "crystal field" model was greeted with enthusiasm by solid-state chemists and physicists between, say, 1950 and 1955. It was the third come-back of the electrostatic model of Jöns-Jacob Berzelius ["chemical bonding is electric"], and Magnus, and later Van Arkel, argued after 1920 that the major facts about chemical bonding between distinctly differently electro-negative elements [this concept existed even before Linus Pauling and Mulliken] might be understood qualitatively with the Madelung potential maintained by the (smaller or larger) electric charges on the cationic and anionic constituents. This paradigm was rendered more quantitative in the book by Rabinowitz and Thilo [13] in 1930 by reliable Madelung constants, as well as ionization energies of many gaseous atoms, and in quite a few cases, gaseous ions with charges $+1$, $+2$, and $+3$ (pushed hard by astrophysicists interested in Fraunhofer lines of the Sun, mainly of Z below 30). For reviews of these two subjects, see [3, 14].

The two most direct applications of the Bethe electrostatic model [9] were short papers by Van Santen and Van Wieringen [15] on the variation of $3d$ group "ionic radii" as a function of the number of electrons a among the various q $3d$-like electrons in the higher sub-shell $(3z^2 - r^2)$ and $(x^2 - y^2)$ directed in a regular octahedral complex MX_6 having all six X nuclei on the Cartesian axes at equal distances R from the M at origin. In 1951, the late Fridrich Ilse and Hermann Hartmann [16, 17] published two papers on the detailed absorption spectra of octahedral titanium(III), vanadium(III), chromium(III) aqua ions (in acidic solution, and in crystalline alums) and Leslie Orgel four [18] in 1955. Being myself a post-graduate student of Jannik Bjerrum in Copenhagen, I published 21 [19, 20] papers on the absorption spectra of Ti(III), V(II), V(III), V(IV), Cr(III), Mn(II), Mn(IV), Fe(II), Co(III), in particular Ni(II), Cu(II), Mo(V), Rh(III), and iridium(III), in the majority of cases having as ligands ammonia, ethylenediamine, multidentate amines and biological or synthetic amino-poly-carboxylates, halides, nitrite, etc. This, rather eclectic combinations of ligands and of d-group ions in different oxidation numbers, established unexpected regularities, as a function of the "ligator atoms" and of M with a definite oxidation state, belonging to the $3d$, the $4d$ or the $5d$ group. Already these results show beyond any doubt that the ligand field parameters, such as the Δ in regular octahedral MX_6 are more than unlikely to have any causal relation with the tiny non-spherical part $U(x, y, z) = U(r) - V(x, y, z)$ of the Madelung potential has any Coulomb attraction. This opinion was growing up within a few years, and even H. Hartmann and H.L. Schläfer [21] gave in 1954 the "coup de grâce" by finding that the aqua ions have lower Δ than the ammonia complexes of, e.g., chromium(III), cobalt(III) and rhodium(III) have *higher* Δ than any known *anion* complexes of these d group Werner-S.M. Jørgensen complexes (this was later confirmed for fluoride ligation). The pale yellow to white color of the corresponding hexacyanide complexes were also then rationalized by very high Δ of CN^-, not having a well-defined spectrochemical position [22, 23].

2 The Angular Overlap Model as Common Paradigm for $3d, 4d, 5d, 5f$ and $4f$ Groups

It is frequently said that the contemporary treatment of energy levels of all the complexes of the five transition groups mentioned is E.H.T. [Extended Hückel

Treatment] originally introduced with the aim of rationalizing the excited electronic levels of conjugated organic compounds [binary such as benzene and tetracene; ternary, such as pyridine and phenanthroline; quaternary such as amino-benzaldehyde]. This is historically true, but the strongest link between the E.H.T. and the A.O.M. (Angular Overlap Model) is the value

$$E_{12} = [E_{11} + E_{22}]^2/(E_1 - E_2) \tag{2}$$

of the non-diagonal element of the "effective one-electron operator" [rather ask a lawyer what the precise definition actually is].

This semi-qualitative argument has been adopted by inorganic chemists along a line of thought: The strongest contribution to the total bonding of atoms 1 and 2 comes from covalency, when the denominator of Eq. (2) is very small (loosely "the two atoms have roughly the same electronegativity" which may easily provide a sixth definition of this elusive concept). It means that two one-electron functions (*orbitals*) practically have the same "one-electron energy". This is a small gnat to swallow for organic chemists, but it became a hot and nagging paradox, when photo-electron spectra of (gaseous or solid) samples show coinciding energies of the measured ionization energy I of two orbitals, one typically being a $3d$ or $4d$ or $4f$ or $5f$ orbital, and the other a normal "valence" orbital.

This paradox first showed up in "ligand field" theory, when gaseous cobalt(III) and iron(III) acetylacetonate molecules had their most loosely bound "organic" orbital a few eV [300 kJ/mol] more difficult to ionize than the Co $3d$ or the Fe $3d$ shell. Energy levels of several monatomic entities [3, 24, 25] show this problem (in absence of any chemical bonding at all). In the gaseous nickel atom, the 20 states of $3d^9 4s$ are, on the average, 0.17 eV above the ground level ($J = 4$) representing 9 of the 45 states of $3d^8 4s^2$. The unique state ($J = 0$) of $3d^{10}$ occurs 1.82 eV above this groundstate. The 44 states (one is still lacking) identified of $3d^8 4s^2$ are, on the average, 1.16 eV above the ($J = 4$) groundstates.

The paradox is far more active in the lanthanide compounds and alloys [2, 3, 4] where, contrary to all concepts in the "electron energy band theory", the filled Sb $5p$ shell in the cubic NaCl type lanthanide monantimonides LnSb almost invariantly is *less* difficult to ionize [lower I_b] than the Ln(III) $4f$ shell, being ionized by the incoming 1253.6 eV photons from a magnesium, or 1486.6 eV from an aluminium anti-cathode. In purely monatomic experiments [3] gaseous Sc^0 has its groundstate belonging to $3d4s^2$ and loses one $4s$ electron when ionized to the groundstate of Sc^+ $3d4s$ and two to the Sc^{++} retaining only one $3d$ electron beyond the 18 electrons in closed shells.

In a Chapter in Gmelin [2], the severe non-additivity of one-electron energies in monatomic systems retaining two or more electrons, is analyzed. The concepts of one-electron energies among inorganic chemists are, largely, imported from atomic spectra [1, 3] serving plausible model paradigms. In one sense, the deviations from additive one-electron energies are even stronger in molecules and other compounds, than in monatomic entities, possibly because the occurrence of comparable orbital radii is more frequent in the former case.

Of course, it is the basis for chemical analysis that matter shows additive values of *rest-mass* [14] but orbitals are not so close to auto spare-parts, as we might perhaps wish.

There is a certain advantage that it is energy *difference* that we study. From this point of view, we enter the same problematic as zero-points for electric potentials. A hybrid of these difficulties appears [3, 26] in the demand for "chemical ionization energies "relative to the prevailing vacuum. Fortunately, atomic spectroscopists determining consecutive ionization energies were not hamstrung by this textbook argument. It is quite interesting how different sciences show wide or very little tolerance. For instance, physicists discuss quite readily a local dielectric constant in solids; among other they start modifying the Rydberg (1895) and Balmer (1885) parameters to describe excitons in many semi-conductors, but never use them for describing red-shifts of lines in cool argon.

3 Studies of Cerium(III) Absorption Bands 1930–1980

Just as many industrial glasses absorb ultraviolet radiation (due to the presence of iron(III), lead(II), ...), cerium(III) both in many (otherwise translucent) minerals or glasses show broad-band absorption increasing toward lower wavelengths. In solid mixed oxides [27], air and other oxidants may give cerium(IV) usually giving yellowish hues. Violet coloration may also occur due to the simultaneous presence of Ce(III) and Ce(IV), comparable to the dark green [to almost black] oxidation product of weakly coloured iron(II) in $Fe(OH)_2$ precipitated in alkaline solution. Violet $Ce(OH)_{3+x}$ has been thoroughly studied [28]. Since the molar extinction coefficient ε of Ce(III) aqua ions is at least 100 times that of green Pr(III) between 280 and 220 nm, the Ce(III) bands of 0.5 percent Ce in the lanthanide content has several times been ascribed to praseodymium (III). As already pointed out by the German-Norwegian crystallographer and geochemist Goldschmidt, cubic dioxides MO_2 [fluorite-type] can readily accommodate Pr(IV), Tb(IV), U(IV); and MO_2 is recently now known for the nine elements from thorium $[Z = 90]$ through californium $[Z = 98]$. Although lanthanides Ln can enter these fluorites, creating oxide vacancies [29] $M_{1-x}Ln_xO_{2-0.5x}$ the cubic oxide with M = Zr tends to favor Ln(III) giving greyish samples and the cubic lattice parameter a_0 in Debye powder-diagrams [27] showing comparatively much less Ln(IV) than Ln(III), whereas it is the opposite for the heavier M(IV) having shorter internuclear distances, containing greater amounts of Ln(IV).

It is evident that the broad bands with oscillator strengths P around 0.01 [for comparison, the strong band with vibrational structure of permanganate in the green is 0.03, and the strongest known electron transfer band of an inorganic complex with only one M, here platinum(IV) in its tomato-red hexa-iodide complex, has $P = 0.18$] differ strikingly from the narrow internal transitions in a partly filled 4f shell [30, 31, 32] behaving almost like atomic spectral lines, but P rarely above 10^{-5} and hard to observe in absorption, if P is not at least 10^{-7}. There was an instinctive feeling among atomic spectroscopists before the revised paradigm by Broer, Gorter and Hoogschagen [33] in 1945, from which proposal Brian Judd [34] and G.S. Ofelt [35] independently of each other (in 1962) developed a highly successful *band intensity* parametrization [31, 32, 36] having three parameters U for each transition between two J-levels of the Ln(III) or Ln(II) to be described. Those "reduced matrix elements" hardly depend on the ligands coordinated to the lanthanide in the complex, solution, or glass, or crystalline solid. The three t-values 2, 4, 6 had to be even, and not larger than $2l = 6$. The t = 2 represents "pseudo-quadrupolar transitions" factually known as weak transitions, e.g. between the half-occupied 4s orbital and the five 3d orbitals in the gaseous potassium atom [6]. In lanthanide complexes, t = 2 sometimes provide the strongest absorption bands due to the partly filled 4f shell, providing P values up to almost 0.001. The tensor U having three parameters t = 2, 4, and 6 for each transition between two J-levels of a partly filled 4f shell represents, in a metaphoric sense, 16-polar transitions for t = 4, and 64-polar transitions for the highest t value, 6. The wave-functions used are those of the S.C.S. [Slater-Condon-Shortley treatment] derived from a magnificient textbook [6].

Simon Freed [37] from the University of Leyden in 1931, studied the ultraviolet absorption spectra by the usually applied technique of contemporary atomic spectroscopy (a quartz prism spectrometer with sensitive photographic plates) where crystalline syncrystallized cerium(III) ethylsulfate with 10 to 5000 times more lanthanum(III). The paper does not make any direct reference to a crystal structure. Fitzwater and Rundle [38] in 1959, published the crystal structure of three isotypic Ln = Y, Er, Pr [Ln(OH$_2$)$_9$] (C$_2$H$_5$OSO$_3$)$_3$ with a trigonal prism of six Er(III) having Er-O distance 237 pm (1 picometer = 0.01 Å) and an equatorial plane (turned 60° relative to the prism) 252 pm. As early as 1939, L. Helmholz [39] published the crystal structure of the ennea-aqua [Nd(OH$_2$)$_9$] (BrO$_3$)$_3$ with closely similar geometry of the nine Nd–O links. Both the bromate and the ethylsulfate have been prepared from La through Lu, and are externally isotypic, with minor deviations in the Ln–O distances of the two kinds.

The absorption spectra [37] of the ethylsulfates with highly varying ratio x in [La(OH$_2$)$_9$](C$_2$H$_5$OSO$_3$)$_3$ with $(1 - x)$ lanthanum and x cerium looked very consistent, La, the abundant lanthanum, hardly contributing to the spectral absorption. The three, quite narrow absorption bands of the crystals at 256, 238, and 220 nm at 300 K, and 255, 237, and 224 nm at 90 K show very moderature shifts with T, the two first move 0.4 percent toward higher energy, and the third band 2 percent toward lower energy [perhaps due to a modified background close to the vacuum ultraviolet]. Perhaps influenced by the work of Dieke, using the almost isotypic LaCl$_3$ and LaBr$_3$ crystals as matrices for other Ln exhibiting numerous narrow lines, many reviews suggest a strong Ln(III) preference for the coordination number $N = 9$ in both aqua ion salts of bulky anions, and in anhydrous halides of rather large Ln(III). An inorganic chemist may extrapolate that this trigonal ($N = 9$) plays a rôle comparable with the octahedral d-group compounds.

However, a weak point in this analogy is that regular octahedral MX_6 and MX_6^{-z} [X being the four halides, or polyatomic ligands such as water and ammonia, are exceedingly commonplace], but MX_9 scarce, and for the most, quite recently characterized. The most clear-cut example is perhaps the rhenium(VII) hydride complex ReH_9^{-2}. Well-read inorganic generalists almost instinctively detect subliminal advertizing for the Sidgwick 18-electron rule that the favorable N for a central atom, of which q among the ten electrons of a given d shell do not participate in chemical bonding, is $N = 9 - (q/2)$. It does not bring home the full [doubtful?] point that lanthanides have perhaps q = 0. The reviewer has the feeling that a more convincing argument is that lanthanides are bulky, at least to the extent of calcium(II), if not strontium(II), and that favored N is almost as nebulous a propensity as atomic weight going with the Z in an element.

The *Angular Overlap Model* [A.O.M] grew out of an Extended Hückel treatment first applied to transition group complexes by Wolfsberg and Helmholz [40–44] assuming the non-diagonal element of one-electron energy in Eq. (2) to be represented by a constant (not far from one) times a smoothed out average energy of the two diagonal values characterizing the orbitals considered in M_1 and M_2 [such as the square-root $(E_{11} E_{22})^{0.5}$ of the, usually, positive product of the two diagonal energies, each probably negative]. The genuine W.H. model takes the "constant not far from 1" to be multiplied by the overlap between the two symmetry-adapted L.C.A.O. at a given $M_1 - M_2$ internuclear distance, restoring a qualitative acceptable picture if it can be forgotten that the ionization energy of the, say, $3d$ or $4f$ orbitals is a dramatically increasing function of the fractional positive charge of the M ions. A related approach is *Equilibrated Differential Ionization Energies* dI/dz, which can be extended to the dI/dz becoming a potential for electrons moving from one atom to another. This idea was suggested by Iczkowski and Margrave [45] in 1961. The results from dI/dz are quite similar to qualitative opinions about heteronuclear compounds [46–51] but are not further discussed in the present review.

Another approach derived from Wolfsberg-Helmholz concepts is the direct transcription of the Schrödinger one-electron [non-relativistic] equation in the sum of the potential energy of the electron considered as a Hartree expression; and a second-differential quotient able to give the kinetic energy of the electron as the normalized integral of its [real] electron density [4, 46, 48]. This idea was discussed by Lifschitz in an early book, and later by Klaus Ruedenberg [52]. What is called modified energy of atomic orbitals in L.C.A.O. today can be shown to be due to simultaneous consequences of the moderate overlap integral S_{12} and the "local kinetic energy" allowing a conservative integral of the total kinetic energy $\mathcal{P}$ and potential energy V having the same sum as in a Newtonian pendulum, the eigen-value being the sum of the integrated potential energy

P and kinetic energy T satisfying the virial theorem $P + T = E$, with the result $P = -2T$ [$T = -E$; and $P = 2E$, being negative in stationary states]. For our purpose, the most important of the three corollaries of the Wolfsberg-Helmholz is the third; the consideration of non-diagonal elements of A.O. one-electron energies being approximately the product of the overlap integral S_{12} times an energy turning out to be typically of order 100 eV, that is ten times the typical orbital *ionization* energies. From sheer considerations of dimensionality, the only numerical values above 60 [if not 100 eV] are the hidden contributions to the total kinetic energy T from the overlap integral S_{12} [not its square!] to increase the energy of anti-bonding L.C.A.O; if we look for energies derived from normalized orbitals, the influence of the overlap integral would be much smaller, and of order the squared S_{12}. This (post-1963) opinion is that "ligand field effects" are due to minute effects of the local kinetic energy operator. Nevertheless, comparison of experimental A.O. energies are in good agreement with the relevant quantity (S_{12}) times the quite huge T of M_1 shells, [3, 49–51].

The first paper directed toward $3d$ and $4f$ group complexes and their orbital energy by Jørgensen, Pappalardo and Schmidtke [53] in 1963 was clearly based on the Helmholz-Wolfsberg paradigm. Talking with Claus E. Schäffer visiting Geneva, he pointed out that it is significant that the total anti-bonding of 5 orbitals with $l = 2$, or with 7 orbitals having $l = 3$, is N times the value for a single MX bond, and that the product $N(2l + 1)$ has a striking similarity with the perturbation of a field with point-sources, being additive from the point of view of each M nucleus. This is the major origin of the name "Angular Overlap Model", the angular functions of the A.O. modifying S_{12} in a predictable way. Of course, there is another, more or less exponentially decreasing, function of the standard internuclear distance M–X.

Seen in hindsight, it could not be excluded that the electrostatic model of Bethe [9] is compatible with an interaction acting only very close to each source of the field, presumably close to each M or X nucleus. The less likely view would be that the A.O.M. interaction is usually a repulsion from the sources. It cannot be absolutely excluded that some higher-order relativistic interaction between the electrons and the small nuclei might destabilize the complex, but one would have expected it to show up in atomic spectra [3], especially those with large isotope shifts.

This scruple disappeared when Claus Schäffer demonstrated the *equiconsequential effect* of the 1963 A.O.M. considered as anti-bonding or bonding between orbitals of M and X and a σ-effect with Kronecker monopoles situated at the nuclei, and extended the A.O.M. with π-orbitals as carriers of interactions originating in tiny Kronecker electric dipoles and [with much less use until now] δ-bonding effects due to Kronecker quadrupoles. The quantum number $\lambda(\sigma:0), (\pi:1), (\delta:2), (\varphi:3), \ldots$ is used for orbitals in linear symmetries [42]. The non-negative λ value indicates the degree of polynomial from the angular factor without z coordinates [the direction of the linear axis in $C_{\infty v}$ and $D_{\infty h}$], this lambda λ should not be confused with the "λ" used [mainly by organic chemists] for planar systems having all their nuclei in the (x,y) plane. The familiar "π" orbitals have the (x,y) as-node-plane, and are directed in the z-direction. "σ" orbitals do not change sign going from (x, $-$ z) to (x, z) (contrary to "π" orbitals). Spheroids, ellipsoids, and all kinds of "blobs" have "σ symmetry.

One kind of one-electron function that has remained important in the A.O.M. paradigm are "blobs". If their value is closely similar at (x, $-$ z) and (x, z), there may not be much need to know the behavior of the "blob" in greater detail. It also means that roughly spherical densities can be considered without too much soul-searching.

In practice, it is very difficult to detect the seven Kramers doublets of the two only ^2F levels with $J = 5/2$ (lowest) followed by $J = 7/2$. The most important parameter of relativistic effects (i.e. those which disappear, if the reciprocal speed of light (1/c) vanishes) is the Landé parameter ζ_{nl}. It is multiplied by a characteristic constant [6] for each Russell-Saunders coupling (S, L) term involving q electrons in a nl-shell able to accomodate between 1 and $(4l + 2)$. Lang [54] established in 1936 that the lowest configuration of gaseous Ce^{3+} is (neglecting the 54 electrons in the closed xenon shells) $4f$ with the ground level $^2F_{5/2}$ followed by $^2F_{7/2}$ at 2253 cm^{-1} whereas the $5d^1$ shows $^2D_{3/2}$ at 49 737 and $^2D_{5/2}$ at 52 226 cm^{-1}. Hence the Landé parameters are $\zeta_{4f} = 2253/3.5 = 644$ & $\zeta_{5d} = (52 226\text{–}49 737)/2.5 = 996$ cm^{-1}. Since the third configuration $6s^1$ (constituted by $^2S_{1/2}$ at 86 602 cm^{-1} alone) has no positive l, the spin-orbit coupling is absent, but higher-order relativistic effects occur [55] such as the Lamb shift [3] modifying the X-ray level structure of U^{+91} and U^{+90} with $1s$, $1s2s$ and $1s2p$ configurations, as well as the strongest bound nl-shells [42] in Dirac-Fock [55] calculations for atoms having z well above 80.

Mainly because of detailed interest in "ligant field" structure of $4f^q$ (as non-degenerate sub-levels for even q, and as Kramers doublets for odd q), Clyde Morrison and Leavitt [12] compiled $4f^q$ sub-levels of Ln(III) in a selected set of crystals with a broad region of transparency. The absorption lines (bands) are usually measured at one or two cryogenic temperatures below 30 K. If crystals contain considerable concentrations of oxide, hydroxide and/or water, it is almost impossible to detect the weak $4f$ sub-levels of cerium(III) because the low-energy vibrational spectra include bending frequencies of LnO $\cdots$ HO, LnOLn and other bridges; and in the presence of carbonate, phosphate, sulfate, . . anions, the stretching X–O frequencies, their over-tones or other combination lines render the sub-levels below 3000 cm^{-1} far less conspicuous. If the problem is protons, it can be worthwhile to substitute all the hydrogens with deuterium.

Anyhow, the best instances (positions in cm^{-1}) of the seven Ce(III) Kramers doublets of ^2F measured in crystalline materials [12] are:

	$^2F_{5/2}$	$^2F_{7/2}$
$La_{1-x}Ce_xCl_3$	0, 33, 110	2168, 2208, 2282, 2399
$La_{1-x}Ce_xF_3$	0, 151	2160, 2240, 2635, 2845
CeF_3	0, 148	2158, 2238, 2638, 2848

$Y_{1-x}Ce_xAlO_3$	0, –	2085, 2485, 2695, 3250	
$Y_{1-x}Ce_xPO_4$	0, 411	2194, 2232, 2599	
$Y_{3-x}Ce_xGa_5O_{12}$	0, 139, 402	2161, 2176, 2373, 2670	
$Cs_2NaCeCl_6$	0, 562, 577	2160, 2661, 3048	(3)

It may be noted that the average energy of the three ($J = 5/2$) Kramers doublets tend to be slightly more below the four ($J = 7/2$) sub-levels than the difference 2253 cm^{-1} in the gaseous ion. Some of this effect may be due to non-diagonal elements of "ligand field" one-electron energy between two sub-levels of same group-theoretical symmetry type.

The Pauli-equivalent "one-hole" configuration $4f^{13}$ is "inverted" having $^2F_{7/2}$ as groundstate. The other J-level $^2F_{5/2}$ is known [61] to be at 8774 cm^{-1} higher energy in Tm^{+2} and at 10 214 cm^{-1} in Yb^{+3} [to be compared with the distance 10 149 cm^{-1} between the two J-levels of $4f^{13}6s^2$ in gaseous Yb^+]. Sub-levels in Ytterbium(III) compounds [12] can exemplified by:

	$^2F_{7/2}$	$^2F_{5/2}$	
$Y_{1-x}Yb_xAlO_3$	0, 209, 341, 590	10 220, 10 410, 10 730	
$Y_{1-x}Yb_xLiF_4$	0, 216, 371, 479	10 288, 10 409, 10 547	
$Cs_2NaYbCl_6$	0, 225, 573	10 243, 10 708	(4)

The last compound in both Eq. (3) and (4) is a cubic *elpasolite* type A_2NMX with a large alkali-metal ion A^+, a smaller ion N^+ [say, sodium], a trivalent metallic element M, and halide anions X. In a certain analogy to the crystal structure of Prussian Blue, it [62] contains perpetuating axes alternating MNMNMN . . . in the directions of all three Cartesian axes, and the large cation [say, caesium] in cuboctahedral cavities, with 12 X coordinated. In double-cyanides, H_2O can occupy the cavities. The Berlin [or Prussian] Blue also has such large alkali cations in the cuboctahedral cavities. The alternating lines running parallel to the Cartesian axes are Fe ^{II}CNFe ^{III}NCFe ^{II}CNFe ^{III}NCFe ^{II}CN . . having each [$S = 0$] iron(II) surrounded by six carbon-bound cyanide ligands, and each [$S = 5/2$] iron(III) in octahedral coordination to six nitrogen ends of the ambidentate cyanide ligands.

Yttrium(III) and the trivalent lanthanides show rather indifferently the $N = 7, 8, 9, 10, 11$, and 12 as coordination number of immediately ligating atoms. $N = 6$ is by no means frequent in crystals containing lanthanides, and even when it occurs [56, 57] it is not often regularly octahedral. The first spectra in a liquid solvent were the $LnCl_6^{-3}$ in acetonitrile [with or without succinonitrile added] studied by Ryan and C.K.J. [56]. In 1969, the chemically fragile hexa-iodide complexes of Ti(IV), Pr(III), Nd(III), Ho(III), and Er(III) were characterized, also in solids. The orange Sm(III) [24900], green Eu(III) [14800], deep yellow Tm(III) [28000] and purple Yb(III) [17850] show quite strong electron transfer spectra, of which the wave-number of the first maximum is given in rectangular parenthesis (Ryan [57]).

It is interesting to compare these electron transfer spectra with the $3d, 4d, 5d$ group hexahalides [28], other lanthanide compounds [4], and a variety of $5f$ group transthorium compounds [63] including uranyl compounds [64].

4 Anti-bonding Cerium(III) $5d$ Orbitals, Born-Oppenheimer Potential Surfaces, and Luminescence Stokes Shifts

The 3-fold axis of the $N = 9$ neighbourhoods of Ln(III) aqua ions, $LnCl_3$ and $LaBr_3$ crystals is a unique axis C_3 [whereas orthorhombic D_2 symmetry [65] has three mutually perpendicular C_2 axes, none of which is a principal axis]. For our purpose, the advantage of uniaxial symmetries is that the linear symmetry D_∞ gives the same symmetry types (of reasonably low l values) as the D_{3h} symmetry. The first extended treatment [66] along this line of $Ce(OH_2)_9^{+3}$ provided a "ligand field" classification of (λ, ω) basis with the relativistic (positive) $\omega = \lambda \pm 1/2$

	$\sigma(\omega = 1/2)$	$\delta(3/2)$	$\delta(5/2)$	$\pi(1/2)$	$\pi(3/2)$
calculated [66]	38 880	42 250	44 630	48 230	49 300
observed	39 060	41 950	44 740	47 400	50 130	(5)

where the upper line gives the calculated energies in cm^{-1} (relative to the $4f$ ground state) of the five Kramers doublets (as eigen-values) where the anti-bonding character of $d\sigma : d\delta : d\pi$ was estimated to have the ratios $2:3:4$ in the earliest A.O.M. paper [53]. The (purely $\sigma - $) anti-bonding effect on $5d\delta$ is due to the three X ligating atoms in the equatorial plane of MX_9 but would be almost vanishing in a linear XLnX (like uranyl ions). The relativistic separation of $5d\delta$ of Ce(III) into $\omega = 3/2$ and $5/2$ is expected to be closely similar to $2\zeta_{5d}$ or $2000 \ cm^{-1}$.

We encounter here the question whether the A.O.M. effects are exclusively anti-bonding. The oldest [9] paradigm of the non-spherical (tiny) part of the Madelung potential would have considered the average energy $44 660 \ cm^{-1}$ of the five Kramers doublets in Eq. (5) as the appropriate zero-point of the "ligand field" stabilization or destabilization. This zero-point is about $5600 \ cm^{-1}$ above the lowest $5d$-like orbital. In a context of purely [42] σ-anti-bonding, the zero-point has a clear-cut meaning as the σ-non-bonding situation. Both Claus Schäffer [42] and W. Urland [109] discuss the relative effect of π-anti-bonding and σ-anti-bonding in $3d$ and $4f$ complexes.

The two scenarios of (ancient) average energy of all five Kramers doublets; and the zero-point obtained by correcting for all σ-anti-bonding, differ by almost $6000 \ cm^{-1}$ (quite a substantial energy in a trivalent lanthanide). The careful analysis [66] asks the question: Is the $5d$ shell of cerium(III) in any sense the precursor of the central atoms Re(IV), Os(VI), Os(IV), Ir(VI), Ir(IV), Ir(III), Pt(VI), Pt(V), Pt(IV), Pt(II), ... with ground states involving a partly filled $5d$ shell? Since the two (λ) one-electron energy differences in Eq. (5) are $43 440 - 38 880 = 4560$ and $48 760 - 43 440 = 5320 \ cm^{-1}$, they only represent 0.4 to 0.15 times the typical Δ in octahedral $5d$ group complexes [19, 24, 46], so, for whatever reason (long Ce–O distances; being closer to overall spherical symmetry), they are smaller, compared to higher Z values with comparable ligands. By the way, this reminds one about $3d^1$ titanium(III) having apparent

Δ [16, 17, 46] close to $3d^4$ manganese(III) [both systems are Jahn-Teller unstable, as well-known [19, 41, 42, 68–71] in $3d^9$ copper(II) with four out of six Cu–X distances between exceptionally short, at least on a spectroscopic "instantaneous picture", and frequently even on a time-average crystal structure. Going back all the way to Pythagoran regular polyhedra, and to a great extent supported by Alfred Werner, we believe that all N bonds M–X in MX_N *usually* are equivalent (arguing for cubal $Mo(CN)_8^{-4}$) which can be shown (by crystallography and by plausible expressions for inter-ligand repulsion) to be *more* stable as an Archimedean tetragonal anti-prism.

S. Freed [37] was worried as long ago as 1931 about the weak Ce(III) band at 33 800 cm^{-1} in the undiluted, hydrated crystals of ethyl sulfate. The question was taken up by Heidt and Berestecki [72] in 1955, showing a weak band in weakly acidic Ce(III) solutions at 33 700 cm^{-1} with its molar extinction coefficient ε increasing smoothly from 16 to 25 with temperature increasing from 15 to 54 °C. Such behavior is compatible with at least three hypotheses: that the absorption band belongs to a (presumably lower) N than the majority Ce(III) aqua ion likely to have $N = 9$, and that the lower N is (are) formed endothermally; that the symmetry of the rare (3 to 6 percent) species is strikingly different from the majority aqua ion; or that the weak band is due to a chemical trace, (e.g. hydroxo complex or strong ion-pair). The reviewer [73] wanted to compare Ce(III) complexes of one or several anions with cerium(III) aqua ions without coordinated anions, going from a mild choice of unidentate chloride (connected with earlier work [74, 75] on ClNd (solvent)$_x^{+2}$ in nearly anhydrous methanol, and in hydrochloric acid [showing no change in the visible, narrow absorption bands below 5 molar $H_3O^+Cl^-$] up to above 12 molar). Absorption spectra of complexes were also studied [73] in aqueous solutions of: multidentate carboxylates (such as tartrate; citrate), and amino-polycarboxylates (such as nitrilo-triacetate; ethylene diaminetetra-acetate), where the alcohol functions in the two nitrogen-free ligands need a somewhat increased pH (one to four ammonia/cerium) inducing a propensity toward air oxidation, provoking weak yellow color and (slowly) Tyndall effect, presumably due to cerium(IV) colloids. On the other hand, the potentially quadridentate nitrilotriacetate (2 ligands/Ce) remains colorless, with two bands at 33 800 and 36 400 cm^{-1}; and the potentially sexidentate ethylene diaminetetra-acetate (1.1 ligand/cerium) at 35 600 and 38 300 cm^{-1} (ε of all four maxima was 500 within a few percent).

For our purposes, a more informative study was of 0.0004 and 0.002 molar CeCl$_3$ in ethanol containing from 0.3 to 5 molar H_2O. The most water-free solution has $\varepsilon = 700$ of a band at 32 400 cm^{-1} not moving its position a lot (up to 33 500 cm^{-1} with $\varepsilon = 100$ in 5 M water). Aqueous 6 molar $H_3O^+Cl^-$ shows $\varepsilon = 32$ at 33 300 cm^{-1}, and 12 M hydrochloric acid $\varepsilon = 430$ at 32 400 cm^{-1}.

Aqueous Ce(III) solutions below a few molar chloride show $\varepsilon = 710$ at 39 600, 600 at 41 700, 380 at 45 100, and 270 at 47 400 cm^{-1}, very similar to the Heidt and Berestecki [72] results. As far as their weak band at 33 700 cm^{-1}

goes, the spectra [73] fortified the opinion that a small quantity of aqua ions have $N = 8$ or 7 only (providing a remote analogy to the earlier belief that zinc(II) aqua ions can have either $N = 4$ or 6). Recently, experimental techniques have become available (by X-ray diffraction [76], or by neutron diffraction, usually exploiting [77] isotope effects) N-evaluation of aqua ions or other complexes in fluid solutions.

It may be worthwhile drawing the attention of more classical complex chemists to the fact that both calcium(II), strontium(II), cerium(III), and thorium(IV) are firmly known to exhibit $N = 12$, Ce(III), Ce(IV), Nd(III) and Th(IV) as approximately icosahedral-anions $M(O_2NO)_6^{-z}$ with six bidentate nitrate ligands, and also $N = 12$ of Ca, Sr, and Ba on the cuboctahedral site of the (historically prototypic) cubic perovskite $SrTiO_3$ showing octahedral titanium(IV) with $N = 6$. As first pointed out by Alfred Werner [78], this $N = 6$ is predominant in Cr(III), Co(III), Ni(II), Ni(IV), Ru(II), Ru(III), Rh(III), Ir(III), Ir(IV), Pt(IV), Au(V) as halide and aqua complexes, and of ammonia too. The previously almost inconceivable octahedral $N = 6$ of Ln(III) in hexahalides [56, 57] in anhydrous solvents such as acetonitrile, or in crystalline elpasolites Cs_2NaLnX_6 [62] have even allowed [67] orange $CeCl_6^{-2}$ and short-lived, dark purple $CeBr_6^{-2}$ to be recognized.

Among d^q aqua ions, very few cases have been found, where N distinctly changes by an integer (in contrast to the photochemical dissociation [66] of the ninth water ligand). Most instances are a simultaneous change of both symmetry and of N, e.g. d^6 low-spin ($S = 0$) $vs.$ high-spin ($S = 2$) (less regular octahedron related to the Jahn-Teller effect [41, 70, 71]). d^8 ($N = 6$) with high-spin octahedral groundstate $^3A_2$ and $N = 4$ in a square, with totally symmetry groundstate $^1A_1$. With a small stretch of imagination, this can also be described as the Jahn-Teller (σ-type) effect on the first spin-forbidden band $^3A_2 \rightarrow {}^1E$ in octahedral (O_h) nickel(II) complexes [19]. The narrow character of this band is related to the Kamimura-Tanabe stability [65] where two sub-levels ^1E of d^8 do not diverge to the first-order in the deviation from O_h. This situation has more practical relevance in the "ruby line" of ^2E (d^3) remaining surprisingly close to degeneracy between the two orbital configurations available to three $3d$-like electrons.

The weak band of cerium(III) aqua ions at $33\,700\ cm^{-1}$ can now [37, 42, 53, 66, 73] be assigned to a reasonably strong band of one (or two) species $Ce(OH_2)_8^{+3}$ in relative concentrations of the order of 5 percent. The reviewer does not recommend a decrease in energy differences between nl and $n'l'$ shells to be called the "nephelauxetic effect" (originally reserved for chemical variation of Slater-Condon-Shortley (or Racah) parameters [1, 4, 6, 19, 31, 41, 46, 65, 73] of interelectronic repulsion in compounds or complexes containing less than ($4l + 2$) electrons in a shell $3d, 4d, 5d, 4f$, or $5f$), although this analogy can be defined as colloquial usage, since the common aspect is an efficient *central field* $U(r)$ decreasing less rapidly with increasing r than in the gaseous Z-ion having the charge corresponding to the oxidation state [46].

5 Photophysics and Photochemistry of Cerium(III) and Scintillating Materials

Since atmospheric air in laboratories is only slightly transparent below 192 nm (above 52 000 cm^{-1} or 6.5 eV photon energies, mainly because of strong absorption in diatomic oxygen, and gaseous H_2O) many recent studies of photophysics have imperceptibly replaced techniques based on "very soft X-rays" usually as 20 to 200 eV photons. This spectral region is not only important to recent extensive studies of the line emission of highly charged, monatomic ions [3] connected with very high-power lasers, but the whole subject of *photo-electron spectra*, starting with David Turner's work in 1962 [80, 81], originally mostly on gaseous molecules bombarded at low pressure with monochromatic photons such as the strongest line of the helium atom (21.2 eV) and the $2p \rightarrow 1s$ emission (40.8 eV = 3 rydberg) of He$^+$. The ionization energies I (usually as Franck-Condon barycentres close to the maxima of the signals due to liberated electrons, much like a Hertz experiment) have been exceedingly important for comparison with M.O. calculations [80–82] and were connected early [4] with the concept of *optical electronegativities* x_{opt} derived from electron transfer spectra of d^q and f^q complexes [24, 28, 31, 41, 46, 56, 63, 64]. The ionization energies below 40 eV have, to a great extent, monopolized the acronym PES. Vaporized $LnCl_3$ molecules have been carefully studied by Lee, Potts and Bloor [83] joining the treatment of ionization probabilities [mainly in solids; XPS or ESCA (Electron Spectroscopy for Chemical Applications) usually bombarding the sample with 1253.6 eV ($2p \rightarrow 1s$ emission from magnesium anti-cathodes) or 1486.6 eV (in aluminium)] $4f^q(S, L, J)$ to $4f^{q-1}(S \pm 1/2, L', J')$ energy levels, treated by coefficients of fractional parentage [84, 85]. It may be noted that the first rationalization of the specific $4f$-group chemistry [86] was mainly based on X-ray induced photoelectron spectra. The chemist may have difficulty adapting to the less familiar ideas of condensed matter physics.

Przibram [87, 88] studied the ultra-violet induced luminescence of certain samples of fluorite (mainly CaF_2) which can be almost colorless (or pale yellow frequently due to traces of $4f^7$ europium(II) and/or $4f^{14}$ ytterbium(II)). Other samples are strongly colored, partly by a variety of color centers. The word "Farbzentrum" (F center) was introduced by Pohl studying far u-v. spectra, coloration etc. of alkali-metal halides such as the blue $NaCl_{1-x}(e^-)_x$ [having been in contact with radioactive U and Th minerals for millions of years]. Butement [89] measured intense $4f^6 \rightarrow 4f^5 5d$ transitions at 17 900, 21 100, 23 400, and 30 300 cm^{-1} of red samarium(II) aqua ions in solution, whereas bands occur for europium(II) at 31 200 and 40 400 cm^{-1} due to excited configurations $4f^6 5d$, and comparable transitions $4f^{14} \rightarrow 4f^{13} 5d$ of ytterbium(II) aqua ions at 28 400 and 40 600 cm^{-1}. Loh [90] measured cerium dispersed as Ce(III) in fluorites, a single band [$4f$ to non-bonding $5d(e_g)$] in cm^{-1}: (CaF_2: 32 500; SrF_2:33 600; BaF_2:34 200) and a triple maximum (CaF_2: 49 500, 51 100, 53 900; SrF_2: 48 800, 50 300, 53 400) due to the three anti-bonding

Ce $5d$ orbitals having angular dependence proportional to (xy; xz; yz). These transitions from Ce $4f$ may be compared to the transitions to the Ce $5d(t_{2g})$ lower sub-shell in O_h symmetry, $CeCl_6^{-3}$ in CH_3CN at 30 300 and $CeBr_6^{-3}$ at 29 150 cm^{-1}, both showing weak shoulders due to spin-orbit coupling [56]. Although the (expected broad and weaker) transitions to the totally symmetric $6s$-like orbital have not been detected, cryogenic spectra [91] of the elpasolite $Cs_2NaY_{1-x}Ce_xCl_6$ show origins of the $[4f \rightarrow O_h\ (t_{2g})]$, two spin-orbit components at 28 196 and 29 435 cm^{-1}.

The mercury atoms in modern glass tube lamps emit a strong line from $6s6p$ $^3P_1$ at 253.65 nm and $6s6p$ from $^1P_1$ a much stronger line at 184.9 nm (54 069 cm^{-1}). It became a major challenge for the Philips lamp laboratories in Eindhoven to utilize the energy of the two kinds of mercury-emitted photons, and capture it in visible luminescence [in "phosphors", a word derived from the Bologna stone BaS (containing ppm of bismuth) made by strong reduction in crucibles of the $BaSO_4$ mineral]. There has been extensive development of luminescent materials containing Sb(III), Mn(II), and Ce(III) each emitting one or two fairly broad bands in the visible spectrum. A somewhat expensive source of green light is terbium, of which the $4f$ level $^5D_4$ down to the seven levels of ^7F_J (the only term lower than $^5D_4$).

Renata Reisfeld [31, 32, 92] studied many cases of energy transfer in glasses from the metastable (hardly radiating) $^3P_0$ state of Hg^0 at 37 645 cm^{-1}; and transfer from Eu(III) $^5D_0$ $^5D_1$ and $^5D_2$ to Yb(III) $^2F_{5/2}$ and the ground state $^2F_{7/2}$; from Ho(III) $^5S_2$ to Sm(III) $^5I_4$ to Tm(III) $^3H_4$ etc., Marvin Weber [93] studied energy transfers from $3d^3$ Cr(III) (^2E & its adjacent doublet $^2T_1$) in $Y_{1-x}Nd_xAl_{1-y}Cr_yO_3$ [perovskite] to Ho(III), Er(III), and Tm(III) $^3H_4$ at 820 cm^{-1} below the ^2E. An interesting type of energy transfer [94] is from the thallium(I) $6s6p$-like level $^3P_1$ to gadolinium(III) $^6P_{7/2}$ at 312.5 nm in borate glass, and a similar behavior is seen ([31], p. 187) of Ce(III) in borate and in phosphate glasses efficiently transferring energy to thulium(III), monitored in the excitation spectrum at 455 nm, a narrow emission from $^1D_2$ to $^3H_4$ (in old days called $^3F_4$). A comparable study was [95] made of Mn(II) [long-lived, lowest quartet state] to thulium(III) in an essentially oxide-free fluoride glass of (mole percentage) composition: 36 PbF_2 + 24 MnF_2 + 35 GaF_3 + 2 AlF_3 + 3 YF_3 + 2 TmF_3.

Pioneer work was carried out by Blasse and Bril [96] on cerium(III) Stokes shifts and luminescence yields in various mixed oxides, such as the cubic garnet $Y_3Al_5O_{12}$ (the Y site has $N = 8$, but rather irregular, the Al site for two-fifths octahedral $N = 6$ and for three-fifths $N = 4$ tetrahedral. The large Stokes shift for $SrY_{2-x}Ce_xO_4$ about 8000 cm^{-1} provides green light.

The comparative study [97] of Ce(III) emission and absorption spectra in both borax and sodium phosphate glass shows 5 approximately equidistant bands with the oscillator strength not far from 0.001 in phosphate glass, and 0.002 to 0.005 in borate. This is reminiscent of Ce(III) aqua ions in Eq. (5), but it is beyond the reviewer's ability to suggest defined linear combinations of the five orbitals written [42, 46] as real (non-complex) linear combinations on a spherically symmetric basis. At least by low T, an almost general tendency of the lowest-energy emission band was the two-maxima shape ascribed by Kröger and Bakker [98] to the major terminal states being derived, cf. Eq. (3), from three Kramers doublets of $J = 5/2$ and four of $J = 7/2$.

Kröger [99] also published the most trend-setting book on emission spectra of solids in the first half of this century.

Scintillation does not mean in this chapter "to twinkle" as a star in transparent, though turbulent (e.g. temperature inhomogeneities) air, but a rapid time evolution of light intensity (and also the variation of spatial angle) of a scintillator, a luminescent material hit by very rapid "elementary" particles, or gaseous ions (such as ^{4}He nuclei emitted from α-radioactive isotopes originating from uranium or thorium minerals). They were the tool (such as ZnS or Willemite Zn_2SiO_4) by which Ernest Rutherford demonstrated that normal matter has 99.94 to 99.97 percent of its rest-mass [in the sense of Einstein's relativity] tucked up in nuclei with a density of 10^{14} g/cm^3 (founding the perennality of the Lavoisier [100] elements).

Before turning toward the specific applications in (ultra-) high-energy physics, we will briefly discuss the *cathodoluminescence* emitted by the wall of a Crookes cathode-ray tube. Abraham Pais [101] stated that the first "elementary" particle was the *electron* characterized in 1897 by J.J. Thomson (inside cathode-ray tubes) having the ratio (− e/m) between their electric charge and their mass [m = 0.00055 atomic mass units] being independent of the metallic cathode material emitting the electrons. However, shortly after Einstein had published his [comparatively "classical"] special theory of relativity in 1905, direct experiments with β-rays emitted by radioactive nuclei, and by electrons accelerated in high potential differences

$$mc^2 = m_0c^2/(1 - v^2/c^2)^{0.5} \qquad (7)$$

where m is the inertial mass resisting acceleration, and m_0 the rest-mass (very slightly below m), when the velocity v relative to the observer is quite small.

When the Van de Graaff linear accelerator (for protons and heavier nuclei) was perfected, as well as the pulsed magnetic-field cyclotron invented by Lawrence in 1932, Eq. (7) was confirmed for protons, deuterons and helium nuclei. Actually, it had a less appealing consequence that very high mc^2 [i.e., v almost c] is not compatible with a constant frequency of field reversal. The remedy was the *synchrotron* admitting bunches [in most cases, of protons] at non-equidistant time-intervals, such as the PS (1959) 28 GeV, and SPS (1976) 400 GeV, both at CERN (west of Geneva) and the Fermi National Accelerator 1 tera-eV = 1000 GeV, million MeV (1984). It may be noted that m_0c^2 for 1 amu is 0.9315 GeV so the TeV machines impart protons with 1000 times more kinetic energy than m_0c^2.

At a laboratory like CERN, many experimental teams work at the same time with the available large accelerators. Developing complex electronic equipment connected with particle detectors are [somewhat metaphorically called "calorimeters" accounting for the energy (more than the heat) and trajectories of the moving clones of particles].

Even in 1945 nuclear fission reactors, scintillating materials such as transparent organic polymers containing terphenyl were extensively used, applying very sensitive photon counters. Most aromatic molecules have a Stokes shift sufficiently large to be almost transparent to their emitted light.

High-energy physics scintillators are more connected to the inertial frame of the observer, and are preferred as rigid crystals or (at least) hard glasses. Relativistic kinematics (of which Eq. (7) is the first statement) is taken seriously. Most present-day experiments have so little $m_0 c^2$ in their total energy mc^2.

Factually, limpid crystals of bismuth germanate $Bi_4(GeO_4)_3$ isotypic with cubic eulytite $Bi_4(SiO_4)_3$ have been made in multi-ton quantities in China, and sold to CERN for about 1000 US\$/kg (~ 150 cm^3). Its near-u.v. emission is rather similar to other 80-electron systems [31] such as Tl(I), Pb(II) and Bi(III).

At a conference [102] in Chamonix, September 1992, cerium(III) emission from CeF_3 and $La_{1-x}Ce_xF_3$ [102, 103] turned up as an interesting alternative to bismuth germanate. These materials crystallize as the hexagonal mineral *tysonite*, but the $N = 11$ of fluoride ligators [104] have quite differing Ce–F distances (seven between 240 and 246 pm, two at 262 and two at 297 pm). Although the absorption peaks at 40 500, 43 100, 45 900, and 48 300 cm^{-1} and the luminescence (in stoichiometric cerium(III) fluoride) at 32 000 cm^{-1} with the strongest band in the excitation spectrum (P about 1) occurs at 161 nm (62 000 cm^{-1}) are rather similar to the cerium(III) aqua ion [67, 109] with $N = 9$, photodissociating to $N = 8$ and then fluorescing at 31 000 in crystals, and 28 000 cm^{-1} in solution.

It may be useful to read the quite numerous recent reviews about scintillating oxides containing large amounts of Z above 50. Here, we can give four references [104–107]. Mostly studied are mixed oxides [including sulfates, phosphates etc.] but all four halides are well represented, e.g. huge crystals of pure caesium iodide, or doped with thallium. One should not concentrate exclusively on ceramic- and glass-type inorganic scintillators; organocerium(III) complexes in tetrahydrofuran [108] "thf" include fascinating halides such as $Cl_3Ce(thf)_3$ which emit in the near u.v. (350 nm) whereas yellow crystals of $((CH_3)_5C_5)_2CeI...(NCCH_3)_2$ having a total $N = 13$ show emission bands at 560 and 628 nm in the thf solvent, and the crystalline complex at 542, 577, 629 and 662 nm. These are, at face value, very large Stokes shifts. The excited states presumably are not amenable to LCAO description, but have Ce $5d$ and aromatic conjugated orbitals mixed in a way [28, 46] as seen with iron(II) $3d^5 4s$ and copper(I) $3d^9 4s$ states. Cerium(III) is not the only lanthanide showing such behavior; a recent thesis by Salvatore Lizzo on europium(II) $4f^6$ (^7F) $5d$ and on ytterbium(II) giving finely structured $4f^{13}$ (^2F) $5d$ absorption and luminescence spectra.

In view of the conceivably rapidly growing interest in scintillators, and in those fields of high-energy physics where they are practically used, the reviewer may finally cite the general references [4, 14, 32, 100].

6 References

1. Jørgensen CK (1962) Solid State Phys 13: 375
2. Jørgensen CK (1976) Gmelin's Handbuch der anorganische Chemie vol 39 B1 (Seltene Erden) p 17 Springer-Verlag, Berlin
3. Jørgensen CK (1991) Comments Inorganic Chem 12: 139
4. Jørgensen CK (1988) In: Gschneidner KA, Eyring L (eds) Handbook on the physics and chemistry of rare earths, North-Holland, Amsterdam, vol 11, p 197
5. Goldschmidt ZB (1978) In: Gschneidner KA, Eyring L (eds) Handbook on the physics and chemistry of rare earths, North-Holland, Amsterdam, vol 1, p 1
6. Condon EU, Shortley GH (1953) Theory of atomic spectra (2nd edn), Cambridge University Press
7. Meyer B (1970) Science 168: 783
8. Ammeter J, Schlosnagle DC (1973) J Chem Phys 59: 4784
9. Bethe H (1929) Ann Physik 3: 133
10. Hellwege KH (1948) Ann Physik 4: 95; 127; 136; 143; 150; 357
11. Prather JL (1961) Atomic energy levels in crystals. NBS Monograph 19, National Bureau of Standards, Washington DC
12. Morrison CA, Leavitt RP (1982) In: Gschneidner KA, Eyring L (eds) Handbook on the physics and chemistry of rare earths, North-Holland, Amsterdam, vol 5, p 461
13. Rabinowitz E, Thilo E (1930) Periodisches System, Geschichte und Theorie, F. Enke, Stuttgart
14. Jørgensen CK (1993) Comments Astrophysics 17: 49–101
15. Van Santen JH, Van Wieringen JS (1952) Rec trav chim 71: 420
16. Ilse F, Hartmann H (1951) Z physik Chemie (Leipzig) 197: 239
17. Hartmann H, Ilse F (1951) Z Naturforsch 6a: 751
18. Orgel LE (1955) J Chem Phys 23: 1004; 1819; 1824; 1958
19. Jørgensen CK (1954) Acta Chem Scand 8: 175; 1495; 1502, (1955) ibid 9: 405; 540; 710; 1362, (1956) ibid 10: 500; 518; 887; 1503; 1505, (1957) ibid 11: 53; 73; 151; 166; 399; 981, (1958) ibid 12: 903; 1537; 1539; (1959) Acta Chem Scand 13: 196
20. Jørgensen CK (1956) Absorption Spectra of Complexes with Unfilled d-shells. Institut International de Chimie Solvay, Bruxelles, p 335
21. Hartmann H, Schläfer HL (1954) Angew Chem 66: 768
22. Schäffer CE, Jørgensen CK (1958) J Inorg Nucl Chem 8: 143
23. Jørgensen CK (1962) Progress Inorg Chem 4: 73
24. Jørgensen CK (1963) Adv Chem Phys 5: 33
25. Jørgensen CK (1969) Chimia (Zürich) 23: 181, (1971) ibid 25: 109; 213, (1972) ibid 26: 252, (1973) ibid 27: 203, (1974) ibid 28: 6; 605, (1975) ibid 29: 53, (1984) ibid 38: 75
26. Jørgensen CK (1981) Comments Inorg Chem 1: 123–140
27. Jørgensen CK, Rittershaus E (1967) Mat fys Medd Vid Selskab (Copenhagen) vol 35, no 15
28. Jørgensen CK (1970) Progress Inorg Chem 12: 101–158
29. Jørgensen CK, Pappalardo R, Rittershaus E (1964) Z Naturforschung 19a: 424
30. Carnall WT, Fields PR, Rajnak K (1968) J Chem Phys 49: 4412; 4424; 4443; 4447; 4450
31. Reisfeld R, Jørgensen CK (1977) Lasers and excited states of rare earths, Springer, Berlin
32. Reisfeld R, Jørgensen CK (1987) In: Gschneidner KA, Eyring L (eds) Handbook on the physics and chemistry of rare earths, North-Holland, Amsterdam, vol 9, p 1
33. Broer LJF, Gorter CJ, Hoogschagen J (1945) Physica (Utrecht) 11: 231
34. Judd BR (1962) Phys Rev 127: 750
35. Ofelt GS (1962) J Chem Phys 37: 511
36. Peacock RD (1975) Structure and Bonding 22: 83
37. Freed S (1931) Phys Rev 38: 2122
38. Fitzwater DR, Rundle RE (1959) Z Kristallogr 112: 362
39. Helmholz L (1939) J Am Chem Soc 61: 1544
40. Wolfsberg M, Helmholz L (1952) J Chem Phys 20: 837
41. Jørgensen CK (1962) Orbitals in atoms and molecules, Academic Press, London
42. Jørgensen CK (1971) Modern aspects of ligand field theory, Elsevier, Amsterdam
43. Ballhausen CJ, Gray HB (1962) Inorg Chem 1: 111
44. Ferreira R (1976) Structure and Bonding 31: 1
45. Iczkowski RP, Margrave JL (1961) J Am Chem Soc 83: 3547

46. Jørgensen CK (1969) Oxidation numbers and oxidation states, Springer, Berlin
47. Jørgensen CK (1988) Quimica Nova (São Paulo) 11: 10
48. Jørgensen CK, Horner SM, Hatfield WE, Tyree SY (1967) Int J Quantum Chem 1: 191
49. Sanderson RT (1988) J Chem Educ 65: 112; 227
50. Jørgensen CK, Faucher M, Garcia D (1986) Chem Phys Lett 128: 250
51. Faucher M, Garcia D, Jørgensen CK (1986) Chem Phys Lett 129: 387
52. Ruedenberg K (1962) Rev Mod Phys 34: 326
53. Jørgensen CK, Pappalardo R, Schmidtke HH (1963) J Chem Phys 39: 1422
54. Lang RJ (1936) Canad J Res A14: 127
55. Pyykkö P (1988) Chem Rev 88: 563
56. Ryan JL, Jørgensen CK (1966) J Phys Chem 70: 2845
57. Ryan JL (1969) Inorg Chem 8: 2053
58. Jørgensen CK, Reisfeld R (1982) Topics Current Chem 100: 127
59. Jørgensen CK (1984) Topics Current Chem 124: 1
60. Jørgensen CK (1989) Topics Current Chem 150: 1
61. Martin WC, Zalubas R, Hagan L (1978) Atomic energy levels, the rare earth elements NSRDS-NBS vol 60. National Bureau of Standards, Washington DC
62. Morrison CA, Leavitt RP, Wortman DE (1980) J Chem Phys 73: 2580
63. Ryan JL, Jørgensen CK (1963) Mol Phys 7: 17
64. Jørgensen CK, Reisfeld R (1982) Structure and Bonding 50: 121–171
65. Reisfeld R, Jørgensen CK (1988) Structure and Bonding 69: 63–96
66. Okada K, Kaizu Y, Kobayashi H, Tanaka K, Marumo F (1985) Mol Phys 54: 1293
67. Jørgensen CK, Brinen JS (1963) Mol Phys 6: 629
68. Oelkrug D (1971) Structure and Bonding 9: 1
69. Smith DW (1978) Structure and Bonding 35: 87
70. Reinen D, Friebel C (1979) Structure and Bonding 37: 1
71. Warren KD (1984) Structure and Bonding 57: 119
72. Heidt LJ, Berestecki J (1955) J Am Chem Soc 77: 2049
73. Jørgensen CK (1956) Mat fys Medd Dan Vidensk Selskab 30: no 22
74. Bjerrum J, Jørgensen CK (1953) Acta Chem Scand 7: 951
75. Malkova TV, Shutova GA, Yatsimirskii KB (1964) Russ J Inorg Chem 9: 993
76. Johansson G (1992) Adv Inorg Chem 39: 159
77. Neilson GW, Enderby JE (1989) Adv Inorg Chem 34: 195
78. Kauffman GB (ed) (1994) Am Chem Soc Symposium Volume 565: Coordination Chemistry
78 a: Jørgensen CK: p 177; p 226
    b: Bendix J, Brorson M, Schäffer CE: p 213
    c: Reisfeld R, Jørgensen CK: p 439
    d: Schäffer CE: p 96
79. Cotton FA, Meyers MD (1960) J Am Chem Soc 82: 5023
80. Turner DW, Baker C, Baker AD, Brundle CR (1970) Molecular photoelectron spectroscopy, Wiley-Interscience, London
81. Rabalais JW (1977) Principles of ultraviolet photoelectron spectroscopy, Wiley-Interscience, New York
82. Jørgensen CK (1975) Structure and Bonding 22: 49
83. Lee EPF, Potts AW, Bloor JE (1982) Proc R Soc London Ser A381: 373
84. Cox PA, Baer Y, Jørgensen (1973) Chem Phys Lett 22: 433
85. Cox PA (1975) Structure and Bonding 24: 59
86. Jørgensen CK (1973) Structure and Bonding 13: 199
87. Przibram K (1936) Z Physik 102: 331
88. Przibram K (1937) Z Physik 107: 709
89. Butement FDS (1948) Trans Faraday Soc 44: 617
90. Loh E (1967) Phys Rev 154: 270
91. Schwartz RW, Schatz PN (1973) Phys Rev B 8: 3229
92. Reisfeld R, Eckstein Y (1973) J Non-crystalline Solids 11: 261
93. Weber MJ (1973) J Appl Phys 44: 4058
94. Reisfeld R (1976) Structure and Bonding 30: 65–97
95. Eyal M, Reisfeld R, Schiller A, Jacoboni M, Jørgensen CK (1987) Chem Phys Lett 140: 595
96. Blasse G, Bril A (1967) J Chem Phys 47: 5139
97. Reisfeld R, Hormodaly J, Barnett B (1972) Chem Phys Lett 17: 248
98. Kröger FA, Bakker J (1941) Physica (Utrecht) 8: 628

99. Kröger FA (1948) Some aspects of the luminescence of solids, Elsevier, Amsterdam
100. Jørgensen CK, Kauffman GB (1990) Structure and Bonding 73: 227
101. Pais A (1988) Inward Bound-Of matter and forces in the physical world, Clarendon Press, Oxford
102. Reisfeld R (1993) In: Heavy Scintillators for Scientific and Industrial Applications, p 155, Editions, Frontières, Gif-sur-Yvette
103. Pedrini C, Moine B, Gaçon JC, Jacquier B (1992) J Phys Condens Matter 4: 5461
104. Moses WW, Derenzo SE (1989) IEEE Trans Nucl Science NS-36: 173
105. Anderson DF (1990) Nucl Instr Methods A 287: 606
106. Moses WW, Derenzo SE (1990) Nucl Instr Methods A 299: 51
107. Book of Abstracts "SCINT 95", international conference on Inorganic Scintillators and their Applications, Delft University of Technology, end of August 1995
    a: abstract p 56 [TU-A-35] Jørgensen CK, Reisfeld R: Inorganic Scintillation as Born-Oppenheimer Constrained Light Emission Between Multidimensional Surfaces
108. Hazin PN, Lakshminarayan C, Brinen LS, Knee JL, Bruno JW, Streib WE, Folting K (1988) Inorg Chem 27: 1393
109. Jørgensen CK (1994) Journal de Physique Colloques C 4: 333
110. Lizzo S (1995) Luminescence of Yb (II), Eu (II) and Cu(I) in Solids. Doctoral Thesis, University of Utrecht

Lasers Based on Sol-Gel Technology

Renata Reisfeld[1]

Department of Inorganic Chemistry, Hebrew University, 91904 Jerusalem, Israel

A new type of solid state laser, tunable in the visible range, has been developed recently.

Incorporation of perylimide and pyrromethene dyes into glasses prepared by the sol-gel method enables the design of new types of visible, stable solid lasers. These can be prepared either in the form of slabs or rods or as waveguiding active media deposited on glass or polymer supports. The difference between the refractive indices of the film and its support determines whether waveguiding occurs in the film or a leaky waveguide laser is formed. The efficiency of the lasers obtained recently is up to 70% when appropriate optical pumping is designed.

We discuss the chemical and physical aspects of the photostability of the dyes in glasses, phenomena responsible for non-radiative relaxation of their excited state, and energy efficiencies of laser.

---

[1] Enrique Berman, Professor of Solar Energy

Structure and Bonding, Vol. 85
© Springer-Verlag Berlin Heidelberg 1996

1 Introduction

Liquid dye lasers tunable in the visible range have been known since the middle of the sixties [1]. They are based on organic colorants dissolved in various solvents. The light absorption from the pump source brings the dye molecule to its excited singlet state, and the emission to the terminal vibrational state can then be tuned by using appropriate resonant cavities to produce laser emission in the spectral range of fluorescent emission.

Since the introduction of organic dyes in gels ten years ago [2, 3], much research work has been devoted to the application of these doped solids as gain media in laser cavities as they provide many advantages compared to liquid dye lasers [4–30]. Many types of gel matrices have been studied, from "totally" inorganic to mixed organic/inorganic, sometimes impregnated to fill up the porous volume (plastic host matrices have also been studied [31–36]). Doping procedures have been applied to both sol and gel. The most commonly used dyes were first rhodamines and are now pyrromethenes. Over these years, the characteristics of laser emission have been constantly improved, ranging from microjoule to millijoule energy levels, the operating lifetimes increasing from a few shots to several thousand shots. We shall give a short historical survey on how these lasers have been developed, describe the way they operate and give some insight to their future.

2 Historical Development

Until recently, liquid dye lasers were the main systems used to achieve tunability in the visible range, and the only commercial choice for tunable lasers between 400 and 660 nm. However, in the last few years an intensive effort has been devoted to producing embedded organic dyes in various solid matrices, with the goal of achieving solid-state dye laser devices that may replace liquid dye lasers; e.g. laser dyes were incorporated into polymers [37, 38], silica-gels [15, 39], xerogels [9], alumina gels [5, 6], ormosils [19], and composite glasses [8, 17, 40, 41]. A solid-state dye laser has the advantages over a liquid dye laser by not being a volatile, flammable, toxic solvent, and by its compact size and mechanical stability. Still, for applications that require high powers, at either C.W. or pulsed high-repetition-rate operation, the problem of heat dissipation is a serious impediment to their utilization. In liquid dyes, on the other hand, a jet or a flowing solution are handy, practical ways of solving the heat problem. In both cases, photostability is a feature of prime importance in selecting a laser dye.

3 Lasers Based on the Sol-Gel Process

The principles of the sol-gel process are as follows.

The sol-gel process consists of: (i) preparing a homogeneous solution of easily purifiable precursor(s), generally in organic solvent miscible with water or the reagent used in the next step; (ii) converting the solution to the 'sol' form by treatment with a suitable catalyzer; (iii) allowing the sol to change into a 'gel' by polycondensation; (iv) shaping the gel (or viscous sol) to the finally desired form or shape such as thin film, fiber or bulk; (v) converting (sintering) the shaped gel to the desired ceramic material at temperatures generally much below ($\sim 500\,^{\circ}$C) those required in the conventional procedure of melting oxides together [42]:

In the sol-gel method, inorganic materials (glasses) are prepared from solutions containing metal compounds such as sources of cations in the final oxides, water as hydrolysis agent and alcohols as solvents. In the solution, metal compounds undergo hydrolysis and polycondensation, forming polymers or particles, and the solution becomes a sol. Further reaction connects the particles, solidifying the sol into a gel. The dried gels may form the final products. The maximum processing temperature may be lower than 100 or 160 °C and so one does not have to worry about the decomposition of the organic compounds imbedded in the gel which have been added as indispensable components. On the other hand, heating of pertinent dried gels to several hundred degree Celsius produces glasses and ceramics which have prescribed shapes, such as plate, cylinder, fiber and coating film. Thus, in general, inorganic materials can be obtained without powder processing.

Dip-coating or spin-coating on glass, ceramic, metal, and plastic substrates by the sol-gel method is very useful for modifying properties of the substrates with a large or small surface area and provides the substrates with new active functions. In the years around 1970, dip-coating was applied mainly to glasses in order to modify their optical properties. In recent years, the dip-coating technique has again attracted much attention in developing various advanced materials.

The sol-gel method of fabricating films offers potential advantages over traditional techniques, the low processing temperature being particularly important in the application to electronic, optoelectronic, and photonic devices where the substrate and other active elements on the substrate are not necessarily highly heat-resistant. Easy coating of large surfaces makes it possible to apply the sol-gel coating to display panel and windows as substrates. The small thickness may be advantageous for coating films for some optical and electronic devices. For other purposes, however, thicker films may be needed. High optical quality films can also be provided [43].

The question as to whether an element is suitable for sol-gel processes or not cannot be answered very easily. However, there are some rules which can be drawn from the basics of the process. In order to form an inorganic network from a solution, a network-forming step is required. This step depends mainly on the structure of the sol and can be roughly divided into two alternative mechanisms. The first is a mechanism which is based on the growth of molecules, leading to macromolecules which then grow together to an infinite network. This mechanism leads to the so-called polymerized gels and is very common in the acid-catalysed hydrolysis and condensation of tetraalkyl silicates. The other type is based on the aggregation of colloidal particles from a so-called colloidal sol and requires a fairly stable sol as intermediate. Otherwise the whole procedure would end up in a precipitation process with no sol phase to be identified. The network-forming step in these sols is the aggregation of particles to an infinite network. In the case of acid-catalyzed silica from alkoxides, the polymerization process can be easily controlled by the limitation of water which leads to stable sols if enough unhydrolysed $\equiv SiOR$ groups can be maintained to keep the average molecular weight small.

The influence of protons and water on the structure of SiO_2 sols has been intensively studied by Sakka and Yoko [43] who showed, by rheological analysis, that various structures of viscous sols can be obtained just by varying H^+ and H_2O concentration in the starting solution.

The stabilization of sols is of great importance in sol-gel processing because it defines very strongly their processing properties which are of special importance for film formation. These are, for instance, the rheology, the maximum solid content and the particle size and distribution. Rheology is a complex parameter and depends on particle shape, temperature, solvent, concentration and particle interaction. The reduced viscosity η/c of non-interacting particles does not depend on concentration:

$$\frac{\eta}{c} = \frac{k}{\rho} \tag{1}$$

where k represents a constant and ρ the density of the particles. For an organic polymer solution, the intrinsic viscosity η_i is related to the average molecular weight M by

$$i = k \cdot M\alpha \tag{2}$$

where k is a constant depending on temperature, solvent and chemistry of the polymer and α represents a parameter depending mainly on the polymer structure. Thus a rough distinction can be made between "basic macromolecular forms", e.g. $\alpha = 0$ for rigid spherical particles, $\alpha = 0.5–1.0$ for flexible chain-like molecules, and for rigid, rod-like molecules α becomes 1.0–2.0. The determination of α allows one to tailor processing properties if the mechanisms of particle shaping can be controlled, e.g. for the hydrolysis and condensation process of tetraethylorthosilicate, α can be selected for optimal fiber spinning. The process has been industrialised by Ashi Glass Co. for the production of high quality SiO_2 fibers. In this case hydrolysis under acid conditions leads to chain-like flexible polymers suitable for fiber drawing.

So far the SiO_2 system has received most scientific interest. This may be due to the fact that SiO_2 precursors in the form of alkylortho silicates have been readily available for almost 150 years, and, compared to almost all other common alkoxides, they are relatively insensitive to moisture. This means that the hydrolysis and condensation to gels take place pretty slowly (depending on concentration, type of alkoxides, and solvent, between hours and weeks). Therefore it is possible to study the reaction kinetics and structure-forming mechanisms of sols by condensation in detail relatively conveniently compared to other systems. The reaction of tetraethyl orthosilicates as interesting precursors in sol-gel reactions has been intensively discussed [44]. The films based on titanium dioxide have also been studied intensively as will be seen later in this section.

The sol structure also influences the maximum film thickness. For given systems, the "cracking thickness" of coating is defined by the system parameters only. The films can be transformed to dense layers by temperature treatment only if their thickness does not exceed certain limits defined by the intrinsic system parameters. In general, the film thicknesses to be obtained in a one-step coating process (spin- or dip-coating) does not exceed some tenths of one micrometer. If the viscosity of a system is adjusted to obtain thicker films, cracks occur during drying. This problem can be circumvented by using composite materials such as ormocers (see below).

The ability to dope sol-gel derived silica hosts [2, 3] with controlled amounts of a lasing species affords the possibility of developing a new generation of advanced tunable solid state lasers. Such lasers, as well as possessing the tunability common to all liquid lasers, can also take advantage of the superior physical properties of sol-gel derived silica. These properties include low nonlinear refractive index coefficient, low strain birefringence, coefficient of thermal expansion closed to zero, low temperature dependence of expansion coefficient and low impurity levels. However in the last few years intensive effort has been devoted to producing embedded organic dyes in various solid matrices (see above) with the goal being to achieve solid-state dye laser devices that may replace liquid dye lasers [45].

In 1989 we succeeded for the first time in preparing a photostable tunable laser by impregnating the orange perylene derivative (perylimide) dye "BASF-241" dissolved in MMA into a silica-gel [8, 40]. The method of Pope and Mackenzie, which allows polymerization of MMA in the pores of the glass, was applied here. The dye, which is orders of magnitude more stable than the conventional laser dyes impregnated in the glass, provided an efficient solid-state laser material. This laser was tunable in the range 568–583 nm.

Later we followed the success of the previous work to study further the possibility of impregnating different dyes into the silica-gel–PMMA composite glass [17, 18]. We started with the red perylimide dye (RPD). The fluorescence emission peak of this dye is centered at about 613 nm. This wavelength is important for medical photodynamic therapy (PDT) and diagnostics. Human blood and tissue absorption is small in the red, allowing the preferential

absorption of light by a photo-active cancer therapeutic agent, such as hema-
toporphyrin derivative, which concentrates in tumors [46]. The properties of
RPD in this system have been described [47, 48].

Other research groups have also reported the introduction of some known
laser dyes (such as rhodamine dyes) into different solid host materials, e.g. into
polymers [37, 38], silica-gels [15, 39], xerogels [9], alumina gels [6], and ormosils (organic modified silicates) [21] and polymeric hosts developed recently
in Los Alamos [49]. Most of the conventional laser dyes used exhibited severe
photoinstability (bleaching) and their lasing output decreased rapidly with time
of operation [6]. The perylimide dyes on the other hand exhibit outstanding
photostability in glass. Also, the pyrromethene 1,3,5,7,8-pentamethyl-2-6-di-
ethylpyrromethene-BF_2 complex (PM-567) has a quantum efficiency of 99.5%
and a triplet extinction coefficient of one-fifth of that of Rhodamine 560. In
solutions, these pyrromethene-based dye lasers have outperformed the rhodamine and coumarin dye in the same wavelength ranges under flashlamp and
laser pumping [36]. The pyrromethene dyes exhibit reduced triplet-triplet
(T-T) absorption over their fluorescence and lasing spectral region.

Hermes et al. [36] incorporated this dye into a modified acrylic polymer.
When pumped with 532 nm light of frequency doubled Nd:YAG this laser
material was able to provide 30 mJ of energy with a loss of 34% after 20 000
shots at 3.33 Hz. Additional BF_2 complex laser dyes have been synthesized
and their spectroscopic and laser characteristics studied in solution [50]. The
8-cyano-pyrromethene-BF_2 complexes showed the best performance with red
emission and slope efficiencies of 48% in solution using the same pumping.

Recently, pyrromethene 567 has been successfully incorporated into glasses
prepared by hydrolysis and polycondensation of vinyltriethoxysilane using
acetone as a solvent [29]. In a single shot of operation under excitation of
frequency-doubled Q-switched Nd:YAG laser providing 532 nm 8 ns pulses of
energy up to 10 mJ, 30% efficiency was observed. The maximum energy output
was 3 mJ/pulse. When the sample is pumped repeatedly the organic molecules
are slowly degraded and the output energy lowered.

Laser properties of pyrromethene-doped ormosils were also reported by
Dunn et al. [26] using longitudinal pumping with slope efficiencies comparable
with those of the lasers obtained in modified acrylic plastic [38] in which
reported efficiencies of close to 60% were obtained under sophisticated pumping
arrangements. These results are very encouraging for the further development of
solid state visible tunable lasers.

4 Perylimide Doped Composite Glass Lasers

As described above, lasers obtained by impregnation of the perylimide dyes into
sol-gel glasses, where the dyes are enclosed in the pores of the glass, seem to be

by far the most photostable system. Their preparation procedure was described recently at the Conference on Sol Gel Optics (III) in San Diego [51] and comments of photostability of these dyes in glasses have been published [52].

While our main efforts went into the detailed elaboration of the preparation procedure and the understanding of photostability, we have used transverse pumping for excitation of the laser, which is less efficient than the longitudinal pumping performed by our colleagues [36, 50].

4.1 An Outline of the Preparation Procedure for Perylimide Doped Glasses

The perylimide dyes were dissolved in a methylmethacrylate (MMA) monomer to form solutions of different concentrations in the range 10^{-6} to 10^{-3} mol/l. Highly porous silica-gel bulk glasses (density about 0.7 g/cm^3) were prepared by the sol-gel method, and dried by slow heating (100 °C/day) from room temperature to 500 °C. The bulks were then immersed in the dye-doped solution of the MMA monomer, which was simultaneously catalyzed by the addition of 2% benzoyl peroxide. The MMA-dye solution thus diffused into the silica-gel glass pores and polymerized therein. After this process of dye impregnation, the bulks were re-immersed in an MMA-dye solution, which at this stage was catalysed for full polymerization by 0.5% benzoyl peroxide, and kept in a sealed container at 40 °C for about a week. The samples were then withdrawn, cleaned, and polished, to obtain parallelepiped slabs of approximate dimensions $10 \times 10 \times 3$ mm³, with clear smooth surfaces.

The density of the composite glass was d = 1.447 ± 0.005 g/cm^3 and the refractive index n = 1.472 ± 0.003.

4.2 Laser Experiments and Results

The light source used for pumping of the glass samples was a frequency-doubled Nd:YAG laser (Lumonix HY600), emitting light pulses at wavelength 532 nm. Typical light pulses of the Nd:YAG laser were of ~8 ns full width at half maximum (FWHM) duration, 1–50 mJ/pulse energy, and 1–10 Hz repetition rate. The exciting beam was focused to a rectangular shape on the dye doped glass surface by a combination of spherical and cylindrical lenses. The excited surface region dimensions were 9×0.7 mm for excitation with the frequency-doubled Nd:YAG laser. Lasing was obtained in a cavity of either a combination of a ~100% reflecting metallic back mirror and a ~50% reflecting output coupler, or a combination of a grating of 1200 grooves/mm at Littrow configuration as the back reflector, and the same output coupler.

Table 1 presents the laser parameters obtained experimentally in our laboratory for composite glass lasers [8, 17, 41, 72] under excitation of Nd:YAG laser emitting at 532 nm (second harmonic) at a repetition rate of 10 Hz and

Table 1. The laser characteristics of perylimide dyes incorporated into glasses under excitation of second harmonic Nd:YAG laser 50 mJ/pulse with a transverse pumping

Dye	RPD	BASF-241
Absorption coefficient	44 000 $M^{-1} cm^{-1}$	85 000 $M^{-1} cm^{-1}$
Lifetime radiative	5.74 ns	3.64 ns
Lifetime measured	5.3 ns	3.5 ns
Quantum yield	0.93	0.96
Tunability range	605–630 nm	575–590 nm
Threshold energy	90 μJ/pulse	60 μJ/pulse
Maximum output energy obtained	0.4 mJ/pulse	2.75 mJ/pulse
Energy flux	2 mJ/cm^2 per pulse	13.75 mJ/cm^2 per pulse
Peak power	75 kW per pulse	790 kW per pulse
Stability	500 000 pulses	long time, more than 1 000 000 pulses

lifetime of 8 ns and energies ranging 1–50 mJ/pulse using transverse pumping. The results in the table correspond to our first work. At present the energies are much higher (patent pending).

4.3 Photostability Measurements

The dye BASF 241 did not show appreciable photodegradation under excitation of one million pulses and thus the photostability could be measured only with red perylimide dye RPD.

The lasing photostability of the red perylimide dye (RPD) in various solid matrices was measured under frequency-doubled Nd:YAG laser excitation. The RPD:composite glass laser intensity decayed to 50% of its initial value after approximately 20 000 pump pulses of 13 mJ/pulse. The output of RPD:ormosil glass and RPD:PMMA glass lasers decayed to 50% of their initial value after 1200 and 1000 pump pulses of the same energy, respectively. For Rhodamine-6G:silica-gel and Rhodamine-6G:ormosil glass lasers, the 50% decay had occurred already after 1000 and 300 pulses, respectively. The decay was non-exponential, indicating that dye bleaching is not a single-photon process. The average laser output decay rates increased linearly with the pump energy.

Figures 1 and 2 show the comparison of the decay rate of the laser action of RPD 300 and Rh 6G in composite glass, ormocers and PMMA under the laser excitation of 13 mJ/pulse of the second harmonic output of Nd:YAG laser.

Singlet-singlet excited state absorption of the RPD dye in solid matrices was also measured between 550 and 730 nm. At ∼600 nm the cross section was ∼2×10^{-16} cm^2/molecule. The excited-state absorption competes with the lasing, and is a main factor limiting laser efficiency.

It should be noted that the decay of laser output intensity as a function of the number of pulses is not exponential. This again suggests that dye bleaching is not a single-photon process.

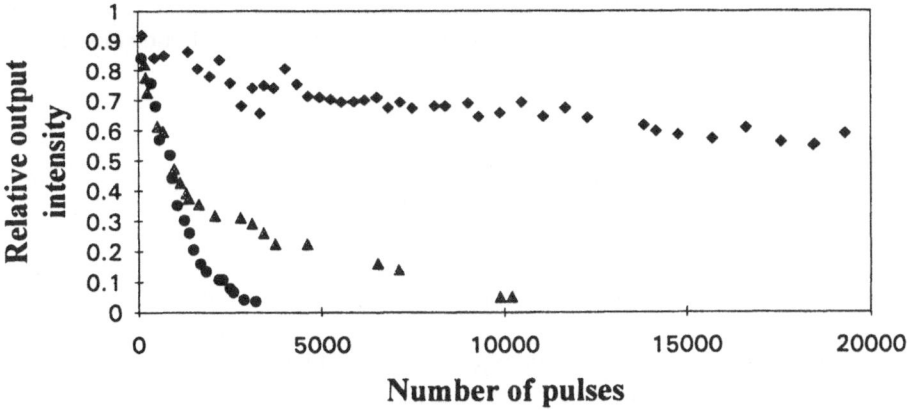

Fig. 1. Relative dye laser output of RPD 300 as function of number of pump pulses of 13 mJ each, for various matrices. *Rhomboids* – composite glass, *triangles* – ormosil, *circles* – PMMA [47]

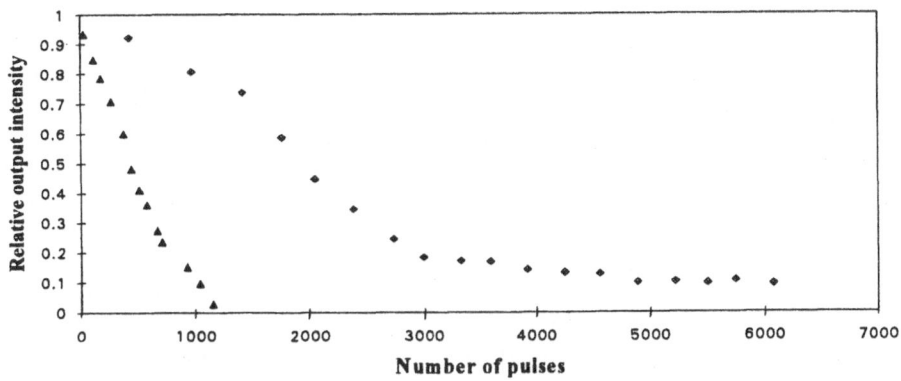

Fig. 2. Relative laser output of rhodamine-6G as function of the number of pump pulses of 13 mJ each, for various matrices. *Rhomboids* – silica-gel glass, *triangles* – ormosil glass [47]

To shed some more light on the dye bleaching mechanism we have measured the laser output decay for different pump light intensities [47]. Obviously the output decay becomes faster as the pump light intensity increases. The results show a linear dependence of bleaching rate on pump pulse energy, with a slope of 1.75×10^{-6} mJ^{-1}. The pump energy density is probably a more meaningful factor than the total pump energy concerning the bleaching effect. The corresponding slope would be 1.25×10^{-7} cm^2/mJ.

The nearest sphere surrounding the perylimide dye molecules is a skeleton consisting of both PMMA and the composite glass. The silicate skeleton of the microphase appears only as a second neighbor. Thus, the effect of this skeleton on the dye photostability is strikingly large. In our opinion the incapsulation of the organic entity in the rigid glass phase allows a back transfer of the excited photoelectron which would otherwise be irreversibly detached and cause photo-decomposition.

5 New "Organic Dye/Xerogel Matrix" Couples

Recently doped xerogels were obtained by hydrolysis-condensation of vinyl-triethoxysilane or methyltriethoxysilane [28, 53], under acid-catalyzed conditions with acetone as common solvent. The molar ratios of silicon alkoxide:water:acetone were respectively 1:3:3. HCl acidified water-acetone mixture was added to the alkoxide-acetone mixture. The solution was stirred vigorously for several hours at room temperature to ensure homogeneous mixing. After complete hydrolysis, a small amount (10^{-3} mol/l) of functionalized alkoxysilane was added to catalyze the condensation reaction. Dye solution was then added in an appropriate solvent. Pyrromethene 567, pyrromethene 580, pyrromethene 597, perylene orange, perylene red, neon red, rhodamine 6G, rhodamine B and xanthylium salt are readily soluble in a variety of solvents including acetone. The resulting clear sols were cast into cylindrical-shaded moulds which were then hermetically closed for gelation, in a drying oven, at 40 °C. A slow gelation process was necessary to prevent possible cracking of the samples. After one week, the containers were opened and finally left to dry for a further three weeks at the same temperature to reach an atmospherically stable condition. Such dried samples had a typical size of 25 mm diameter by 15 mm thickness. Bulk density, measured by Archimedes method in water, was around 1.3 g/cm^3. The xerogels used for laser experiments were polished to obtain parallel surfaces with good optical quality (about 4 nm roughness). Xerogel thin layers (0.5 mm thick) were cut with a diamond saw and polished for optical absorption measurements.

The pyrromethene dyes in xerogels gave the best results with generally more than 50% slope efficiency. The efficiency is strongly dependent on the porosity of the matrix. This dependence is still not completely understood and is the subject of our current investigations.

Preliminary experiments show that it takes much longer to bleach a perylene red doped sample than a pyrromethene 597 one when submitted to a similar C.W. laser flux (531 nm).

A solution often proposed to increase the stability of the doping molecules is to bond the dye covalently to the solid matrix. Unfortunately this, to our knowledge, has always led to a dramatic decrease of fluorescence intensity preventing any laser action with such a sample. However, this method used with a relatively long spacer in order to retain fluorescent characteristics was applied to classical rhodamine B as it is relatively easier to graft it on the matrix than the perylene of pyrromethene dyes [53].

The active molecules (rhodamine B) were attached to the inorganic polymer backbone via a flexible spacer. This approach allows a high concentration of the incorporated active side groups without segregation. The dye containing monomer was prepared from rhodamine B isothiocyanate and (3-aminopropyl) triethoxysilane (APTES) [53]. The synthetic route was as follows. Under inert atmosphere the amino groups (10% excess) of the silicon precursor was reacted with the isothiocyanate groups of the dye in anhydrous ethanol. A small amount of triethylamine was used as catalyst. The reaction mixture was stirred overnight at room temperature. Ethanol was removed under reduced pressure. The crude product was washed with ether to eliminate excess APTES through three or four repeated decantations and dried in vacuo (yield: 50%).

6 Waveguide Lasers

Numerous modern applications such as recording, communication printing, display etc. demand compact wave guiding lasers that can be tuned in the visible spectral range [54]. Such a system was proposed by us about a decade ago, theoretically suggesting an introduction of laser dyes into glass films [55, 56]. Since then, tunable lasers as waveguide amplifiers based on deposition of doped films on glasses have been reported by us [51, 54, 57–63].

7 Resonators for Waveguided Lasers

Distinct from conventional lasers based on bulk optics, waveguide lasers offer the option of using *distributed feedback* (DFB) or *distributed reflectivity* (DBR). These configurations have several advantages as compared to resonators based on localized reflectors. The advantages were demonstrated most clearly in the case of semiconductor lasers [64–66]. The use of reflecting facets in that case results in high power concentration on the facets and limits the output power due to the danger of damaging the emitting surfaces. Moreover, the wide gain band of a semiconductor laser medium and the weak frequency selectivity of a Fabri-Perot resonator cause the excitation (even for a slight excess over the threshold) of many modes whose spectral envelope amounts to several nanometers. These problems are avoided in DFB structures which naturally provide high selectivity of wavelength. In addition, a periodic modulation of refractive index and/or width provides the option of *distributed surface extraction* of radiation through the grating. These advantages of DFB layouts are even more apparent in the case of dye based medium. The bandwidth of gain here amounts to tens of nanometers, so that wavelength control is needed.

The first demonstration of DFB took place more than two decades ago by Shank et al. [64]. Periodic modulation of the refractive index was achieved by several methods, including modulation of gain in the substrate medium by the use of optical interference pumping. Their results showed a drastic reduction of bandwidth, well below 1 nm. In the case of interferometric pumping, tunability was achieved by the change of the intersecting angle of the pump beams. The utilization of DFB in semiconductor lasers is now a matter of routine, and the drastic reduction in linewidth achieved by this method makes it very widely used in commercial lasers.

Turning to very recent reported work, Shamrakov and Reisfeld [60] and He et al. [67] reported the operation of a leaky waveguide laser based on a sol-gel glass film and polymeric films respectively. Leaky-mode waveguided lasers are based on a guiding layer with refractive index lower than the supporting substrate. In a gain medium, such lossy modes can be steadily supported. Since the losses due to Fresnel reflection increase with the value of incident angle, the mode losses effectively discriminate the lowest order transverse mode in the case of a thick film structure. The leaky wave mechanism here provided a natural way of outcoupling the radiation. The resonator cavity here had external mirrors.

Different types of planar resonators based on surface gratings can be designed. First or second order Bragg reflection can be used depending on whether the periodic structure should or should not provide a means of outcoupling the radiation. The resonator structures can be further divided into linear or ring types. Ring resonator types have several advantages as follows.

1. The broader area of light generation enlarges naturally the volume of the amplifying material. The enlarged area releases the need for high concentration of pump radiation and reduces the danger of optical damage to the films.

2. The lithography resolution requirements are released in the case of ring type resonators since the period of the gratings is inversely proportional to $\sin \theta$ where θ is the angle of incidence of light on each grating section.

3. A narrow ray of light obliquely incident at a grating structure will be spatially spread out by the successive reflections, reducing the possibility of self-focusing, filamentation or other sources of light concentration.

The theory of distributed feedback lasers was originally postulated by Shank et al. [64], and further developed subsequently [68–70].

8 Tunability in Thin-Film Lasers

The large amplification bandwidths encountered in transparent media doped with organic dyes makes it an ideal source for the development of tunable sources, although conventional ways of achieving tunability by means of external resonators are also viable.

A more attractive way of achieving tunability, which is proposed here for the first time, is to use the interaction of light propagating in the thin film with surface acoustic waves (SAW). Fortunately, *we were able to demonstrate the growth of sol-gel films on quartz substrates*, a material which has well known piezo-electric properties, and found applications in diverse SAW devices. The basic principles involved in the interaction between guided light waves and SAWs are very well known. Basically, they were proved to function as TE-TM converters, wavelength filters, optical correlators and RF spectrum analyzers. In all cases light-SAW interaction provided a way of controlling the periodic perturbation of both frequency and amplitude. We proposed to use this principle in order to tune the film waveguided lasers.

The possibility of producing thin films of high optical quality by the sol-gel method has many attractive features. We have concentrated on the development of waveguided lasers based on organic-dye doping. Nevertheless, very significant preliminary results were obtained. Optical waveguides of high quality were demonstrated, and guided modes and their propagation constants and losses were characterized. In a subsequent study, organic Rhodamine B dye was introduced into the sol-gel solution and the thin films were optically pumped by frequency doubled Nd:YAG laser light. An amplification factor of the order of 50 dB/cm was measured followed by superradiant emission. These outcomes have been published [57–59, 63]. Furthermore, the high gain factor being available allowed the achievement of laser action in a resonator produced simply by cleaving two facets of the glass support. The graphs in Fig. 3, which are very recent measurements, are presented here for the first time and show a very typical lasing characteristic. Lasing was also indicated by a narrowing of the resulting pulse duration.

Here several sections of sol-gel thin films are deposited, each of them doped with a different laser dye. Grating structures are inscribed on each part with

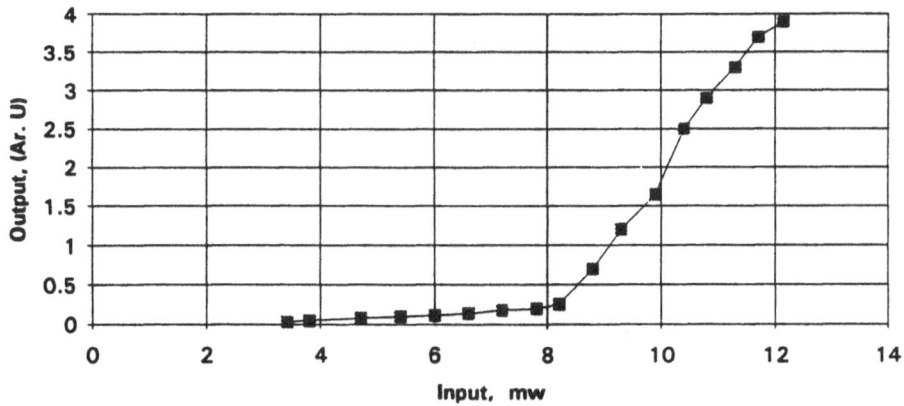

Fig. 3. Output of waveguide laser as a function of pump power showing lasing threshold. From: Finkelstein I, Ruschin S, Sorek Y, Reisfeld R (submitted 1995) Waveguide visible lasing effects in dye-doped sol-gel films. Appl. Phys. Letters

different grating periods, selecting a given wavelength from each dye medium by a distributed feedback (DFB) mechanism. By steering the pump beam, each laser can be individually addressed, and practically the entire visible spectrum can be spanned. The applicability range of such a device would be very large indeed; it includes color displays, color printers, spectroscopy and sensing. The waveguided character of the generated emission makes it suitable for coupling into optical fibers, a property that, by itself, brings with it a host of applications.

The high refractive index glass films were prepared at room temperature as follows [54, 59, 63]. The glass synthesis was performed by hydrolysis and subsequent copolymerization of titanium tetraethoxide $Ti(OEt)_4$ or titanium tetraisopropoxide $Ti(OPr)_4$ with organically modified silane ORMOSIL, namely glycidyloxipropyltrimethoxysilane (GLYMO):

$$CH_2-CH-CH_2O(CH_2)_3-Si(OCH_3)_3 \, .$$

Glass films with refractive index up to 1.66 were thus obtained. In order to prepare films of lower refractive index of 1.48 and density of 1.68 gr/cm^3 we used a procedure based on azeotropic distillation with benzene or toluene solution, including the laser dye, tetraethoxysilane $Si(OC_2H_5)_4$ (TEOS), triethoxyvinylsilane $CH_2=CHSi(OC_2H_5)_3$ (TEVS), and low molecular weight polymethylmethacrylate (PMMA). The block diagram of this procedure is given in Fig. 4 [58, 60].

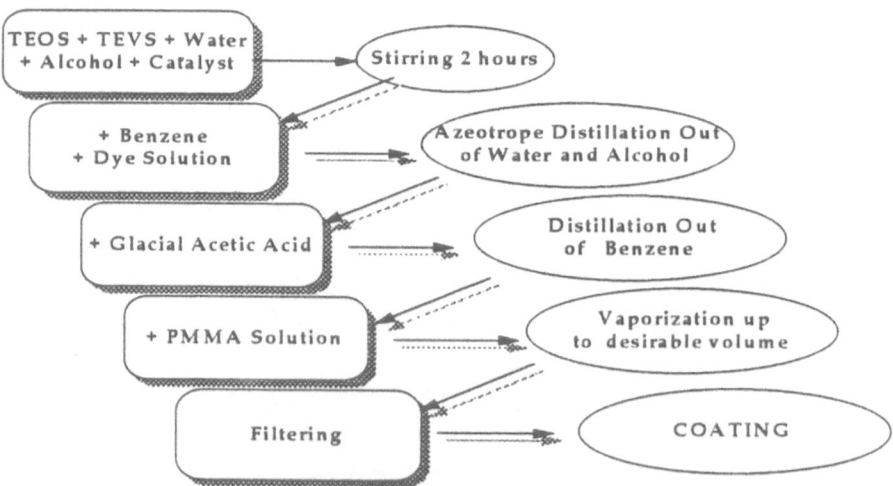

Fig. 4. Procedure for preparation of composite glass

Table 2 presents examples of laser dyes introduced into the films (prepared in our laboratory), the lasing range, and the threshold of laser operation.

R = alkyl

R^x = O-aryl

Structural formula of LFR 300

Figure 5 illustrates laser emission, as observed by narrowing of the fluorescence at threshold energy, of the Lumogene LFR 300 provided by BASF. The spectroscopy of this dye in various solvents has been discussed in detail [47].

Glass films from alumina, prepared by peptization of aluminum hydroxide doped by Rhodamine 6G, Rhodamine B and Oxazine 4 showed laser action with calculated conversion efficiency of 21%. The intensity decreased linearly with the shot number of exciting N_2 laser.

Fluorescence intensity, photodegradation and kinetics of degradation of excited states of R6G in sol-gel films was reported recently [72]. Photodegradation of the dye increased with the concentration as a result of interaction of the monomers with higher aggregates.

The coupling of leaky modes from doped film with a lower refractive index than that of the support has been calculated theoretically [73], and a laser configuration based on active dielectric thin film, acting as an optical amplifier designed.

It has been predicted that the resonant beam width is dominated by the active medium width of the film, not by the resonator-mirror curvature. The calculations were based on the leaky-mode concept [74].

An investigation of oblique plane wave scattering in active dielectric films reveals the existence of anomalously large resonance that occurs at discrete plane wave angles of incidence. This fact allows application of the active films with lower refractive index than the support as an efficient amplifier.

Table 2. Spectral characteristics of the dyes used for glass films doping

Dye	Abs. max. (nm)	Tuning range (nm)	Spontaneous spec. width (nm)	Laser Spec. width (nm)	Threshold (µJ/pulse)
Lumogen LFR 300	578	605–630	70	9	30
Lumogen LFO 240	525	568–583	30	3	50
R6G	546	560–610	44	11	50
DCM	472, 496	595–650	79	19	100
Rhodamine 610	560	585–635	45	10	40

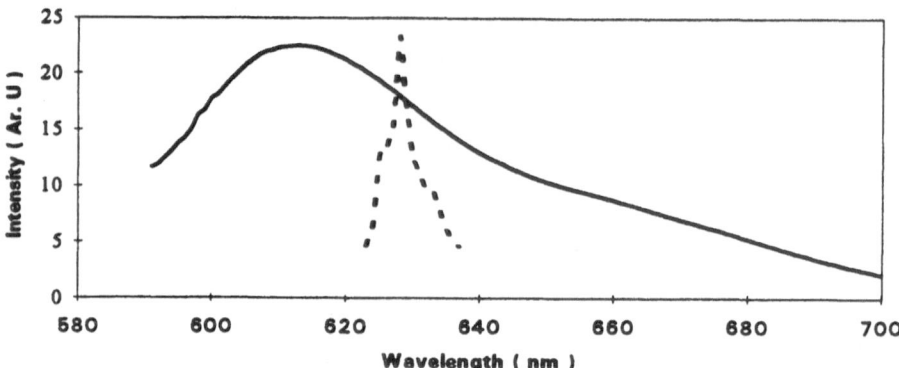

Fig. 5. Fluorescence and laser emission of LFR 300 in a composite glass

Very recently, new blue heptamer laser dyes didecyl *p*-polyphenyl heptamer (DDPPH), didecyloxy *p*-polyphenyl heptamer (DDOPPH), and bisbenzothiazole 3,4-didecyloxy thiophene (BBTDOT) were incorporated into a composite glass. Under excitation with a frequency-doubled dye laser (300 nm), the DDPPH lased at 377 nm. The DDOPPH lased at 425 nm and the BBTDOT lased at ~450 nm when excited by the third harmonic of a Nd:YAG laser (355 nm). The output from the second heptamer in tetrahydrofuran was photostable (less than 10% decrease) for more than 900 000 pulses and with a slope efficiency of approximately 20% [75].

9 Conclusion and Perspective

The state of the art of tunable lasers has today reached such an advanced level that a large workshop was held recently [71] at which a number of commercial companies participated, in addition to the scientific community, indicating not only potential but also existing application.

The solid state dye laser is a rapidly expanding field. Much progress has been made concerning the mechanical, thermal and optical properties of the matrix, and more suitable dyes have been identified, especially pyrromethene (597) for efficiency and perylene (red) for lifetime. Efficiency (several tens of percents) and tunability (several tens of nanometers) have now reached expected values for short operating times. The lifetime attained with a single area of the doped solid has been considerably increased (several tens of thousand pulses), yet this is certainly where future further progress can be made. Grafting the dye to the host matrix could provide a solution. The results with the (photounstable) rhodamine B are preliminary indications as to increased stability, but further study is needed. The recent results with rhodamine B show that the dye stability may be increased by covalent bonding to the silica network while retaining a strong fluorescent emission, therefore making it utilizable for tunable solid state dye lasers. Attempts to graft more suitable dyes such as perylenes or pyrromethenes are in progress.

Acknowledgements. This work was supported by the US Army European Research Office and Night Vision Laboratory contract DAJA 45-90-C-055 and by the Israeli Ministry of Science. Dr. Al Pinto of Night Vision has been extremely helpful in providing important relevant information and discussing the current results.

We also thank D. Brusilovsky, R. Gvishi, H. Minti, D. Shamrakov, Y. Sorek and Mrs. I. Finkelstein for help in the experiments, and Prof. C.K. Jørgensen, Z. Burshtein and S. Ruschin for helpful discussions.

10 References

1. Schäfer FP (ed) (1990) Dye lasers. Springer, Berlin Heidelberg New York
2. Avnir D, Levy D, Reisfeld R (1984) J Phys Chem 88: 5956
3. Avnir D, Kaufman VR, Reisfeld R (1985) J Non-Crystalline Solids 74: 395
4. Altshuler GB, Bakhanov VA, Dulneva EG, Erofeev AV, Mazurin OV, Roskova VP, Tzekhom-skaya TS (1987). Op spectrosc (USSR) 62: 1201
5. Kobayashi Y, Kurokawa T, Imai Y, Muto S (1988) J Non-Crystalline Solids 105: 198
6. Mckiernan JM, Yamanaka SA, Dunn B, Zink JI (1990) J Phys Chem 94: 5652
7. Pouxviel JC, Dunn B, Zink JI (1989) J Phys Chem 93: 2134
8. Reisfeld R, Brusilovsky D, Eyal M, Miron E, Burstein Z, Ivri J (1989) Chemical Physics Letters 160: 43
9. Salin F, Lesaux G, Georges P, Brun A, Bagnall C, Zarzycki J (1989) Optics Letters 14: 785
10. Altshuler GB, Bakhanov VA, Dulneva EG, Mazurin OV, Roskova GP (1990) SPIE 1328: 89
11. Dunn B, Mackenzie JD, Zink JI, Stafsudd OM (1990) SPIE 1328: 174
12. Knobbe ET, Dunn B, Fuqua PD, Nishida F (1990) Applied Optics 29: 2729
13. McKiernan JM, Yamanaka SA, Dunn B, Zink JI (1990) J Phys Chem 94: 5652
14. Reisfeld R (1990) SPIE 1328: 29
15. Whitehurst C, Shaw DJ, King TA (1990) SPIE 1328: 183
16. Canva M, Georges P, Brun A, Larrue d, Zarzycki Z, "Impregnated SiO₂ gels used as dye laser matrix hosts". Proceedings VI International workshop on glasses and ceramics from gels, Seville, 6–11 October 1991
17. Gvishi R, Reisfeld R (1991) J Physique (Colloques C7), 1: 199
18. Reisfeld R (1991) J Physique (Colloques C7) 1: 415
19. Altman JC, Stone RE, Nishida F, Dunn B (1992) SPIE 1758: 507
20. Dunn B, Nishida F, Altman JC, Stone RE (1992). Chemical Processing of Advanced Materials 84: 941
21. Hsin-Tah Lin, Bescher E, Mackenzie JD, Hongxing Dai, Stafsudd OM (1992) Journal of Materials Science 27: 5523
22. King TA (1992) Chemical Processing of Advanced Materials 90: 997
23. Larrue D, Zarzycki J, Canva M, Georges P, Brun A (1992) SPIE 1758: 420
24. Liu S, Hench LL (1992) Chemical Processing of Advanced Materials 85: 953
25. Dai H, Lin HT, Stafsudd OM (1993) SPIE 1864: 50
26. Dunn B, Nishida F, Toda R, Zink JJ, Allik TH, Chandra S, Hutchinson JA (1994) Mat Res Soc Symp Proc 329: 267
27. Lo D, Parris JE, Lawless JL (1993) Appl Phys B 56: 385
28. Canva M, Georges P, Perelgritz JF, Brun A, Chaput F, Boilot JP (1994) "Agile solid state dye lasers", Supplement au Journal de Physique III, Colloque 4, volume 4: 369

29. Canva M, Georges P, Perelgritz JF, Brun A, Chaput F, Boilot JP (1994) Mat Res Soc Symp Proc 329: 279
30. Rahn MD, King TA (1994) "Solid state dye doped sol-gel glass composite lasers", CLEO '94 proceedings, p 389
31. Peterson OG, Snavely BB (1968) Appl Phys Lett 12: 238
32. Itoh U, Takakusa M, Moriya T, Saito S (1977) Japan J Appl Phys 16: 1059
33. Gromov DA, Dyumaev KM, Manenkov AA, Maslyukov AP, Matyushin GA, Nechitailo VS, Prokhorov AM (1985) J Opt Soc Am B 2: 1028
34. Allik TH, Chandra S, Hermes RE, Hutchinson JA, Soong ML, Boyer JH (1993) "Efficient and Robust Solid-State Dye Laser", OSA Proceedings on Advanced Solid-State Lasers, Pinto AA, Fan TY (Eds) (OSA Washington, DC, 1993) 15: 271
35. Allik TH, Chandra S, Robinson TR, Hutchinson JA, Sathyamoorthi G, Boyer JH (1994) Mat Res Soc Symp Proc 329: 291
36. Hermes, RE, Allik TH, Chandra S, Hutchinson JA (1993) Appl Phys Lett 63(7): 877
37. Wang HHL, Gampel L (1976) Optics Comm 18: 4
38. Allik TH, Chandra S, Robinson TR, Hutchinson JA, Sathyamoorthi G, Boyer JH (1994) Mat Res Soc Symp Proc 329: 291
39. Altshuler GB, Bakhanov VA, Dulneva EG, Erofeev AV, Mazurin OV, Roshova GP, Tsekhomskaya TS (1987) Opt. Spectroscopy 62: 709
40. Reisfeld R, Brusilovsky D, Eyal M, Miron E, Burshtein Z, Ivri J (1989) "Perylene dye in a composite sol-gel glass:- a new solid-state tunable laser in the visible range", Binational French-Israeli Workshop on Solid-State Lasers, George Boulon, Jørgensen CK, Reisfeld R (Eds) SPIE 1182: 230
41. Gvishi R, Reisfeld R, Burshtein Z, Miron E (1993) "New Stable Tunable Solid-State Dye Laser in the Red", 8th meeting on Optical Engineering in Israel, Tel-Aviv Israel: Optoelectronics and Applications in Industry and Medicine, Oron M, Shlodev I, Weissman I (Eds) Proc SPIE 1972: 390
42. Mehrotra RC (1992) Structure and Bonding 77: 1
43. Sakka S, Yoko T (1992) Structure and Bonding 77: 89
44. Schmidt H (1992) Structure and Bonding 77: 119
45. Reisfeld R, Jørgensen CK (1992) Structure and Bonding 77: 207
46. Dougherty TJ (1987) Photochem and Photobiol 45(6): 879
47. Reisfeld R, Gvishi R, Burshtein Z (1995) J Sol-Gel Science and Technology 4: 49
48. Gvishi R, Reisfeld R, Burshtein Z (1993) Chem Phys Lett 212: 463
49. Hermes RE, McGrew JD, Wiswall CE, Monroe S, Kushina M (1992) Appl Phys Comm 11(1): 1
50. Allik TH, Hermes RE, Sathyamoorti G, Boyer JH (1994) SPIE Proceedings, on Visible and UV Lasers 2115: 240
51. Reisfeld R (1994) SPIE 2288: 563
52. Jørgensen CK, Reisfeld R (1994) SPIE 2288: 208
53. Canva M, Dubois A, George P, Brun A (1994) Sol-Gel Optics III, Mackenzie JD (Ed) SPIE 2288: 298
54. Sorek Y, Reisfeld R, Finkelstein I, Rushin S (1993) Appl Phys Lett 63: 3256
55. Reisfeld R (1983) Chem Phys Lett 95: 95
56. Reisfeld R (1985) Chem Phys Lett 114: 306
57. Reisfeld R (1994) J de Physique 4: 281
58. Reisfeld R, Shamrakov D, Sorek Y (1994) J de Physique 4: 487
59. Sorek Y, Reisfeld R, Finkelstein I, Ruschin R (1994) Optical Materials 4: 99
60. Shamrakov D, Reisfeld R (1993) Chem Phys Letters 213: 47
61. Shamrakov D, Reisfeld R (1994) J Optical Materials 4: 103
62. Sorek Y, Reisfeld R, Tenne R (1994) Chem Phys Letters 227: 235
63. Sorek Y, Reisfeld R, Finkelstein I, Ruschin R (1995) Appl Phys Lett 66: 1169
64. Shank CV, Bjorkholm JE, Kogelnik H (1971) Appl Phys Lett 18: 395
65. Kogelnik H, Shank CV (1972) J Appl Phys 43: 2327
66. Yariv A (1991) Optical electronics, 4th. edn. Saunders
67. He GS, Zhao CF, Park C-K, Prasad PN (1994) Optics Commun 111: 82
68. Streifer W, Scifers DR, Burnham RD (1976) IEEE J of Quant Electron QE-12: 422
69. Yamamoto Y, Kamiya T, Yanai H (1978) IEEE J of Quant Electron QE-14: 245
70. Hardy A, Welch DF, Streifer W (1989) IEEE J of Quant Electron QE-25: 2096

71. Proceedings of the Solid-State Dye Lasers Technology Workshop, 4 August 1994. Sponsored by Night Vision and Electronic Sensors Directorate and Science Applications Int Co edited by Allik TH
72. Narang U, Bright FV, Prasad PN (1993) Appl Spectroscopy 47: 229
73. Halmos MJ, Fletcher TM, Stafsudd OM (1992) Appl Opt 31(21), 4132
74. Yeh P (1988) Optical waves in layered media, Wiley, New York, Chap 11
75. Gvishi R, He GS, Prasad PN, Narang U, Li M, Bright FV, Reinhardt BA, Bhatt JC, Dillard AG (1995) Appl Spectroscopy 49: 834

Author Index Volumes 1-85